Superconductivity

Library of Congress Cataloging-in-Publication Data
Dahl, Per F., 1932–
Superconductivity: its historical roots and development from mercury to the ceramic oxides / Per Fridtjof Dahl.
p. cm.
Includes index.
ISBN 0–88318–848–1
1. Superconductivity. I. Title.
QC611.92.D34 1992
537.6′23--dc20 91–36487
CIP

American Institute of Physics
335 East 45th Street
New York, NY 10017-3483

Library of Congress Cataloging-in-Publication Data

Dahl, Per F., 1932–
Superconductivity: its historical roots and development from mercury to the ceramic oxides / Per Fridtjof Dahl.
p. cm.
Includes index.
ISBN 0–88318–848–1
1. Superconductivity. I. Title.
QC611.92.D34 1992
537.6′23--dc20 91–36487
CIP

Contents

Illustrations

Preface

The history of superconductivity spans a number of relatively distinct phases, and with the recent worldwide flurry of attention over high-T_c superconductivity in the ceramic oxides, its history is still being made. Undeniably the explosive growth in critical temperature arriving on the heels of Bednorz and Müller's epochal discovery in 1986 is wondrous to contemplate from, say, the perspective of the late Bernd Matthias. Nevertheless, equally impressive has been the growth in the scale of high-field, high-current superconducting technology over barely three decades: from centimeter solenoid bores to tens of kilometers of accelerator dipoles in the latest generation of hadron colliders. Moreover, since the near-term significance of these latest—albeit exciting—developments in oxide superconductivity from the standpoint of large-scale applications is still in question, the focus of the present volume is on "conventional" (metallic) superconductivity.

The original intention had been to limit the volume to the largely undocumented prehistory, discovery, and development of superconductivity up to about 1940, relegating, possibly, treatment of the multifarious and extensively documented development of the subject in the post-World War II years to a subsequent volume. In the event, I was persuaded that some coverage of the more recent period would be desirable as well. Accordingly, the material is organized around the following principal phases: nineteenth and early twentieth-century developments (principally liquid helium and erroneous notions) that led to Kamerlingh Onnes's discovery of superconductivity in 1911; events surrounding the actual discovery and its immediate aftermath, including the critical current and field, and attempts to explain superconductivity within the contemporary theoretical framework; the "golden years" between the first and second world wars, culminating in the Meissner effect and marking the elucidation of type I (elemental) superconductivity as a true thermodynamic state; the phenomenological theory from Gorter and Casimir to Ginzburg and Landau, London's new electrodynamics, and the experimental and theoretical basis for the BCS theory; the advent of type II superconductivity in alloys and compounds, particularly the materials development of the 1950s that gave rise to high-current density conductors and paved the way for their exploitation in high-field technology; the fruits of the technology, ranging from Kunzler's first stabilized solenoid to the brink of the Super Collider. I finish with the ceramic oxides circa 1989, although on a guarded note. The emphasis is unabashedly on experimental and applied aspects, becoming increasingly selective in post-World War II developments and reflecting my own interests, but with enough theoretical detail and basic minutiae thrown in to offer a balanced, coherent account.

The treatment from 1911 to 1933 (the Meissner effect) is based as much as possible on primary sources—for example, letters, laboratory notebooks, and original publications. Foremost among these are the archives of the Kamerlingh Onnes Laboratory, and particularly the papers of Kamerlingh Onnes himself; nowadays these are preserved at the Museum Boerhaave in Leiden, along with those of W. Keesom and other Leiden giants. Invaluable as well are the collections in the Deutsches Museum, Munich, and the library of the Eidgenössische Technische Hochschule, Zürich, both of which contain considerable correspondence between de Haas, von Laue, Mendelssohn, Kronig, McLennan, Meissner, and others. The bulk of the Meissner papers, including notebooks and correspondence, has until fairly recently been retained by the family, and left unavailable for examination; however, in 1986 the papers were deposited in the Sondersammlungen of the Deutsches Museum where I was privileged to gain access to them, though in their rather unedited state.

The Archive for History of Quantum Physics proved to be an excellent source, and frequent use was made of its depositories at the Office of History of Science and Technology, University of California, Berkeley, and the Niels Bohr Library of the American Institute of Physics, New York.

A word about units and conventions. No systematic effort has been made to adhere to a uniform set of technical units, for example gaussian or Système International units. Rather, I follow the conventional usage at any particular time. Thus, both gauss and teslas are used where appropriate, as is the older form of 0 °K when the period demands it.

I am indebted to the following individuals and institutions for archival assistance: Steven Engelsmann, Peter de Clercq, and H. J. F. Leechburch Auers (Museum Boerhaave, Leiden), Ms. van der Werf (Instituut Lorentz, Leiden), Jurgen Teichmann and Rudolf Heinrich (Deutsches Museum, Munich), Maria Hopfer (Bayerische Akademie der Wissenschaften, Munich), Beat Glaus (History of Science Collection, Library, Swiss Federal Institute of Technology, Zürich), Ms. Irena M. McCabe (The Royal Institution, London), Robin Rider and Sheila O'Neill (History of Science and Technology Division, The Bancroft Library, University of California, Berkeley), John Heilbron and Diane Wear (Office of History of Science and Technology, University of California, Berkeley), Richard L. Robinson (Research Library, Lawrence Berkeley Laboratory, Berkeley), Research Library staff, (Brookhaven National Laboratory, Upton, New York), Pat Kreitz and Library staff (Superconducting Super Collider Laboratory, Dallas), Jean Hrichus and Ann Kottner (American Institute of Physics, New York). Similar thanks go to the following individuals for permission to quote correspondence or reproduce material from the indicated sources or collections: Lieselotte K. Templeton and Robin E. Rider (quotations from letters in Otto Stern collection, Bancroft Library); John Heilbron (use of excerpts from interviews in the Archives for History of Quantum Physics); Ralph Kronig (quotations from letters in Archive Hs 1045, ETH library); Rudolf Heinrich (quotations from Meissner papers, Deutsches Museum); G. A. C. Veeneman (photos from Museum Boerhaave); J. E. Kunzler (figure from journal article of 1960). I also thank R. de Bryun Ouboter and Georgio Frossati, of the Kamerlingh Onnes Laboratory, Leiden, for their warm hospitality. The volume had its origins in articles prepared for John Heilbron's fine journal *Historical Studies in the Physical Sciences*; as such, I am most obliged to Heilbron for his indirect role in launching an expanded project, and for subsequent hospitality during my wanderings in and out of Berkeley. It is also a pleasure to thank Peter VanderArend, Martin Berz, and certainly Richard Beth, my dear mentor, for help in deciphering obscure Dutch and German handwriting.

The volume owes much to Diana Barkan for her methodical reading of the penultimate draft of the manuscript, for her historical insight, constructive criticism, and many helpful suggestions. Equally important was the spendid support and cooperation in all stages of manuscript processing by the editorial staff, including Timothy W. Taylor and Maria Taylor of the Books Division, American Institute of Physics, and Rhona B. Johnson and Helen Wheeler—my editor and copyeditor, respectively.

Especially, though, I am indebted to Eleanor, my wife and collaborator, for her lack of intimidation with software, for her meticulous attention to details, for her patient typing and expert word processing of the manuscript, and for her companionship and steadfast support during the unsettled life of recent years on the road between Long Island, Berkeley, and Dallas.

CHAPTER 1

The Convergence of Classical Concepts circa 1900

1.1 The Nadir of Temperature

Nineteen hundred and eleven was an auspicious year in the transition from classical to modern physics. Max Planck's quantum of action had proved to be sufficiently pervasive to justify a summit conference on its far-reaching implications; the Solvay Council held that same year, while focusing on the quantum theory of radiation, also provided a forum to air the latest developments on the kinetic theory of matter. The crowning achievement of classical atomism was Ernest Rutherford's seminal 1911 paper presenting his celebrated interpretation of the experiments of Hans Geiger and Ernest Marsden. Rutherford's atomic model would soon be profoundly revised in light of the quantum theory. Niels Bohr, the quite unknown young man responsible for that major revision, was even then on his way from Copenhagen to Cambridge to study under J. J. Thomson; late in 1911 Bohr was introduced to Rutherford at Manchester. By 1911 another area of the physical sciences was on the threshold of a new age. In that same year the following admonition appeared in the 11th edition of the *Encyclopedia Britannica*:

> Though *Ultima Thule* may continue to mock the physicist's efforts, he will find ample scope for his energies in the investigation of the properties of matter at the temperatures placed at his command by liquid air and liquid and solid hydrogen. Indeed, great as is the sentimental interest attached to the liquefaction of these

refractory gases, the importance of the achievement lies rather in the fact that it opens out new fields of research and enormously widens the horizon of physical science.[1]

The author of these views was none other than Sir James Dewar, the well-known low-temperature physicist and doyen of low-temperature technology. That very year, by a quirk of fate, a startling discovery in low-temperature physics was made by Dewar's nemesis Heike Kamerlingh Onnes of Leiden, Holland. The discovery was virtually assured by a final liquefaction step reached at Leiden 3 years earlier which, to Dewar's bitter consternation, he himself had failed to achieve in a barely contested race with Onnes.

These two intertwined milestones, liquid helium and superconductivity, brought many honors to Kamerlingh Onnes, including a Nobel Prize. They were poignantly, if less formally, juxtaposed in the international Jubilee on 11 November 1922, which honored Onnes on the 40th anniversary of his professorate at the University of Leiden. This was also the occasion for the presentation of a well-deserved festschrift from colleagues at home and abroad.[2] Among the photographs taken at this festive occasion,[3] is one depicting a room in Onnes's laboratory cluttered with cryogenic apparatus and on the wall of which was a plainly lettered placard in Dutch, erected for the day to mark the location of a historically important spot:

SUPRAGELEIDERS

The cryogenic feat making the discovery of superconductivity possible was commemorated that same day, in an adjacent spot, with the unveiling of a more lasting memorial stone of grey marble. Translated from the Dutch, it reads: "On this spot, on the 10th of July, 1908, helium was liquefied for the first time by Dr. Heike Kamerlingh Onnes."[4]

Onnes's successful liquefaction spelled the end of Dewar's own low-temperature work at the Royal Institution. Virtually overnight, Dewar's world-famous scientific program was reduced to the study of soap bubbles and thin films. Mainly as a consequence of his infamous temper, he left no school in low-temperature science and technology. Although British laboratories would in time play a vital historical role in unraveling the subtleties of superconductivity, Dewar's laboratory was not to be one of them.

The events that set the stage for the experiments of 1911 have often been told. Even a superficial recapitulation of them must take account of the two dominant personalities primarily responsible for their shaping.[5] They had much in common, yet they were poles apart.

James Dewar (1842–1923), the senior of the two men, was initially steered into physical science at Edinburgh University by the physicist (and geologist) David Forbes. Dewar soon came under the tutelage of the chemist Lyon Playfair, followed by Playfair's successor Crum Brown. Some of Dewar's mannerisms may possibly be blamed on his exposure to the pompous Playfair, who was widely disliked. In some specific heat measurements performed by Dewar

Figure 1-1. James Dewar and his famous flask. Courtesy AIP Niels Bohr Library and W. F. Meggers Collection.

at this time, he used a crude vacuum-jacketed calorimeter, the forerunner of the famous dewar vessel. In 1869, Dewar became lecturer in chemistry at the Royal Veterinary College of Edinburgh, and subsequently served as assistant chemist to the Highland and Agricultural Society. He was appointed Jacksonian Professor of natural experimental philosophy at Cambridge in 1875, and Fullerian Professor of chemistry at the Royal Institution in 1877. Dewar held both chairs until his death, in spite of the fact that he and Cambridge barely got along.

Dewar's early work was mainly in organic chemistry, but encompassed physics and physiology as well. At Cambridge, he concentrated on atomic and molecular spectroscopy. Upon arriving in London, he began a new series of high-temperature and spectroscopic studies. However, a sudden and ultimately overriding passion for cryogenic investigations developed within a year of his taking residence at the Royal Institution; perhaps he was inspired by the spirit of Michael Faraday, which was said to have permeated the institution on Albemarle Street. (Faraday, during his tenure there, liquefied all the known gases

Figure 1-2. Heike Kamerlingh Onnes. Courtesy AIP Niels Bohr Library and Burndy Library.

except the six "permanent" ones.) More likely, Dewar's immediate inspiration was a significant scientific development in Paris that broke a 30-year hiatus in cryogenic progress. But first, before recounting the chronology of these events, we must devote a few words to Heike Kamerlingh Onnes.

Kamerlingh Onnes was born in 1853, the oldest son of a prominent family in Groningen, Holland. His formal schooling began in 1865 at the Hoogere Burgerschool in Groningen where, according to the school records, the boy was an assiduous student, and eager for knowledge.[6] During these years, the principal of the school was van Bemmelen, later Professor of Chemistry at the University of Leiden, who appears to have exerted strong influence on him. Onnes passed his final examination in 1870, matriculated at the University of Groningen, and received his bachelor's degree in physics and mathematics in 1871. That year, when Onnes was only 18 years old, he received a gold medal in a competition sponsored by the University of Utrecht. His prizewinning essay provides a strong hint of the scientific subjects and slightly cumbersome style of expression that one identifies with the mature Onnes: "A critical investigation of the methods of determining vapor density and of the results obtained thereby, with respect to the relation of the nature of the chemical compounds and the density of their vapours."[7]

Like Dewar, Onnes was mainly attracted to chemistry early in his career. With that in mind, he then received a stipend to study under Robert Bunsen and Gustav R. Kirchhoff at Heidelberg, the birthplace of spectrum analysis. Onnes's own fascination with calorimetric measurements was kindled at this time; Bunsen's low-temperature investigations led, among other things, to the rediscovery of the ice calorimeter. Seizing the subject of calorimetry, as Dewar had slightly earlier, Onnes dashed off another prizewinning essay, this time sponsored by the State University at Groningen, entitled "A critical survey of the methods of determining the quantities of heat which are set free by chemical reactions and dissociations ... " Following a stint with Bunsen, the guru of chemistry (who was, in contrast to Playfair, also famous for his great wit), Onnes now inclined toward the study of physics, and he finished his *Wanderjahr* with a start on his *Seminarpreis* under Kirchhoff's guidance. (The *Seminarpreis* involved experiments with Foucault's pendulum.) In 1873, he returned to Groningen to complete his graduate studies.

Onnes passed his "doctoraal" (preparatory) examination in 1876, and defended his doctoral dissertation on "New proofs for the axial rotation of the Earth" in 1879 (following his *Seminarpreis*). The year before he obtained his degree he secured a position as assistant to Johannes J. Bosscha, who was then the director of the Polytechnic School at Delft. The 4 years Onnes worked under Bosscha proved to be an excellent training period in designing and handling experimental apparatus.

Equally as important for Onnes was the lasting influence of a paper that he came across by Diderik van der Waals, then a professor at the University of Amsterdam. This was van der Waals's second paper to gain prominent attention and influence in low-temperature science. Johannes Diderik van der Waals received his doctorate at Leiden the same year that Onnes returned to Groningen. His dissertation "On the continuity of the gaseous and liquid states"—albeit published in Dutch—had nevertheless elicited instant public notice by James C. Maxwell (who learned to read Dutch in order to read the thesis), among others. Initially the main importance for low-temperature physics was that the paper, which contained van der Waals's famous equation of state based on molecular forces, came to the attention of Thomas Andrews of Queens College, Belfast—the father of "real" isotherms and the thermodynamic critical point. (The critical temperature, we recall, is that temperature below which a gas can be liquefied by application of sufficient pressure.) Van der Waals's impact on Onnes through his second influential paper came later, upon his arrival at the University of Leiden, as we shall see shortly. In 1882, Onnes was appointed successor to Pieter L. Rijke there—a professorship he was to hold for 42 years, or nearly as long as Dewar's tenure in London.

Dewar and Onnes's formative years therefore involved rather similar professional aspirations that were quite typical of budding academicians of the time. No claim for a similarity of their personalities can be made, however. Dewar's temperament was not only autocratic but utterly irascible, leaving him with no friends and no pupils. Onnes, to be sure, was paternalistic, opinionated, and a man of strong principles—traits not uncommon among the moguls of late nineteenth-century science. But he proved to be a benevolent leader, kind and scrupulously fair in his relations with friends and pupils alike—behavior that was certainly within the norms of his time. Intellectually, Dewar was a pertinaciously enthusiastic and brilliant experimentalist. Onnes was a more disciplined experimenter and had more theoretical flair. The preface to his doctoral thesis is telling in this respect. He quotes Helmholtz in his memorial lecture on Gustav Magnus:

> It seems to me that nowadays the conviction gains ground that in the present advanced stage of scientific investigation only *that* man can experiment with success who has a wide knowledge of theory and knows how to apply it; on the other hand, only *that* man can theorize with success who has a great experience in practical laboratory work.[8]

The chronology of the milestones leading to liquid helium, briefly, begins the year that Dewar came to the Royal Institution in London.[9] On Christmas Eve, 1877, the first liquefaction of a "permanent" gas, oxygen, although only as a mist of liquid, was announced by Louis Cailletet before the Academy of Sciences in Paris. His method, discovered by accident while attempting to liquefy the gas by simply applying pressure, was that of expanding precooled, compressed oxygen against a column of mercury acting as a piston. The event had actually taken place three weeks earlier, but Cailletet delayed its announcement so as not to bias his bid for corresponding membership to the academy. The delay proved costly: Two days prior to his announcement Raoul Pictet claimed the same achievement by telegram from Geneva. Pictet's approach was based on pumping on the vapor in several stages, though also in his case retrospective scrutiny of the procedure attributes the end result, a jet of liquid, to expansion.

In 1882, Sigmund von Wroblewski assumed the chair of physics at Jagiellonian University, Cracow (an institution under the Hapsburg aegis, since Cracow was then part of Austria). He took advantage of the occasion and imported a Cailletet apparatus from the renowned instrument maker E. Dukretet in Paris where he had recently studied under a stipend from the Cracow Academy. Wasting no time, he teamed up with Karol S. Olszewski of the university's Chemistry Department. The two soon succeeded in achieving the static liquefaction of oxygen, producing liquid "boiling quietly in a test tube" in 1883.[10]

Their method combined those of Cailletet and Pictet, expanding gas into a container cooled by pumping off the vapor above the liquid ethylene coolant bath. Boiling nitrogen followed almost immediately. Unfortunately, the teamwork proved to be short-lived because of personal differences, the nature of which remain obscure to this day. Working separately, both turned to liquefying hydrogen with equally marginal success; neither achieved more than a cloud of fog in the midst of escaping hydrogen gas. Noteworthy, however, are the basic low-temperature resistance measurements extending down to the temperature of solid nitrogen, taken by Wroblewski in the course of these efforts. His contributions to the question of the variation of specific resistance with temperature, already then a matter of some academic importance, came to an abrupt halt by his untimely death. In 1888, working late at night, he perished in the flames of an overturned kerosene lamp.

The Royal Institution developed a tradition in gas liquefaction much earlier, dating from 1823 when Michael Faraday liquefied chlorine gas while he was an assistant to Humphry Davy. Thus it was natural that Dewar felt drawn to the subject when he assumed his professorship in chemistry there. After the announcement of liquid oxygen was made in Paris, he too lost no time in procuring a Cailletet apparatus from Dukretet. Its use, although limited, was most notable in the first of Dewar's legendary Friday Evening Discourses in 1878. However, an improved version of the Cracow plant was operational in London by 1886. Curiously, it is described in an appendix to Dewar's paper on meteorites,[11] a circumstance that became a factor in later friction between the cantankerous Dewar and the incautious Olszewski.

Dewar's chemical laboratory, housed in the basement of the Royal Institution in these years, has been vividly described by Lord Rayleigh. Rayleigh's own domain at the time was the physical laboratory upstairs. "When I first saw it, probably about 1889," he recalls, "the most conspicuous object in it was a tube 18-metres long, which was filled with compressed oxygen for observations on the absorption spectrum. As his work on the liquefaction of gases progressed, and as the scale of operations increased, it gradually took on more and more of the aspect of a factory, full of machinery run by means of shafting from a large gas engine."[12]

Dewar's program had two objectives: further lowering of temperature, particularly with the aid of liquid hydrogen, and a method for producing cryogenic liquids in quantities adequate for low-temperature research purposes generally (and for his demonstration lectures). The latter objective also demanded the ability to retain volatile liquids for long periods. The key to large-scale liquefaction—and to liquid hydrogen—was the application announced in 1895 by Carl von Linde in Germany and William Hampson in England of the principle of free (Joule–Thomson) expansion for producing liquid air on an industrial scale. Although the instantaneous temperature drop obtained by "external" work (e.g., expanding a gas against a piston) is greater, cooling by internal work (expanding gaseous molecules against their own attractive forces or into a region of lower pressure) with the aid of a Joule–Thomson expansion valve is a *continuous* process allowing a cumulative cooling effect. The availability of copious quantities of liquid air, furthermore, provided the precooling necessary in the case of hydrogen which, unlike oxygen and air, cannot be cooled by free expansion from room temperature. Dewar's problem of storing and transporting his cryogens had been solved by his own ingenuity considerably earlier, during his Edinburgh researches, as previously noted. His famous vacuum-jacketed, silvered flask was first demonstrated in its mature form in an Evening Discourse in January of 1893. (Double-walled cryostats had been in use from Cailletet's time, but only water vapor had been excluded from the interspace.)

With a good supply of liquid air on hand, Dewar's overall laboratory program on the behavior of materials at cryogenic temperatures made steady progress in the years after 1895. Great importance was attached to low-temperature resistance measurements, both in calibrating gas thermometers and for their intrinsic interest. At the same time, preparations continued with a heightened sense of urgency for what was then regarded as the final cryogenic push. Dewar was well aware that Olszewski had not given up on liquid hydrogen. More ominous in Dewar's ever-pessimistic view, Kamerlingh Onnes was looming as a formidable contender across the Channel. But Olszewski was frustrated by the low critical temperature of hydrogen and by a lack of funds for acquiring equipment of Linde's cryogenic caliber. Onnes, in his own methodical way, though he was now in possession of a cryogenic plant superior to that of the Royal Institution, was biding his time. His goal was a more ambitious liquid hydrogen *facility* for exploiting the new scientific opportunities to the fullest, not merely to beat Dewar to the new temperature domain. In addition, however, the cryogenic work at Leiden was interrupted at this critical juncture by an alarm called by the Leiden Town Council over the dangerous work performed in the laboratory. (Both Dewar and Olszewski gallantly testified on Onnes's behalf in the ensuing inquiry, warning of "a terrible disaster for science" if the work in the laboratory were curtailed.[13])

While Onnes was obliged to move slowly, Dewar forged ahead, and achieved a jet of partially liquefied hydrogen in 1896. His triumphant goal, in the form of 20 cubic centimeters of boiling liquid, was reached in May of 1898—the year that Onnes obtained permission from the Leiden authorities to press forward. The boiling point was estimated, after considerable thermometric confusion, to be about 20 °K. The next year Dewar also produced solid hydrogen by pumping off the vapor above the liquid and, finally, by pumping the vapor from solid hydrogen, reached a lowest temperature of 12 °K.

Onnes, in his inaugural address in 1882 (the year of Wroblewski's appointment at Cracow), had sketched out a program of quantitative research "in establishing the universal laws of nature and increasing our insight into the unity of natural phenomena."[14] His program was inspired by van der Waals's call in 1873 for an experimental test of the equation of state and, as noted, particularly by van der Waals's law of corresponding states, which was published in 1880. Put briefly, the law results from introducing ratios of temperature, pressure, and volume to their critical values in the equation of state, thereby casting the equation in a form approximately valid for all gases and liquids. This paper was the aforesaid second publication by van der Waals to influence low-temperature science, this time by attracting Onnes's attention, and held the key to quantitative low-temperature experimentation.

> This law had a special charm for me [he emphasized in later years], because I was of the opinion that it was based upon the stationary mechanical similitude of the substances considered. From this point of view, the study of the divergences in substances of simple chemical structure with low-critical temperature seemed to me of great importance. Precision measurements at low temperature must be very attractive in my opinion. For this purpose, it was necessary to dispose of large apparatus with which the measurements could be made at a constant temperature, and it was indispensable to construct suitable temperature baths, baths which could also be used for numerous other investigations.[15]

Above all, his famous dictum *Door meten tot weten* ("By measurement to knowledge") set the style for a "modern physical laboratory ... modelled upon astronomical lines."[16]

The first order of business in furnishing the new laboratory[17] was, as usual, the purchase of a Cailletet compressor, augmented somewhat later by a set of Pictet pumps. However, a decade elapsed before the first phase of installing the laboratory equipment was successfully concluded. In 1892, at long last, Onnes's liquefier not only produced 20 cubic centimeters of liquid oxygen, but the liquid could be poured into a separate container, albeit long after this operation became standard practice in Cracow and London. And then the apparatus promptly broke.

This setback proved to be but a mere perturbation of a program that was by then in full swing. By 1893 a large-scale liquefaction plant was fully operational, providing a cascade of low-temperature cycles with methyl chloride, ethylene, oxygen and, somewhat later, air, successively reaching slightly below −200 °C. Thermodynamic measurements in connection with the equation of state had initial priority, and were particularly aimed at corroborating the law of corresponding states—as always Onnes's guiding principle. The scientific program soon embraced magnetic, magneto-optic, radioactive, and electrical subjects, most of which demanded investigations at low temperatures, as evidenced from the earliest issues of the famous *Communications*—Onnes's unprecedented in-house laboratory journal, volume 1 of which dates from 1885.

Originally, some of this work was initiated by quite separate teams or individuals within the laboratory. One member of the staff was Pieter Zeeman, who had studied under Onnes and later became assistant to Lorentz at Leiden. His discovery of the effect that bears his name, the splitting of spectral lines in a magnetic field, was made in 1897—the year of the Leiden Town Council flap, and also the year Joseph J. (J. J.) Thomson obtained firm evidence linking cathode rays to negatively charged particles or *electrons* (G. Johnstone Stoney's term coined in 1894 but a term Thomson shunned even in his Nobel lecture). Zeeman's discovery, inspired by Faraday's discovery half a century earlier of the action of magnets on light (rotation of the plane of polarization), and augmented by Lorentz's insight, yielded the magnitude and sign of the ratio of charge-to-mass (e/m) of the particles involved, bound as they were within atoms but otherwise unmistakenly identical to J. J. Thomson's *free* electrons. This discovery also propelled Zeeman into the scientific limelight, and shortly thereafter he was enticed to accept an appointment at the University of Amsterdam. After Zeeman's departure, the Leiden laboratory's scope was narrowed somewhat and its investigations were consolidated into the mainstream of Onnes's program, one that now exclusively embraced low-temperature phenomena.[18] Thus, of the programmatic legacies of the two great theorists who had crossed his early path, van der Waals and H. A. Lorentz, Onnes had decisively chosen for his own pursual van der Waals's thermodynamic line over Lorentz's magneto-optics that were continued by Zeeman. Onnes's personal, explicit involvement also becomes more evident in the shifting authorship in the *Communications.*

With oxygen conquered, attention shifted to hydrogen. Again, the law of corresponding states was to be the guide. The critical parameters, temperature (T_c) and pressure (P_c), first had to be derived from measurements on the equation of state for hydrogen down to the lowest temperature attainable with oxygen. Hydrogen was liquefied on 5 June 1905, but it was decided to proceed to a larger liquefier without delay, the goal being 1.5-liter baths with the temperature held constant to 0.01 degrees. The hydrogen plant became operational in February of 1906. In May Onnes, taking a cue from Dewar, demonstrated liquid hydrogen before the Royal Dutch Academy of Sciences in Amsterdam.

Even before liquid hydrogen was a fait accompli at Leiden, the race was on again between Dewar, Olszewski, and Onnes. This time the goal was liquid helium, taking advantage of the latest addition to the roster of permanent gases. William Ramsay, who had detected helium in terrestrial sources in 1895, eventually joined in, aided almost immediately by Morris Travers, who devised his own hydrogen liquefier in a remarkably short time. But now Onnes's seeming procrastination paid off handsomely. By the time liquid hydrogen was available at Leiden, the technical machinery for liquefying helium—pumps, compressors, all sorts of apparatus—was largely in place, with a new room in the laboratory dedicated to the task. It was clear by then, from unsuccessful attempts by all four parties to liquefy helium by simple expansion, that success would again have to depend on regenerative expansion—that is, on a refrigeration cycle in which the refrigerant, after being cooled by expansion through a nozzle, is then passed through a heat exchanger where it further cools the incoming compressed gas. This was a formidable undertaking, but one already utilized in the liquefaction of hydrogen. Thermodynamically, moreover, the prospects for successful liquefaction of helium by this route looked good to Onnes. At first, he believed that the critical temperature of helium lay extremely low, perhaps even at absolute zero. But by 1907, 5 to 6 °K seemed more likely. And determinations of isotherms of helium were "rendering it very probable that the Joule–Kelvin effect might not only give a decided cooling at the melting point of hydrogen, but that this would even be considerable enough to make a Linde–Hampson process succeed."[19]

One major difficulty remained: obtaining helium gas of sufficient quantity and purity. Here Onnes's brother, O. Kamerlingh Onnes, head of the Office of

Commercial Intelligence at Amsterdam, saved the day with monazite sand procured from North Carolina. Monazite contains (because of its radioactive origin) from 1 to 2 cm^3 of helium gas per gram, and occurs in the United States in the Carolinas, Idaho, and Florida, as well as in India, Brazil, and South Africa. "The monazite being inexpensive, the preparation of pure helium in large quantities [by heating the monazite] became chiefly a matter of perseverance and care," reads Onnes's report.[20] Indeed! Four chemists labored in succession on the purification for 3 years.[21] The minimum requirement was 200 liters, with 160 further liters needed as backup reserve. Dewar's defeat, by the same token, can be attributed to his decision to use gas from the Bath Springs as his source of helium. Onnes, in fact, had proposed to Dewar in 1905 to share the expenses for a plant to extract helium from that particular gas; luckily for Onnes, that proposal fell through. Dewar had installed a large-scale collector at Bath, and laboriously transported the crude gas to London where the helium was extracted. Unfortunately, it proved to be contaminated with neon that could not be easily removed by the purification methods then available. Dewar might have done better if he had collaborated with Ramsay, the authority on rare gases, but—as usual—a long-standing feud between the two ruled that out.

A footnote to Dewar's paper in 1908, "The nadir of temperature," is a short but sad epilogue to his final effort: "Helium was liquefied by Professor Dr. Kamerlingh Onnes, of Leiden University, on 9th July, 1908." Actually, the memorable day was July 10. Preparations for the attempt began at 5:45 that morning. The critical stage came about 6:00 p.m. "At first the fall of the helium thermometer which indicated the temperature under the expansion cock, was so insignificant that we feared that it had got defect ... after a long time, however, the at first insignificant fall began to be appreciable, and then to accelerate. Not before 6:35 p.m. an accelerated expansion was applied, on which the pressure in the coil decreased from 95 to 40 atms, the temperature of the thermometer fell below that of the hydrogen."

> In the meantime the last bottle of the store of liquid hydrogen was connected with the apparatus: and still nothing had as yet been observed but some slight waving distortions of images near the cock. The thermometer indicated first even an increase of temperature with accelerated expansion from 100 atms, which was an indication for us to lower the circulation pressure to 75 atms. Nothing was observed in the helium space then either, but the thermometer began to be remarkably constant from this moment with an indication of less than 5 °K. When once more accelerated expansion from 100 atms was tried, the temperature first rose, and returned then to the same constant point.
>
> It was, as Prof. Schreinemakers, who was present at this part of the experiment, observed, as if the thermometer was placed in a liquid. This proved really to be the case. In the construction of the apparatus it had been foreseen that it might fill with liquid, without our observing the increase of the liquid. And the first time the appearance of the liquid had really escaped our observation. Perhaps the observation of the liquid surface, which is difficult for the first time under any circumstances, had become the more difficult as it had hidden at the thermometer reservoir. However this may be, later on we clearly saw the liquid level get hollow by the blowing of the gas from the valve and rise in consequence of influx of

Figure 1-3. Kamerlingh Onnes and Flim with the helium liquefier. Courtesy Deutsches Museum, Munich.

liquid on applying accelerated expansion, which even continued when the pressure descended to 8 atms. . . .

The surface of the liquid was soon made clearly visible by reflection of light from below, and that unmistakably because it was clearly pierced by the two wires of the thermoelement.

This was at 7:30 p.m. When the surface had once been seen, it was no more lost sight of. It stood out sharply defined like the edge of a knife against the glass wall.[22]

When the experiment was ended at 9:40 p.m., "not only had the apparatus been taxed to the uttermost during this experiment and its preparation, but the utmost had also been demanded from my assistants," continues Onnes's report. Onnes particularly acknowledged Mr. G. J. Flim, Chief of the Technical Department of the Cryogenic Laboratory (Figure 1-3), who had superintended the construction of the cryogenic apparatus.

Most laboratories have their indispensable technical factotum; Flim is a superb example. He joined the laboratory in 1901, and remained there long after Onnes passed away. (Flim's son was shot during the German occupation in World War II.) His skill in the laboratory arts and crafts would be equally important in the superconductivity programs in later years. Superconductivity owes much to the venerable Flim.

In Dewar's case, Robert Lennox played a somewhat similar role, although he was a man of more formal training. Dewar was, like Onnes, very careful to acknowledge the contribution of his assistant, but the mutual relationship between the two was often strained, as a result of his temper. The climax came soon after Dewar's crushing blow in 1908, instigated by a quarrel with Lennox over the performance of Lennox's separator of helium from the Bath gas. Lennox promptly left, and he did not come near the Royal Institution again until well after Dewar's death in 1923.[23]

1.2 From Resistance Thermometry to Basic Questions

The electrical resistance of pure metals and alloys proved to be an important addition to the low-temperature developments we have already sketched. It is relatively easy to measure and, being strongly dependent on temperature, offered a highly convenient everyday substitute for the cumbersome gas thermometers. Hand-in-hand with its utilitarian application as a thermometric aid in the various temperature ranges opening up, grew a purely empirical preoccupation with the temperature dependence of electrical resistance. In addition, however, fledgling theoretical models of metallic conduction sprang up in academic circles mostly outside the ongoing cryogenic programs; these studies of electron behavior in metals gradually developed into a major subject area of late nineteenth-century theoretical physics. The last decade of the nineteenth century saw gas liquefaction, resistance thermometry, and the electron theory embroiled in a climactic dispute over the electric properties of pure metals at the lowest temperatures. Before coming to grips with the resolution of this classic controversy—a precursor to the experiments of 1911—we will review briefly in this and the following section the background for the experimental and theoretical understanding, respectively, of the electrical conduction in metals circa 1900.

The first significant experiments on the relative "conducting power" of various substances were made by Henry Cavendish late in his celebrated electrical researches.[24] Humphry Davy systematized the subject with the observation that the electrical conductivity of a wire falls with increasing temperature.

In 1821 (perhaps again assisted by Faraday), Davy compared the conductivity of a wire when heated a dull red from a voltaic pile with a wire brought to a state of incadescence by a lamp.[25] Credit for attempting the first quantitative determination of the fall in conductivity belongs to Emil Lenz.[26] His measurements on silver, brass, iron, and platinum, read before the St. Petersburg Academy of Sciences in 1833, were among the earliest of his electromagnetic investigations conducted at St. Petersburg, which spanned nearly three decades. Lenz's analysis of his results is remarkably sophisticated for its time: He fitted his results to a quadratic expression for the conductivity as a function of temperature n (using his notation), of the form

$$\gamma_n = X + Yn + Zn^2 , \tag{1-1}$$

with the aid of Gauss's method of least squares, where X is the conductivity at the freezing point. Unfortunately, expressing the formula in terms of conductivity rather than resistance, and postulating its validity beyond the range of his highest observed temperature (200 °C), led him to the awkward conclusion that all metals share the property of a *minimum* in electrical conductivity at some relatively modest temperature, typically 300 °C.

Closer to the mark was the Norwegian physicist Adam Arndtsen. In 1857, when he was Adjunct and keeper of the physical cabinet at the University of Christiania, he established the general trend in the temperature variation of resistance of metals and alloys at "ordinary" and higher temperatures.[27] Suspecting that the dubious quadratic term in Lenz's expression was influenced by the fact that Lenz, and later Edmund Becquerel and F. Muller[28] (neither of whom actually found any evidence for a minimum) used bare wire samples immersed in heated oil baths, Arndtsen insulated his wires with silk. He then wound them around a glass former mounted in oil or water kept at a constant temperature with a spirit or gas lamp. The temperature variation of the resistance, he found, could be expressed generally by:

$$W_t = W_0 \, (1 + at + bt^2) \tag{1-2}$$

where W_0 is the resistance at $t = 0$ °C. The quadratic term, he surmised, is only significant for alloys and iron. Arndtsen argued that the (linear) temperature coefficient a is actually the same for perfectly pure metals. His paper, published in 1858, immediately caught the attention of Rudolf Clausius. Clausius was then serving his first professorship at the new Polytechnicum in Zürich. He noted that, neglecting the quadratic term, Arndtsen's average coefficient was very close to the coefficient of cubical expansion of a permanent gas (which is given by the reciprocal of the absolute temperature). That is, to a very good approximation,

$$W_t = W_0 \, (1 + 0.00366 \cdot t) \tag{1-3}$$

or in Clausius's words "the resistance of a simple metal in the solid state is closely proportional to the absolute temperature."[29]

By 1873 further measurements bridging the temperature range from 0° to 200 °C, particularly those made by August Matthiessen at the University of London in the 1860s (and those of the German industrial pioneer Werner Siemens), had cast doubt on Clausius's simple linear rule.[30] That year Rene Benoit,[31] on the basis of his own measurements on the resistance of the more common metals from the temperature of boiling water up to 860 °C (boiling cadmium), resurrected the quadratic formula for resistance, and published a careful tabulation of the *specific resistance* (or "resistivity") of each metal, reduced to 0 °C.[32]

The variation of resistance with temperature took on a practical significance as a thermometric tool after 1877, the year of Cailletet and Pictet's "mist," and especially after Wroblewski and Olszewski's static liquefaction in 1883. Two years later, several important papers appeared on resistance measurements in the novel temperature range that was now becoming accessible. One paper, written by Cailletet himself with F. Bouty, reported measurements on various metals, including mercury, down to the temperature of boiling liquid ethylene (about −100 °C).[33] They summarized their results in terms of a "mean coefficient of resistance change," that is, the linear coefficient in the resistance formula.[34] Wroblewski extended resistance measurements on electrolytic copper down to the temperature of solid nitrogen, −200 °C, finding that in this temperature region the resistance diminishes much faster than the absolute temperature. His tables show typical values for copper of 0.004 at a mean temperature of 0 °C, in good agreement with Cailletet and Bounty, but then falling sharply, approaching 0.007 near −200 °C.[35]

The same year, 1885, Hugh Longbourne Callendar began systematic measurements at the Cavendish Laboratory, the home of electrical standards in Great Britain. A personage of some note in these unfolding developments, Callendar got his start as an undergraduate of no special promise under J. J. Thomson. In retrospect, Thomson considered Callendar's development at the laboratory "in some respects the most interesting of all [those encountered in his] career. ... [Callendar] had not been in the laboratory long ... when I saw that he possessed to an exceptional degree some of the qualifications which make for success in experimental research. The problem was to find a subject for his research which would give full play to his strong points. ... It seemed to me that the most suitable research would be one which centered on the accurate measurements of electrical resistance."[36] Within months J. J. Thomson's perceptive hunch paid off, as Callendar, on Thomson's suggestion, focused on precision platinum resistance thermometry (something begun by Siemens in 1861 but not followed up), including a method for reducing the observed "resistance temperatures" to those of an air thermometer,[37] since the standard of thermometry was then Regnault's air thermometer. Platinum was selected for having a roughly linear response over a wide range of temperatures, for its stable resistivity, and for its attainability in very pure form.

By way of a postscript on Callendar, it may be noted that after a stint as professor of physics at the Royal Holloway College, Egham, Callendar accepted an appointment in 1893 as professor of physics at McGill University in Montreal, where he introduced, with the help of his protégé Howard T. Barnes, his platinum thermometer—to which we will return shortly—as a new standard in a succession of noteworthy physical and engineering measurements. Callendar's reputation soared, and when he left McGill for University College, London, in 1898, his colleagues were sorely pressed to find a replacement for him. His successor, in fact, proved to be an altogether different type, being none other than Ernest Rutherford, who accepted the professorship with some ambivalence toward his predecessor.[38]

> On the one hand, Rutherford was gratified that Callendar should have been, and been esteemed to be, so excellent. ... On the other hand, Rutherford objected to being measured against the great measurer, whose old-fashioned physics, with its tedious exactness and proximity to engineering, he did not admire.[39]

Despite Rutherford's belittling of Callendar's style, at the time of Callendar's tutelage under J. J. Thomson, Thomson was continuing in the proud footsteps of James Clerk Maxwell and Lord Rayleigh at Cambridge on the determination of exact electrical standards. But as for low-temperature measurements per se, the center of excellence now shifted to London.

By this time Dewar's team at the Royal Institution Laboratories, possessing copious quantities of liquid oxygen and the workhorse of cryogens, liquid ethylene, turned to the effect of low temperatures on various physical phenomena per se. (Olszewski's group, following the "dissolution of collaboration" with Wroblewski, as Olszewski put it,[40] continued to concentrate on the thermodynamic properties of the cryogenic fluids themselves.) Joining forces with John A. Fleming of University College, London, Dewar began a systematic charting of the specific resistance of metals, alloys, and nonmetals from the boiling point of water to the lowest point within reach, boiling liquid oxygen at 30 mm Hg, or −197 °C. The actual experiments were performed largely by Fleming and his assistants. The resistances were prepared as wires wound in flat coils on mica formers, and mounted in double-walled flasks containing the cryogenic liquids. The insulating properties of liquid oxygen obviated Arndtsen's old concern for electrical insulation, as well as the problem of obtaining good thermal contact.

The preliminary trend in Dewar and Fleming's data published in 1892, "if plotted as specific resistance vs. absolute temperature as abscissæ,"[41] revealed that: (1) lines of resistance are more or less curved lines that tend downward in such a way as to show that if prolonged beyond −200 °C they would probably pass through or near the origin or absolute zero; (2) the curves divide into three classes, those that are concave upward (e.g., iron, nickel, tin), those that are concave downward (e.g., gold, platinum, silver), or those that are essentially straight (e.g., aluminum); and (3) whereas perfectly pure metals show an enormous decrease in specific resistance at the lowest temperatures, the decrease is extremely sensitive to the smallest impurities. Alloys and impure metals exhibited qualitatively similar departures from this trend, with resistance lines nearly straight and of much reduced slope compared to those for pure metals, or slopes "never ... in such a manner as to indicate that if prolonged they would pass through the absolute zero."[42]

This behavior agrees qualitatively with August Matthiessen's original "rule" of 1860 (then based on alloys at room temperature) "that the increase of resistance due to a small concentration of another metal in solid solution is in general independent of temperature."[43] (The more common version attributed to Matthiessen, that the measured resistivity is the sum of a temperature-independent impurity contribution and an intrinsic temperature-dependent contribution, is actually of more recent origin.) Be that as it may, the essential question now was whether pure metals approach "a *minimum* specific electric resistance in proportion as the absolute zero ... is approached."[44] This exciting possibility was strengthened by the fact that pure nonmetals showed a decrease in resistance with a rise in temperature. Dewar's immediate task, then, was clear: "to complete the examination of the change of conductivity with diminished temperature for all the metals in a state of the greatest chemical purity."[45]

Unfortunately, nearly a decade elapsed before Dewar again dealt with the fundamental issue in question, fully 3 years after liquid hydrogen became available to him, not withstanding his follow-up paper with Fleming in 1893 carrying the measurements, with greater accuracy, to −200 °C.[46] The interruption in this fundamental inquiry was, no doubt, largely the result of Dewar's preoccupation with preparations to liquefy hydrogen. Another distraction was a nasty priority exchange over cryogenic matters in general that erupted between Dewar and Olszewski—a debate that soon focused on resistance thermometry in particular. The dispute began in 1895 with a paper by Olszewski that angrily and unwisely accused Dewar of scientific improprieties.[47] Olszewski argued that the platinum thermometer was first proposed by Cailletet and Collardeau in 1881, a fact ignored by Dewar in print; moreover, "practically" he claimed, "it was used for the first time by Witkowski [a colleague of his] at Cracow in 1891."[48]

In the midst of these distractions another laboratory entered the debate on resistance thermometry when Ludwig Holborn of the Physikalisch-Technische Reichsanstalt at Berlin-Charlottenburg took up the subject in 1895, initially in collaboration with Wilhelm Wien. Wien himself would address the subject again from a different perspective nearly two decades later.

A good summary of the trend in specific resistance with temperature as perceived or as actually measured by the various parties is found in an interesting plot prepared by Callendar in 1899 and reproduced here as Figure 1-4. (Resistance values are in C. G. S. units.) The spurious peak claimed by Lenz and Matthiessen around 250 °C is quite conspicuous—as is the curious curve attributed to Holborn and Dickson! The more reasonable temperature functions for various metals over the much more restricted temperature range of relevance bordering on absolute zero, published by Dewar and Fleming in 1893, are shown in Figure 1-5.

Despite a rash of publications on platinum thermometry from London, Cracow,

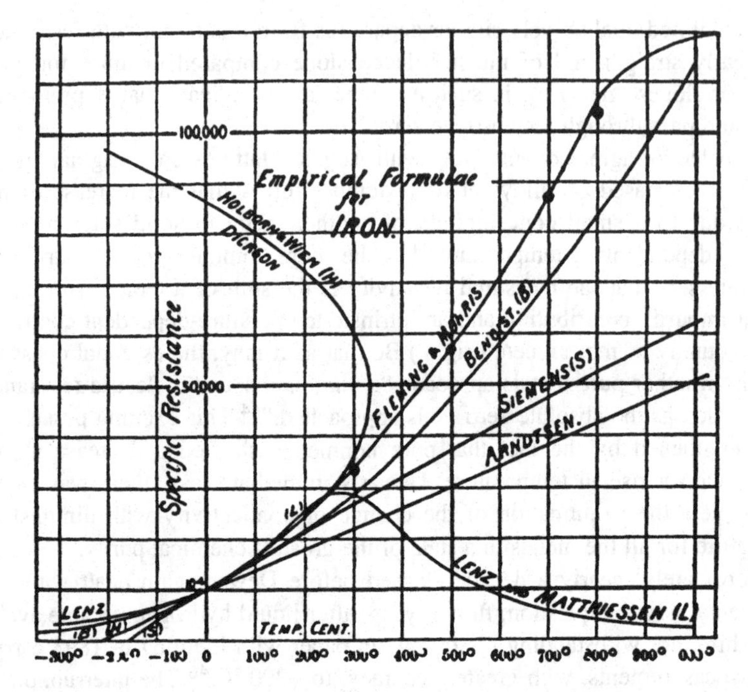

Figure 1-4. Callendar's summary of resistance data by various investigators, starting with those of Emil Lenz. Callendar, PM, 47 (1899), 193.

and Charlottenburg during 1895–1896, interest in the question of the resistance behavior at the lowest temperatures then within reach subsided, with one exception. It was taken up by Callendar, the principal expert on electrical thermometry in Great Britain (and elsewhere), making use of Dewar's multifarious measurements now extending to liquid hydrogen (–240 °C). (Olszewski's measurements in 1895 reached the same temperature through nonstatic, fleeting liquefaction efforts.) Callendar's major paper of 1899,[49] unfortunately published long after he had left the Cavendish, starts with a concise discussion of methods for reducing platinum temperatures (*pt*) to those of the air scale (*t*) with the aid of a parabolic difference formula of the type:

$$t - pt = d\,(t/100 - 1)\,t/100\,, \tag{1-4}$$

in which *d* is a constant which depends on the purity of the platinum sample employed. This formula can be equally expressed in the form

$$t - pt = a\,(pt) + b\,(pt)^2 + c\,(pt)^3 + \ldots$$

with the coefficients *a*, *b*, *c*, and so forth, determined by calibration at various fixed points (e.g., air, steam, sulfur). Eschewing the "widespread tendency among non-mathematical observers to regard with almost superstitious reverence the value of results obtained by the method of least squares," he determines the coefficients by graphical curve-fitting instead. His rationale for favor-

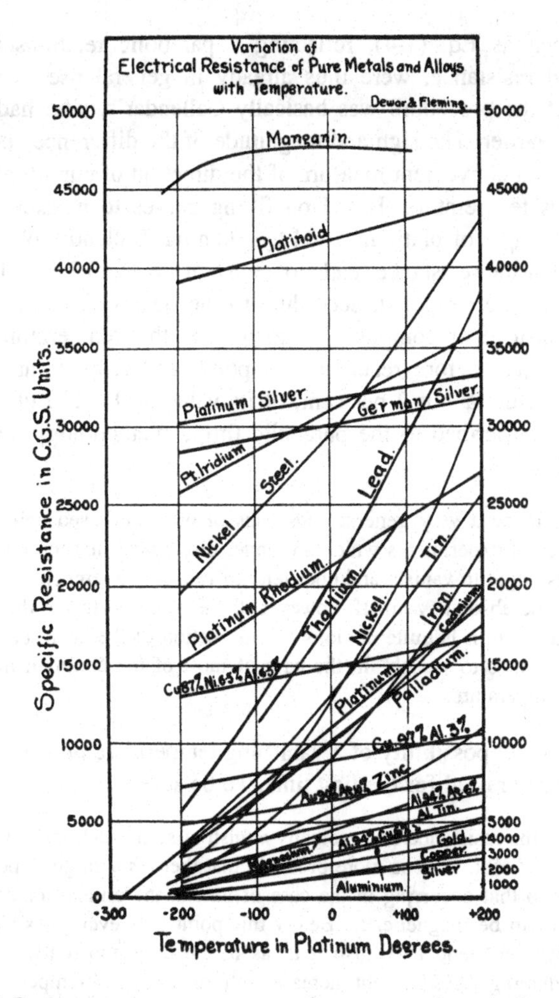

*Figure 1-5. Resistance versus temperature for pure metals and alloys: Dewar and Fleming, 1893. Dewar and Fleming, PM, **36** (1893), Plate I.*

ing the graphic method over that of least squares being that "... it is not easy to decide on the appropriate weights to be attached to the different observations," he offers as a case in point the contemporary analysis of available resistance data by J. D. Hamilton Dickson of Cambridge, fellow practitioner of platinum thermometry. "It is a mistake [warns Callendar], in reducing a series of observations of this kind, to put all the observations, including the fixed points, on the same footing, and then apply the method of least squares, as Mr. Dickson has applied it in his reduction of the results of various observers [including those of Callendar himself] with platinum thermometers. ... In order to make his formula fit my observations ... he is compelled to admit an error of no less than 0°.80 on the fundamental interval itself, which is quite out of the question, the probable error of observations on this interval being of the order of 0°.01 only."[50]

Formulae such as Eq. (1-4), reflecting a parabolic relationship between temperature and resistance, were thus already in general use in the various laboratories, but the procedure was basically Callendar's, who had devised it nearly a decade earlier. The sign and magnitude of the difference coefficient, d, (Dewar uses δ) is a convenient measure of the direction of curvature in plots of resistance versus temperature. Based on fitting curves to measurements on a particular pure sample of platinum used by Fleming, Callendar observes in his paper of 1899 that the resistance tends to vanish at $t = -240.2$ °C. "It seems not unlikely, however [he cautions], according to the observations of Dewar, that the resistance, instead of completely vanishing at this temperature, ceases to diminish rapidly just before reaching this point, and remains at a small but nearly constant value, about 2 per cent, of its value at 0 °C." But the physical significance to be ascribed to the parabolic fitting methodology was to him a troubling point.

> There appears to be a very general consensus of opinion, based chiefly on the particular series of experiments which are under discussion, that the resistance of all pure metals ought to vanish, and does tend to vanish at a temperature which is no other than the absolute zero. If, however, there is any virtue in the parabolic method of reduction, it is quite obvious [on inspection of the enclosed data] that the resistance "tends to vanish" in the case of most of the common metals at a much higher temperature.

Here, then, the novel possibility of a vanishing temperature *above* absolute zero seems to have been raised for the first time. To be sure,

> whether or not the resistance does actually vanish at some such temperature may well be open to doubt. ... It is more probable that there is a singular point of the curve, similar to that occurring in the case of iron at the critical temperature, at which it ceases to be magnetic. ... Below this point it is even possible that the resistance might not tend to vanish, but, as in the case apparently of bismuth [Dewar and Fleming, 1895], might increase with further fall of temperature. It has been suggested that at very low temperatures all metals might become magnetic. It is very probable that the change of electrical structure here indicated would be accompanied by remarkable changes in the magnetic properties.[51]

The question was belatedly raised anew by Dewar himself in his Bakerian lecture in 1901, where he summarized a study of platinum, gold, silver, copper, and iron resistances extending down to the boiling point of hydrogen.[52] The results had been reduced by Callendar's method and, as noted, by a somewhat similar method employed by his associate Dickson. They were ultimately published, with some additional data, in Dewar's seminal paper read before the Royal Society in March of 1904. On that occasion, he began by stressing two salient points:

> The resistance of an unalloyed metal continuously diminishes with temperature and in each case appears to a definite asymptotic value below which no further lowering of the temperature seems to reduce it,

and

the parabolic connecting between temperature and resistance is no longer tenable at very low temperatures.[53]

Moreover, the temperature at which the resistance would vanish (with some exceptions) but as a general rule, rises with the purity. (That is, the temperature coefficient α increases with purity, while the difference coefficient decreases.)

In fact, Dewar continues, "it is ... remarkable that in the cases of all the purest metals examined, their resistances calculated by either method of reduction vanish at temperatures above −273 °C." From these data, Dewar infers "that the curves have taken a more or less quick turn in the neighborhood of the boiling point of hydrogen." While the curves for most metals are concave toward the temperature axis down to the boiling point of oxygen, below that temperature they are convex, reminiscent of the behavior hinted by Fleming and Dewar in 1886, which showed that "the curve connecting the resistance of mercury with temperature, throughout this range, ... was somewhat like the disused old English ʃ." Callendar had been struck by the same tendency in reviewing the older Dewar-Fleming experiments, suggestive of a point of inflection in the case of platinum. At the very least, one conclusion was no longer in doubt: "in no case can anything parabolic connect resistance and temperature ranging from the boiling point of water to that of hydrogen."[54]

At Leiden, too, where the cryogenic facilities were expanding along with the thermodynamic, thermomagnetic, and galvanic investigations,[55] there arose in these years a growing need for electrical thermometers to augment gas thermometers,[56] particularly in anticipation of a vigorous liquid hydrogen program and the growing expectation for liquid helium. The task of implementing such a measurement system was assigned to Onnes's student B. Meilink shortly after his arrival at Leiden in 1900. The program had a twofold objective. The first was to express as accurately as possible the resistance as a function of temperature for some standard metal. Platinum seemed the best candidate for an absolute determination, based on the experience at Cracow, Berlin, and London.[57] Meilink's goal was a new standard of accuracy: 0.03% at the boiling point of oxygen, 0.01% at room temperature. His second objective was to express as accurately as possible the temperature variation in the ratio between the resistance of pure or deliberately impure metals and that of the standard metal. For the express purpose of thermometry, absolute values were judged to be of "less moment" than ratios.

Meilink's first *Communication*, published in 1902[58] describes the preparation of resistances with the usual precautions of Callendar, and techniques for Wheatstone bridge measurements and zero-point determination. However, a full 3 more years elapsed before another progress report finally appeared with actual measurements.[59] Meilink attributes the excessive delay to "one error or another," and the need to gain confidence in the pronounced deviations from the simple parabolic resistance behavior they were observing in subcooled liquid nitrogen. So far this merely confirmed the onset of behavior which, "as follows from Dewar's experiments, appears so strongly at the temperature of liquid hydrogen."[60] Still, even in liquid oxygen the quadratic representation was grossly inadequate for the accuracy that was demanded. At lower temperatures, the shortcomings of the platinum thermometer were such as to require an altogether different approach. Concerning resistance *ratios*, in particular gold to platinum, Meilink, on Onnes's prodding, devised a method for measuring ratios directly using the relatively new differential galvanometer, sidestepping the cumbersome normal procedure requiring separate determinations.[61] In so doing, Meilink again confirmed what Dewar already knew, that gold exhibits a temperature function of considerably less curvature than platinum. Since gold can also be obtained in purer form, it seemed a more promising wire at the lowest temperatures.

Two more years passed before further progress was reported. When it finally was, authorship of the *Communication* had passed to Onnes himself and a new student, Jacob Clay.[62] The reason Onnes attached heightened importance to these measurements at this particular juncture appears to be tied to his singular preoccupation with "corresponding states" at that time, as we shall see. Onnes was also keenly aware of the widening search in laboratories everywhere for the ideal resistance standard. Meanwhile, he and Clay noted, the English team of Travers and Gwyer had also joined the effort to standardize platinum thermometers,[63] but they had little of substance to contribute, as they lacked the formidable cryostat technology now in place at Leiden.

This time the delay in the Leiden program was deliberate, however. Hydrogen had been liquefied at Leiden in June of 1905, albeit no more than 15 cm^3. But as a result of Onnes's grave decision to delay exploiting his success and proceed immediately with the construction of a much larger liquefier, liquid hydrogen in baths of the size of a liter became available the following spring, making temperatures in the range 252 to 259 °C possible.

In spite of this important technological milestone at Leiden, no definitive progress was made on the resistance behavior itself. Even the question of a *point of inflection*, raised by Dewar, had proved difficult to resolve. And the more tantalizing possibility of a *minimum* at very low temperatures, with the resistance increasing at still lower temperatures and becoming infinite at absolute zero, remained entirely speculative. About the only conclusion to be drawn was that even a cubic representation of resistance was quite inadequate in the new temperature region. For a lack of better ideas, Onnes and Clay resorted to more complicated formulae involving reciprocal powers of the absolute temperature—no less than ad hoc representations, conveniently and unjustifiably forcing infinite resistance at $T = 0$.

1.3 The Electron Theory

While the experimental investigation of electrical resistance at low temperatures shifted from the Continent to England and by 1900 was dominated by Dewar and his associates, the theoretical development of the subject remained a German monopoly. That subfield began with Wilhelm Weber.

The "atomistic" nature of electricity seems to have been first proposed by

Weber in 1846 during his relatively brief tenure at Leipzig, where he assumed the position vacated at the onset of Gustav Theodor Fechner's long neurotic illness. Fechner, who first adumbrated Weber's law of electrical force in a semi-quantitative treatment in 1846, was an ardent atomist, and he and Weber remained very close. Weber expounded on the similarity of metallic and electrolytic conduction, despite the latter being accompanied by chemical activity, and a theory of electrical conduction in metals occupied him off and on from the time of his return to Göttingen in 1849 until his retirement in the 1870s. Although he is best remembered for his famous law, Weber's lasting contribution lay in stimulating the development of a model for the transport of electricity, particularly by motivating his student and successor at Göttingen, Eduard Riecke.

Weber's theory, elaborated in his major article of 1875[64] and revisited in an unpublished, final manuscript,[65] treats a metal as a lattice of fixed positive charges surrounded by negative particles rotating in elliptical orbits. The system is inherently unstable; in the presence of an electrical potential, the negative particles move in ever widening spirals until they are captured by neighboring atoms. This process is repeated, and the particles propagate randomly between ponderable molecules of the metal. Weber also identified the *vis viva* of the moving electrical particles with the conduction of heat, and with thermoelectricity. His attempts to formulate these particulate views quantitatively in terms of molecular properties were not very successful, however. They were jolted by the frenzy of *Naturphilosophie* in the 1860s and 1870s, and by the view that rays in the ubiquitous discharge tubes in use since Faraday's time are an aether propagation independent of ponderable matter—a view championed by Gustav Hertz and in tune with the Maxwellian concept of displacement. Weber's particulate views on electricity were exonerated in his last decade, largely on the authority of Hermann von Helmholtz. In spite of a prolonged and rather bitter feud between the Weber and Helmholtz camps over criteria for electrodynamic forces (whether velocity-dependent forces are conservative), Helmholtz himself gradually came to accept "atoms of electricity."[66]

Weber's model of electrical conduction was extended by Wilhelm Giese of Berlin. Giese applied the increasingly popular analogy of gaseous conduction to electrolysis in order to account for the conductivity of the hot gases of flames; later he turned to metallic conduction as well. He argued that conductivity must be associated with the *ions* of a metal, not with charged molecules,[67] which originate when the molecules dissociate by collision. Since, moreover, in a metal there is no evident net transport of mass, electrical conduction, as in Weber's model for example, is a transitory process with the individual ions only traveling in very short paths. Somewhat similar ideas were expounded by Arthur Schuster of Manchester, a protagonist of J. J. Thomson and one whom Giese bitterly accused of having stolen his thunder; Schuster declared that "one molecule cannot communicate electricity in an encounter in which both molecules remain intact."[68] Schuster, incidentally, had been a fellow student with Onnes during his *Wanderjahr* under Bunsen and Kirchhoff at Heidelberg, and had worked with Weber at Göttingen and in Helmholtz's laboratory in

Berlin, as well. Like J. J. Thomson's, Schuster's research centered on the discharge of electricity through gases but, with his Continental background, his approach was predicated on the particulate theory advocated by Helmholtz and the German school in general, while Thomson's was based on the Maxwellian aether dynamics.[69]

Both Giese (by acknowledgment) and no doubt Schuster and nearly everybody else, were influenced by the increasingly accepted hypothesis of ionic dissociation propounded by the Swedish physicist Svante Arrhenius, the Dutchman J. H. Van't Hoff, and others in line when it became fashionable to view the space between the neutral molecules or atoms of a metal as filled with electric charges, discharged by collisions. But it was Weber's student Eduard Riecke who constructed what proved to be the forerunner of a sounder model of metallic conduction, in terms of charged ions drifting between molecules bound in a lattice. He thereby laid the definitive groundwork for the classical electron theory of metals—a subject to be completed by Paul Drude, another figure from the Göttingen school. The versatile Riecke inherited both his mentor's experimental and theoretical program. On the experimental side, among other things, his work with Geissler tubes (a particular variety of discharge tubes mass produced by Heinrich Geissler of Bonn) helped to establish the identity of negatively charged particles emanating from the cathode, particles which, he argued, are none other than the free negative particles in a metal. Although his theoretical contributions to electrolytic and metallic conduction were of more lasting importance and mainly of interest in our present chronicle, a short digression on the intense experimentation on cathode rays taking place in Riecke's time would be worthwhile.[70] Strictly speaking, the simultaneous experimental work on electrolytic dissociation and conduction in liquids deserves equal notice (we have referred to it already and will have occasion to do so again), but pursuing this would carry this narrative too far astray.

Be that as it may, we will start with Julius Plücker who, in 1859, accidentally (it seems) discovered that the fluorescence seen in a Geissler tube (highly evacuated for that time with the mercury pump—Geissler's invention as well) is influenced by a magnetic field, hinting of a current of electricity emanating from the cathode and striking the wall of the tube. Further experiments by Johann Wilhelm Hittorf (one of Plücker's pupils) and Eugen Goldstein confirmed that the *Kathodenstrahlen* (Goldstein's term) proceed from the cathode in rectilinear paths, similar to light rays in their shadow-casting effects. In 1879, Sir William Crookes published the results of an extensive investigation of cathode radiation, on the basis of which he concluded that the rays possess mass (and thus are not aether waves as maintained by some) and electric charge. Unfortunately, Crookes had rather antiquated ideas about the nature of cathode rays ("negatively charged atoms"). A further complication was the prevailing view of the current flow in a Geissler tube that linked a flow of negative electricity from the cathode toward the anode with an equal and simultaneous positive flow in the opposite direction. The positive rays were nowhere to be found, until Goldstein had the inspiration of piercing holes in the cathode, thereby revealing a positive current component—his *Kanalstrahlen*.

Further work by Heinrich Hertz (1892) indicated that the cathode rays were much smaller than Crookes's negative atoms. This was soon confirmed by his pupil Philipp Lenard in an experiment in which the cathode rays were passed through a thin window of aluminum foil and a considerable distance in the external air—something quite impossible for even the smallest atoms according to the prevailing dynamic theory of gases. Any lingering doubt as to whether the rays were indeed material particles was dispelled in 1895 when Jean Perrin demonstrated their negatively charged particulate nature by collecting them in a "Faraday cup" (open-ended metal cylinder). Two years later, and coinciding with Zeeman's discovery in Leiden, J. J. Thomson nailed down the electron in a series of experiments that subjected the cathode rays not only to magnetic deflection but also, for the first time, to electric deflection. (This was now possible with a vacuum in his cathode tube that was good enough to maintain a static electric field by eliminating the neutralizing action of the ions.) This combination allowed Thomson to determine both the ratio of charge to mass, e/m, and the velocity of the particles. The ratio was ~1000 times greater than for hydrogen ions, something he attributed solely to the smallness of m, and it was in close agreement with Zeeman's value. Not only that; the value was the same regardless of the composition of the cathode—a fact that Riecke confirmed independently in 1899. The nature of the positive canal rays, it must be added, was not sorted out until well into the second decade of the twentieth century.

Eduard Riecke's first major theoretical contribution to the subject was published in 1898.[71] In this paper, he derives rather elaborate formulae for both the thermal and electrical conductivity of metals in terms of the number of carriers discharged by the metallic atoms, their charge, mass, velocity, and mean-free path, and for ionic velocities; he utilized these formulae to investigate both thermoelectric and thermomagnetic effects. Unfortunately, he furnished no numerical estimates for the various physical parameters involved and, consequently, produced no quantitative predictions.

Whereas Riecke's treatment of metallic conduction was a direct extension of Weber's electrodynamics, Drude was a pupil of the magneto- and electro-optical school of Franz Neuman and Wilhelm Voigt. Drude completed his dissertation in physical optics under Voigt at Göttingen, but his classic papers on electron theory,[72] all dated 1900, were written at Leipzig. Even at this late stage Drude, like Riecke, found it necessary to assume the simultaneous presence of positive and negative free particles in a metal (partly by analogy with the coexistence of cathode rays and canal rays in Geissler tubes), though Drude purposely used the term *electron*, a term also introduced in 1900.[73] He reserved the term ion "for the aggregate electrical particles and ponderable masses encountered in electrolytes."[74] But his major advance over Riecke lay in explicitly adopting the viewpoint of the kinetic theory of gases. Drude assumed the electrons of the "gas" move freely and are in thermal equilibrium with their surroundings through the collisional exchange of energy with the atoms; thus he ignored mutual collisions between the electrons. This view implied a kinetic energy of the electrons equal to that of a gas molecule at the same temperature, or proportional to the absolute temperature. (Riecke was not far behind; he too assumed that the mean kinetic energy of the free charge carriers is proportional to the temperature but he again fell prey to a flawed tendency of his: he failed to derive the constant of proportionality.)

Drude's expression for electrical conductivity is a minor watershed in nineteenth-century physics. Assuming that the electron drift velocity u under the influence of the electrical field E is much less than the average electron velocity v of thermal motion (the gas velocity) and that the electron loses all of its original momentum in colliding with a metal molecule, his derivation went as follows (ignoring his cumbersome subscripts denoting electrons of opposite polarity): The acceleration of the electron, of charge e and mass m, is given by $m\,du/dt = eE$. If τ is the average time between successive collisions, we may write:

$$u = \frac{eE}{2m}\tau .$$

The current density is $J = neu = nEe^2\tau/2m$, with n the number of free electrons in unit volume. The time can be replaced by λ/v, where λ(Drude uses l) is the mean-free path, to give Ohm's law for the conductivity:

$$\sigma = \frac{J}{E} = \frac{e^2 n\lambda}{2mv} . \qquad (1\text{–}5)$$

Now comes the essential point of the argument: equating the kinetic energy of the electron to the mean translational kinetic energy of a gas molecule at the same temperature, or $mv^2/2 = 3kT/2$ (Drude and his contemporaries use Loschmidt's number α for $3k/2$). On substitution for v in Eq. (1-5),

$$\sigma = \frac{1}{6}\frac{e^2 n\lambda v}{kT} . \qquad (1\text{–}6)$$

This is Drude's celebrated expression for σ. It differs slightly in the numerical constant (being larger by a factor of 3/4) from the expression published post-haste by Riecke,[75] when he too adopted Drude's assumption for the kinetic energy. The difference stems from slight differences in kinematic treatment.[76]

A temperature gradient in a metal will also cause an electron current to flow. Drude's expression for the coefficient of thermal conductivity is:

$$K = \frac{nv\lambda k}{2} , \qquad (1\text{–}7)$$

making the ratio between the thermal and electric conductivities the same for all metals at the same temperature, in agreement with the old rule of Wiedemann and Franz,[77] or

$$\frac{K}{\sigma} = 3\left(\frac{k}{e}\right)^2 T . \qquad (1\text{–}8)$$

Again, Riecke's and Drude's expressions differ slightly in the numerical constants; Drude's value for the constant is 8/9 times Riecke's value.

The agreement with Wiedemann and Franz, in Drude's opinion, suggests that only a single type of free particle is operative, or mainly operative, in a metal. More than that, the agreement constituted a major success of the electron theory at this stage, as Drude was quick to point out—albeit its only success. Max Reinganum of Leiden immediately noted that although the physical constants in Eq. (1-8) (or α and the electron charge) were then poorly known, their *ratio* could be determined with considerable accuracy from two other known constants.[78] These are the number of molecules of a gas (e.g., hydrogen) per unit volume at a certain temperature and pressure, and the amount of electricity required to liberate a unit quantity of hydrogen at this temperature and pressure in an electrolyte (the so-called electrochemical equivalent of hydrogen). Reinganum's ratio agreed remarkably well in order of magnitude with measurements already made by Jäger and Diesselhorst.[79] This agreement is, in fact, fortuitous, due to an inconsistent definition of mean-free path in Drude's calculation of σ and K.[80] Curiously, the agreement with the results of H. A. Lorentz's more rigorous calculation made in 1905 is worse.[81]

Lorentz's model assumes, for the first time, a single species of free (negative) electrons, though troubling aspects of the Hall-effect still did not firmly rule out positive free electrons as well.[82] More important, though, Lorentz treated the velocity of the electrons as a *statistical* quantity in accordance with Maxwell's velocity distribution (in fairness, a refinement deliberately neglected by Drude). His value for the numerical coefficient for K/σ, defined in terms of k, is 2, instead of Drude's value 3 (Riecke's value was 27/8), yielding an agreement "not quite satisfactory," though close enough to give reassurance of the "sound basis" of Drude's theory.[83] The poor agreement must have irked Lorentz—it was an ominous hint of fundamental flaws in the basic theory. Still, the capital conclusion that K/σ does not depend on the nature of the metal and varies with the absolute temperature could only mean confirmation of another essential assumption in the underlying model: a preponderance of electron collisions with the metallic atoms compared with collisions between electrons. Mutual collisions between electrons would diminish K, but not affect σ, since the motion of the center-of-mass is unchanged in such collisions.[84]

Lorentz's contributions were conceptually important for their statistical treatment of the kinetic theory of gases; in practice his results differed little from Drude's. In particular, they retained the basic weakness of the electron gas theory in general, by failing to provide a basis for calculating a priori the values of n and λ as a function of temperature.

Sidestepping the questions of the detailed shape of the curve of resistance versus temperature, the simplest of analytic expressions for the dependence had, at the very least, to reflect its general trend: an approximately linear dependence of resistance with increasing temperature, at least upward from the boiling point of hydrogen. Thus, the product of $n\lambda v$ in Eq. (1-6) must be independent of temperature. However, by assuming $v \propto \sqrt{T}$, the free-electron theory requires $n\lambda \propto 1/\sqrt{T}$.

Returning to the problem in terms of Lorentz's "unitary" formulation in 1909, Riecke suggested that n is independent of temperature[85]—an assumption that he readily admitted was dubious at best. If so, λ must be proportional to $1/\sqrt{T}$. This latter conclusion was already implicit in his paper of 1898, but he now justified it by a calculational model based on closely packed, cubically arranged atoms at the melting point. Lorentz had reached the same shaky conclusion in 1905, but to his chagrin was "unable to explain why n should vary in this way"[86] (i.e., why n should *not* vary with temperature to ensure $\lambda \propto 1/\sqrt{T}$).

A method for estimating n directly from optical measurements had been proposed by Drude in his original paper on the subject.[87] It was based on the fact that in the reflection of radiation of long wavelength by metals, the portion of the radiation absorbed depends on the conductivity of the metal. The idea was followed up in 1904 by Arthur Schuster.[88] Schuster's analysis of available data indicated that the number of free electrons in a metal is of the same order as the number of atoms in the same volume; this would imply that the total number of electrons in a metal is constant. Yet, it seemed much more plausible that $n\lambda$ should increase with temperature due to increasing molecular dissociations and more space for the freely moving electrons. In particular, the Thomson effect, the flow of heat caused by an electric current in a nonuniformly heated conductor, requires that n vary as $T^{3/2}$ or possibly as $T^{1/2}$.[89] This would call for a temperature dependence of λ no faster than $1/T^2$—still too slow to account for the ongoing resistance measurements.

Alternatively, attempts were made by Johann Koeningsberger and his co-workers to treat conductivity as a dissociation equilibrium between free and bound electrons,[90] assuming on very general thermodynamic principles that

$$n = n_0\, e^{-Q/RT}\,. \qquad (1\text{-}9)$$

Here n_0 is the number of bound electrons within the metallic atom, and Q is a heat of dissociation (work required to remove the electron). But to obtain a quantitative fit to the low-temperature data Koeningsberger, too, was forced to make additional, quite ad hoc assumptions about the temperature variation of λ, invoking some kind of complicated temperature function, either exponential or perhaps quadratic as in Eq. (1-2). In the event, fresh insight into the whole problem came from another direction, with the resolution of a much more serious difficulty confronting the classic electron theory: the specific heats of metals.

The upshot was that the electron theory, despite all of its variations, refinements, and specific assumptions, proved unable to account even qualitatively for the resistance data becoming available at the lowest attainable temperatures. Nor did it matter very much, since, in any case, the Leiden resistance program was increasingly driven by an altogether different notion: the possibility of a *minimum* in the curve of resistance versus temperature for sufficiently pure metals. The idea of electrons condensing as T approaches zero followed naturally from Onnes's favorite notion of electrons, not as an ideal "gas," but as a substance characterized by an equation of state. Perhaps this explains why he now took a personal hand in the investigation. In fact, the concept of the density of the free electrons diminishing by cooling, approaching zero at absolute zero with the electrons condensing onto the atoms, was not Onnes's alone. It was first promulgated largely on the authority of the omnipotent Lord Kelvin, whose opinion counted. Kelvin introduced his views on this in the 1901 jubilee volume to Professor Bosscha.[91] Bosscha had, after all, figured prominently in Onnes's development in the years immediately preceding his appointment at Leiden, and the volume must have ranked high among Onnes's favorite private documents.

A solid substance, according to Kelvin, should be a perfect electric insulator at $T = 0$, under the influence of electric forces sufficiently moderate not to "pluck electrions out of the atoms."[92] (*Electrion* was Kelvin's favorite term for atoms of resinous electricity; entities much smaller than atoms of ponderable matter and permeating all of space.) As the temperature rises, some of the electrions are "shot" out of the atoms. If an electric field is present, the body acquires some degree of electrical conductivity, the greater the higher the temperature. Such behavior is exhibited by "bad" conductors, for example, glass, or guttapercha. Metals, however, are so crowded with electrions that some are "spilt out" by thermal agitation even as low as 16° above absolute zero, as evidenced by the high electrical conductivity Dewar reported at this low temperature in his Bakerian lecture. A metal, like a nonmetal, should still be a near-perfect insulator below, for example, 1° of absolute temperature, but somewhere between 2° and 16° the conductivity ought to exhibit a maximum. *Some* thermal motion is necessary to sustain conductivity, but *too much* thermal motion "must mar the freedom with which an electrion can thread its way through the crowds of atoms to perform the function of electric conduction."[93]

Onnes voiced similar views in his *Rector Magnificus* address on the occasion of the 329th anniversary of the University of Leiden in 1904, entitled "The importance of accurate measurements at very low temperatures."

> We may account for [the supposition that the resistance of metals near absolute zero should increase infinitely] by observing the actions which the dynamids [the electrical dipoles that constitute atoms in Lenard's model] exercise on the electrons, especially on those which are deprived of their satellites. It seems as if the vapour of electrons which fills the space of the metal at a low temperature condenses more and more on the atoms. Accordingly, the conductivity, as Kelvin has first expressed it, will at a very low temperature reach a maximum and then diminish again till absolute zero is reached, at which point a metal would not conduct at all, any more than glass. The temperature of the maxima of conductivity probably [lies] some times lower than that of liquid hydrogen. At a much lower temperature still, there would not be any free electrons left, the electricity would be congealed, as it were, in the metal. As yet it does not seem possible to attain the temperature at which conductivity reaches a maximum, but the very great importance attached to observations of the conductivity of metals at extremely low temperatures—where also the investigation of the influence of small admixtures on the conductivity play[s] a part—urges on to numerous researches which may be said to lead to the equation of state of the electrons.[94]

These ideas received quantitative support in 1908 from analogous observations on the absorption spectra of the compounds of the rare earths in the liquid hydrogen temperature region, observations reported jointly by Onnes and Jean Becquerel.[95] (Becquerel was the son of Henri Becquerel and, with his father, a frequent guest collaborator at Leiden.) On cooling to liquid air temperatures, most spectral bands become narrower and increase in intensity—a signature, according to the electron theory, of an increase in the number of electrons determining the band. But at still lower temperatures, several bands, including those of certain crystals and sulfides, pass through a maximum intensity with decreasing temperature, then virtually disappear, again suggestive of diminished numbers of partaking electrons.

> An investigation into the connection between what we already know about those functions [for the temperature dependence of the number of electrons which belong to a certain band] and what the change of the electrical resistance of the metals leads us to expect about the action of forces exercised by the ponderable substance on the electrons naturally suggests itself. At very low temperatures we shall no longer be justified in considering the electrons as a perfect gas, but we shall rather have to compare them to a vapour which precipitates on parts of the atoms (dynamides [Lenard]), and solidifies at still lower temperature. When we approach these centres the paths of the electrons are subjected to changes which modify the free length of path in the same way as van der Waals' quantity *b* is subjected to a change by the forces exerted by the molecules on each other [calculated by Reinganum].[96]

Onnes and Becquerel distinguish "three states of aggregation" corresponding to distinct stages of stability of the electronic path length. Adding that it is the *ratio* of absolute temperatures on which the degree of change of the spectra depends, they suggest "as a heuristic image the idea that we may speak of corresponding states according to different units of temperature caused by mechanic similarity of the motion of the electrons round the centres."[97]

The idea of a minimum resistance was central to Koeningsberger's treatment of metallic conduction as a result of recombination of electrons previously freed by dissociation from the atom. Koeningsberger based his analysis partly on the low-temperature measurements of Dewar, Meilink, and others, and partly on the observational evidence that many metals have a negative temperature coefficient at low temperatures and even exhibit *minima* in their resistance curves at rather easily attainable temperatures. These materials include certain metallic oxides and sulfides, or today's semiconductors. However, even though Koeningsberger showed that the transport of electricity in them takes place solely by electrons, not by ions, the conduction mechanism involved here was gradually construed by most physicists to be very different from that operative in pure metals. Koeningsberger's work, while vaguely perceived as supporting the Kelvin notion at the time, played no essential role in the low-temperature developments of interest here.

Meanwhile, in the late spring of 1906 Onnes and Clay had resumed the resistance measurements last reported by Meilink two years earlier. A progress report, with the measurements now extending to the melting point of hydrogen—essentially Dewar's limit in 1901—was presented by Onnes at Amsterdam in June.[98] Comparison of the resistance of gold and platinum now showed that the point of inflection was lower for gold (–220°) than for platinum, and that the elusive minimum of resistance "seems not to be far off at –259°."[99] They noted a quantitative disagreement with Dewar's result for gold at the lowest temperature that was reported in his Bakerian lecture (Dewar's value was 25% lower) but offered no opinion for the discrepancy. And they quite ignored Dewar's more complete paper of 1904.

A definitive report came one year later, in June of 1907. It was preceded by a brief, separate summary of further platinum sample calibrations,[100] with the sensitivity of resistance to purity and metallurgical history also thrown in. The full report constituted Onnes and Clay's last major *Communication* of results in the hydrogen temperature region. Rather than viewing the investigation merely as a calibration of resistances, they now "preferred to consider [it] as part of our more general investigation on the change of resistance with temperature."[101] Although the report emphasized gold and the effects of impurity admixtures, deferring quantitative questions associated with wire drawing, it also covered silver, bismuth, lead, and mercury—with samples purer than Dewar's. The measurements now reached –262 °C, the temperature of solid hydrogen evaporating under 2.5 mm. Among these are the first measurements on mercury at hydrogen temperatures. Mercury was chosen not so much for purity as for its tendency to obviate the influence of drawing and ensuring a uniform distribution of admixtures.[102]

The authors now spoke of a temperature corresponding to the point of *inflection* ($d^2r/dT^2 = 0$), to the point of *proportionality* ($dr/dT = r/T$), and to the point of *minimum* resistance ($dr/dT = 0$). Only rarely was the point of inflection actually observed, and for these samples the purity was open to doubt; nevertheless, the rule seemed to be: the purer the sample, the lower the inflection point. (Evidently Dewar had passed the inflection point for some of his wires, but their purity was also in doubt.) As for the point of proportionality, Onnes and Clay speculated that it might be tied to the melting point; if so, a good candidate might be osmium. And, "until we have reached the proportional point, we need not discuss the question of the minimum point."[103] Indeed, if we judge by Onnes's stronger statement made in 1910, he had already become uneasy with the whole concept of a minimum as applied to perfectly *pure* metals, "when the latest experimental results obtained with extremely pure gold showed that the point of proportionality would always have to be sought at still lower temperatures."[104] Worse, the point appeared to lie lower in temperature, the purer the metal. Although by now Onnes may have entertained growing doubts as to the proposition of electrons condensing onto atoms, the experimental question of a minimum was left unresolved. Significantly lower temperatures were simply unattainable in hydrogen, even under reduced pressure.

CHAPTER 2

Unexpected Guidance

2.1 Into a New Domain

Two and a half years elapsed between Onnes's liquefaction of helium in the summer of 1908 and his first report on resistance measurements in the exciting new temperature domain that opened up as a result of this milestone achievement. Naturally, first priority went to determining the properties of liquid helium. The delay was characteristic of Onnes's methodical preparation for exploiting new scientific opportunities. And this time the new domain was his alone.

The very day helium was seen as a liquid, Onnes's team attempted to solidify it by evaporation under reduced pressure. Although an extemporized arrangement of powerful vacuum pumps ensured "that the pressure could decrease even to one centimetre, the helium still [did not cease] to be a liquid."[1] In 1909 they tried again. Even though the liquid still refused to solidify with the vapor pressure over the liquid reduced to 2 mm, the attempt yielded one positive result: a minimum temperature of 1.38 °K. The next year a further lowering of the vapor pressure by a factor of 10 (to 0.2 mm, corresponding to 1.15 °K, later reevaluated as 1.04 °K) again failed to achieve the primary goal. To be sure, "with regard to the obtaining of solid helium [Onnes reflected in retrospect], this result was disappointing. However, it appeared from this result that the region of temperatures in which the properties of substances could be investigated by means of baths of liquid helium extended further than might have been hoped for ..."[2]

With the nadir of temperature apparently reached, Onnes decided the time had come to press on with a program of systematic physical measurements in liquid helium. In particular, he felt that extending the old measurements with Clay held promise for finally resolving the lingering controversy over the resistance behavior near absolute zero. First, however, some further improvements in the cryogenic apparatus were necessary. Most urgently needed was a method for transferring the liquid helium from the liquefier proper to a cryostat still physically attached but with more room for measuring apparatus. An attempt at such a transfer in 1909 was a qualified success at best, achieved more by accident than by design. A substantial overhaul of this end of the liquefier was in the planning stage, however. In the meantime, Onnes went ahead with preliminary measurements, altering the liquefier as little as possible in other respects. The only significant modification made was enlarging the liquid helium volume somewhat. The earlier cryostat had been blocked at the top by a regeneration spiral that left little room for instrumentation other than a small helium thermometer.

The new resistance measurements began in the late fall of 1910, and Onnes presented the first results at meetings of the Amsterdam Academy on 24 December and again on 25 February in 1911. He was the sole author of the printed report,[3] but, as was his custom, he did not fail to acknowledge the assistance of Cornelis Dorsmann and Gilles Holst—Dorsmann for general assistance during the whole investigation, and Holst specifically for the resistance determinations with the Wheatstone bridge. Holst, an enigmatic participant in the experiments now under way, had recently returned to Holland from Zürich's prestigious Eidgenössische Technische Hochschule (ETH) with an intermediate degree in physics and mathematics. He joined the Leiden team on a temporary basis with the objective of conducting an experimental investigation under Onnes's guidance in support of a doctoral thesis at ETH. To get him started, Onnes suggested he lend a hand to the resistance measurements; his thesis work could afford to wait.

Credit for the liquefier and experimental apparatus again belonged mainly to Flim, with assistance from O. Kesselring. Kesselring was *Glasbläsermeister* of the laboratory—a master craftsman Onnes imported from Jena, the burgeoning seat of optical technology. Their liquefier-cryostat is depicted on the left of Figure 2-1. As noted, it differed little from the 1908 arrangement except for a somewhat roomier cryostat. The letters in the figure are mostly those of Plate III of *Communication 108* announcing liquid helium. The helium gas delivered from the compressor was cooled in stages by routing it through the following successive circuits: (1) through a tube (in upper part of far-left dewar) which at its lower end was cooled by vapor of liquid air where the helium was dried, (2) through a tube divided in two parts where it was first cooled by hydrogen abducted (pumped off) as a vapor (D_a) and then by abducted helium (D_b), (3) through a liquid air-cooled charcoal air trap (lower part of far-left dewar), (4) through refrigerating tubes cooled by liquid air and evaporated liquid hydrogen evaporating under reduced pressure (further down, main dewar). Here the compressed helium was cooled to 15 °K. Next, it passed through a regenerating coil (A), wound fourfold in a Hampson spiral (which still interfered somewhat with the measuring apparatus in the modified cryostat), and the gas was finally expanded through a Joule–Thomson cock (M_1). The liquid was collected in the inner silvered-glass dewar of the multiple-walled cryostat (E_A) only partially

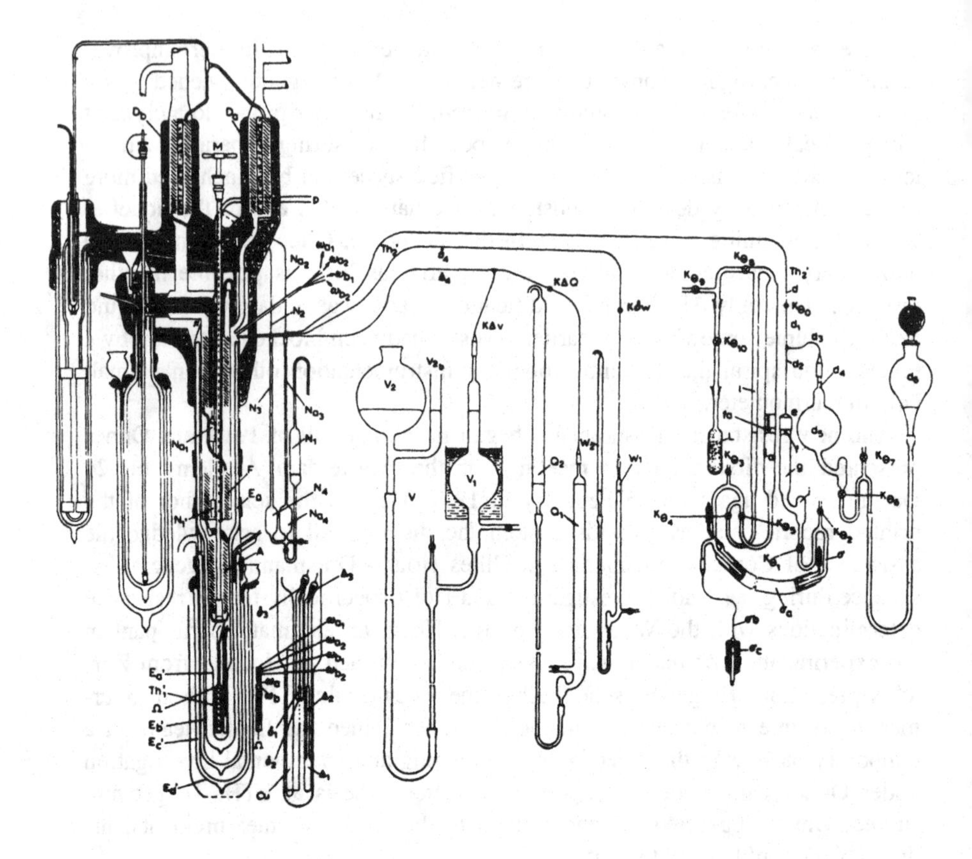

Figure 2-1. Liquefier-cryostat, Leiden, circa 1910. Kamerlingh Onnes, RN, 13 (1911), Plate I.

shown at the bottom of Figure 2-1. The innermost vessel, also depicted to a larger scale, was thermally insulated, in turn, by liquid hydrogen, hydrogen gas, liquid air, and alcohol. In addition to serving as the liquid helium reservoir, its lower part—the cryostat space—contained a helium thermometer for determining the bath temperature, the resistance (denoted by Ω), a dilatometer (Δ) for determining the mass of helium volumetrically, and the control apparatus for the dilatometer (δ). Since stirring the liquid was not possible, a thick copper rod (Cu) conducted heat away to insure that the bath temperature remained as constant as possible. The wall of the cryostat was pierced by three capillaries soldered to the new silver wall of the liquid air vessel: one leading to the thermometer, the second to the dilatometer, and the third to the control apparatus of the dilatometer. In addition, four insulated wires passing through the wall, enclosed in new silver tubes, led the current to and from the resistance.

"Practice in the various operations and the improvement made by introducing the second helium thermometer [indicating the quantity of liquid hydrogen present, replacing earlier thermocouples] rendered it possible to save a fair amount of liquid hydrogen so that, as a part of the necessary hydrogen had been liquefied the previous day, it was possible to begin at half past seven in

TABLE 2-1 Resistance of platinum wire, Pt_B

T	W_t/W_0
273.09 °K	1.0
20.2	0.0171
14.2	0.0135
4.3	0.0119
2.3	0.0119
1.5	0.0119

the morning and have the cryostat part of the apparatus full of liquid helium by a quarter to two in the afternoon."[4] Once it was going, the liquefier produced about 0.28 liters of liquid per hour.

The vapor pressure, a measure of the temperature at the surface of the liquid in the cryostat, could be regulated quite accurately by auxiliary apparatus not shown in Figure 2-1. The bath temperature, however, was less assured, since stirring was not possible. The arrangement allowed a maximum vapor-pressure gradient corresponding, at the lowest temperatures, to a temperature difference of 0.06 degree—an uncertainty not greater than uncertainties arising from other causes. The helium thermometer, a conventional constant-volume gas thermometer, was accurate to about 0.1 degree. Its limit was set by the low pressure at helium temperatures (typically 1 mm), and by the lack of corrections for deviations from the perfect gas laws (pending further experiments). The apparatus at the far right in Figure 2-1 was part of the helium thermometer, used for adjusting the constant volume and for reading the pressure by a double mercury manometer. The apparatus in the center belonged to the dilatometer; its principal components were a small reservoir Δ_1 in the helium bath, the bulb V_1 used for determining the mass of helium present in the reservoir, and a volumenometer Q for density determinations.[5]

The resistance Ω was a fine platinum wire, denoted technically by sample designation Pt_B. It had been previously calibrated at hydrogen and higher temperatures against a resistance standard Pt_I established by Onnes and Clay in 1907.[6] The wire was wound tightly on a glass cylinder while hot (not a recommended procedure, since it strains the wire by its thermal contraction during cooldown), and the thicker platinum ends were fused to the glass. To these ends, double platinum leads were tin-soldered (Wa_1, Wa_2, Wb_1, Wb_2, near the top of the cryostat).

The resistance values were measured with a Wheatstone bridge. The results for Pt_B are listed in Table 2-1 as ratios of W_t, the resistance at the temperature of the observation, to W_0, the resistance at 0 °C (assumed to be 273.09 °K) versus absolute temperature.

Far from approaching zero at absolute zero, much less rising after reaching a minimum, the resistance reached a constant but finite value extending from somewhere above or near the boiling point of helium down to at least 1.5 °K.

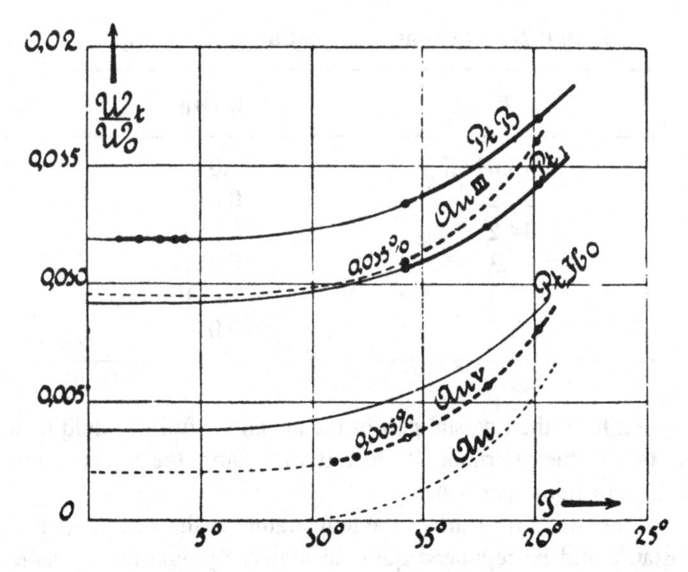

Figure 2-2. Resistance versus temperature for platinum and gold. Lightly drawn curves are extrapolations of the data. Kamerlingh Onnes, RN, 13 (1911), Plate III.

The results are also plotted as the curve Pt_B in Figure 2-2, "which shows well the asymptotical approach of the resistance to a constant value at 4.3 °K."[7]

The influence of inevitable impurities in the platinum specimen had to be evaluated before much significance could be attached to the results. In view of the close resemblance of the deviations of the platinum resistances from a linear function of temperature to those observed in 1907 for gold by Onnes and Clay,[8] Onnes plotted results from the earlier measurements on several gold samples of different degrees of purity (Au_{III} and Au_V) as well.[9] The estimated percentage of impurities in the samples, furnished by C. Hoitsema, Master of the Royal Mint in Utrecht, are indicated in the figure. Curve Au_V, the measured points of which extended down to 11 °K (heavy-dashed curve), was, in view of the platinum behavior, confidently extrapolated to a constant plateau at helium temperatures. Since, moreover, according to Onnes and Clay's earlier investigation, the influence of admixtures can be represented with rough approximation even down to hydrogen temperatures by an additive resistance that is independent of temperature (an extension of Matthiessen's rule), curve Au_{III} was similarly extrapolated. But, again according to Onnes and Clay, the effect of admixtures should be proportional to the quantity of admixture present. If so, these data would lead to an implausibly negative resistance for *pure* gold. Still, there were many experimental uncertainties, among them the degree of contamination of the samples, the effects of strain in the wire and of drawing, and the simplistic extrapolation to zero impurity. (The 1907 measurements had shown clear deviations from Matthiessen's rule already at hydrogen temperatures.) In view of the uncertainties, Onnes settled for the most plausible conclusion: "*that within limits of experimental error (the degree of purity attainable) the resistance of pure gold is already zero at helium temperatures*"[10] (lowest dashed curve in Figure 2-2).

A similar extrapolation for platinum was less clear-cut. Comparing the resistance ratio of Pt_B with that of the old reliable standard Pt_I in the hydrogen region[11] suggested Pt_B was less pure than Pt_I; the method of attaching the wire to the glass former was worrisome as well. Nevertheless, assuming a constant additive resistance, as before, gave the light solid curve for Pt_I, Figure 2-2. Even so, the indicated constant residual resistance was not necessarily representative of pure platinum, since both Holborn (at the Reichsanstalt) and Onnes and Clay[12] had shown that thinner wire decreased less in resistance with temperature than thicker wire, at least down to the hydrogen region (reflecting presumably the effects of strain from drawing). But, assuming a resistance difference inversely proportional to wire thickness (Onnes does not give the wire size for Pt_B) in order to deduce a value for pure platinum compensated for the influence of drawing "would lead ... too far into the region of pure conjecture."[13] With that disclaimer, Onnes concluded that, like gold, the resistance of pure platinum ought to reach zero at helium temperatures.

The possibility that the resistance of pure metals tends to vanish *above* absolute zero was, strictly speaking, not a novel concept; as noted earlier, it had been suggested by Callendar as early as 1899,[14] but only Onnes was in position to act on it.

2.2 A Clue from Berlin

Two days before Onnes gave his second report to the Amsterdam Academy on the preliminary resistance measurements with Holst, Walther Nernst submitted a progress report to the Prussian Academy of Science in Berlin; his investigations were highly pertinent to Onnes's ongoing program. Nernst's cryogenic group was barely 6 years old, having been established soon after his appointment to a professorship in physical chemistry at Berlin in 1905. That was also the year he announced his heat theorem,[15] deduced while attempting to calculate chemical equilibria from first principles. Later generalized in several versions as the third law of thermodynamics, the law in its initial and rather mundane form holds that curves for the total energy and free energy of a reacting system merge asymptotically near absolute zero. Henceforth, practically all of Nernst's energy until the onset of the war would be devoted to testing the theory experimentally. The key test involved measuring integral specific heats of solids down to the lowest temperatures attainable—measurements needed for determining the temperature variation of the total energy of substances. By chance, the specific heats of metals were proving to be a major embarrassment for the electron theory of Riecke, Drude, and Lorentz just then. Nernst's extensive measurements, finally under way in 1910 after 5 exhausting years of preparatory effort, would throw light on the troubling questions confronting the classic theory. More than that, they would furnish early experimental support for the revolution in physics that was then fermenting. From a more limited

Figure 2-3. Walther Nernst, circa 1938. Courtesy Francis Simon Collection, AIP Niels Bohr Library.

perspective, Nernst's measurements would provide an important clue to the resistance behavior near absolute zero. The clue had been offered to Dewar but, somewhat surprisingly, Dewar either missed it or somehow failed to follow up on it.

The classical guiding principle for studying the specific heat of metals was the venerable law of Dulong and Petit, established in 1819.[16] Pierre Louis Dulong and Alexis Therese Petit concluded from measurements on various elements that the product of the specific heat of a monatomic solid and its atomic weight, a product known as the atomic heat, is independent of temperature and approximately equal to 6 calories per degree per gram-atom. This behavior was shown by Ludwig Boltzmann, half a century later, to follow from simple considerations underlying the kinetic theory of matter. According to Boltzmann's principle of the "equipartition" of energy (1871), each degree of freedom of a solid atom performing simple harmonic oscillations about some equilibrium position at the absolute temperature T is given by kT where k is a constant.[17] (The energy is equally shared between kinetic and potential contributions, each $1/2kT$, in contrast to gases where only translational motion is involved.) Three independent degrees of freedom give a total mean energy of $3kT$, or for a gram-atom:

$$U = 3N_0 kT = 3RT . \tag{2-1}$$

Here Avogadro's number, N_0, is related to Boltzmann's constant (k) and the gas constant (R) by $R = kN_0$. The atomic heat at constant volume is thus:

$$C_V = \frac{dU}{dT} = 3R\,, \tag{2–2}$$

or about equal to that found by Dulong and Petit.

In spite of its elegantly simple explanation of a well-known experimental fact, equipartition barely survived 12 months. Coincidentally, notable exceptions to the law of Dulong and Petit were reported simultaneously by Dewar before the Royal Society at Edinburgh and by Heinrich Friedrich Weber at a meeting of the Chemical Society in Berlin.[18] Weber had initially found that the atomic heat of carbon, whether diamond or graphite, has at room temperature a much smaller value than that typical of other monovalent substances (about 1.8 calories), but its value triples when the temperature rises to 200 °C. Dewar's investigation extended all the way to the temperature of the oxy-hydrogen blow pipe, some 2000 °C. At this temperature carbon did, indeed, seem to agree with the law of Dulong and Petit. Subsequent measurements by Weber showed, in Dewar's opinion (borrowing a favorite analogy of his own), "that the specific heat of carbon, in both its forms, when plotted to temperature as abscissa, produced a curve with a point of inflection in it, like the Old English ʃ."[19] In fact, the trend with temperature of Weber's data on diamond could be fitted by an expression of an equally familiar form, namely,

$$C_V = a + bt - ct^2\,. \tag{2-3}$$

In 1905 Dewar himself extended the measurements on diamond and graphite to the temperature of liquid hydrogen. These, as well as "observations extracted from laboratory records"[20] supported qualitatively the unavoidable conclusion that the atomic heats of solids generally drop sharply from the classical value with decreasing temperature.

Nor could the classical value be reconciled with the electron theory. According to it, n free electrons per unit volume should have a kinetic energy of $3/2nkT$. If n is about equal to the number of atoms, the electrons should contribute $3/2R$ to the specific heat, giving $C_V = 3R + 3/2R = 9/2R \approx 9$ calories per gram-atom instead of the equipartition value of $3R$. The fact that the electrons do *not* contribute much to C_V was supported by Dewar's observations that at room temperature, the equipartition value appears to hold about equally well for metals and insulators. The contradiction is not resolved by postulating a temperature dependence for n. Any change in the volume density with a rise in temperature is in the direction of heat absorption [since Q in Eq. (1-9) is positive], making the specific heat even larger.

These facts prompted Albert Einstein to reexamine the problem of specific heats in 1906,[21] in light of the new developments engendered by Max Planck. (That 1906 was to be a watershed for classical physics is poignantly underscored by Drude's as well as Boltzmann's suicide that year.) Einstein may have been inspired by Weber himself, whose lectures on the subject he attended at Zürich. Perceiving that "Planck's theory of radiation strikes the heart of the matter,"[22] Einstein boldly replaced the average energy of an atom as given by equipartition, $3kT$, with that deduced from quantum theory for an atom of a monatomic solid vibrating with a frequency ν:

$$U = 3kT\left(\frac{x}{e^x - 1}\right) \tag{2–4}$$

where $x = h\nu/kT$. We now have for the specific heat of a gram-atom of solid:

$$C_V = \frac{d}{dT}\left(3RT\frac{x}{e^x - 1}\right) = 3Rx^2\frac{e^x}{(e^x - 1)^2}. \tag{2–5}$$

The trend with temperature of this expression, Einstein found, agreed rather well with available data, in particular with Weber's old data on diamond. At high temperatures ($x \ll 1$), the expression approaches $3R$, the classical value. At low temperatures, however, it has the exponential form x^2e^{-x}, and approaches zero as T approaches zero. The only material parameter in the expression is the frequency ν characteristic of each substance. At "corresponding" temperatures, chosen such that ν/T are the same, C_V is the same for all substances.

Specific heats of solids are much more difficult to measure than electrical resistances, especially at very low temperatures. For one thing, the quantity measured is not the specific heat at constant volume, the quantity of most theoretical relevance, but the specific heat at constant pressure, C_P. From it, one derives C_V by various analytical or empirical formulae involving separately measured values for the coefficient of cubical expansion, compressibility, and atomic (specific) volume. In addition, Nernst needed a method for measuring specific heats at a well-defined temperature, instead of simply averaging it over a range of temperatures as was usually done. Finally, the measurements had to be extended low enough in temperature to allow extrapolation to absolute zero with some confidence. Since Leiden was closer than London, Nernst visited Onnes and consulted with him on this last point; however, on returning to Berlin he decided that a liquid hydrogen plant of the Leiden magnitude was well beyond his resources—or patience. Instead, he followed Morris Travers' early example and personally improvised a small liquefier "for the equivalent of about twenty pounds sterling."[23] Kurt Mendelssohn, a student in Nernst's laboratory much later, recalled "that it could be made to work more or less occasionally and then only by Nernst's trusted old mechanic Alfred Hoenow."[24] Hoenow, it seems, was Onnes's Flim or Dewar's Lennox. But, like Leiden and unlike Dewar's Royal Institution, Nernst's laboratory was also an academic institution, with a cadre of bright research students on hand. During 1906–1909, Nernst developed, principally with the assistance of Arnold Eucken and the colorful English brothers Frederick Alexander and Charles Lindemann (the latter two stand in vivid contrast to the difficult Eucken who eventually came to a sorry end), the necessary calorimetric methods. These included methods for performing the various ancillary measurements for reducing the results. The crux of the apparatus was the famous vacuum calorimeter, whose significance for low-temperature research has been likened to that of the spectrograph for spectroscopy or Wilson's cloud chamber for nuclear physics.[25] The solid specimen, serving as its own calorimeter, was surrounded by a resistance heater that also served as resistance thermometer; the calorimeter block was enclosed in a vacuum-walled jacket to minimize heat loss.[26]

Nernst presented their first results on specific heats of various sundry elements and compounds down to the temperature of liquid air to the Berlin Academy in February of 1910.[27] As expected, they confirmed a marked drop with decreasing temperature. "One gets the impression," Nernst reported, "that the specific heats are converging to zero as required by Einstein's theory."[28] (Nernst's own theory required them to converge on a constant value, but not necessarily zero.) Frederick Lindemann and Alfred Magnus, he added, were preoccupied with obtaining a quantitative fit to Einstein's predictions, while Charles Lindemann concentrated on the tedious corrections for thermal expansion. Good progress was reported by Nernst in a lecture before the French Physical Society later in the spring. Indeed, not only was the agreement corroborating his heat theorem, but the direction of the investigation now took an unexpected turn: The theorem, and specific heats in particular, seemed to acquire a deeper significance in view of the apparent link with quantum theory. Wary of premature speculation, Nernst pushed on with preparations for definitive measurements in liquid hydrogen.

New data were presented to the academy 1 year later, on 23 February 1911.[29] They covered measurements on lead, silver, zinc, copper, aluminum, potassium chloride, and chloride of mercury, at temperatures down to slightly above the boiling point of hydrogen—as low as Hoenow could coax the liquefier. Indeed, the drop in specific heats was dramatic. Whereas the specific heat of lead had only dropped 10% from its room temperature value in liquid air, in boiling hydrogen it fell by 50%. These data are plotted in Figure 2-4 as atomic heat versus absolute temperature (heavy curves). The results for potassium chloride and chloride of mercury fell between the curves for silver and zinc. The thin curves were calculated from Einstein's formula for various $\beta\nu$ values as indicated in the plot ($\beta \equiv h/k$ in Einstein's notation).

From the behavior exhibited in Figure 2-4, Nernst drew a number of conclusions: (1) The trend in the atomic heats at low temperature was similar for all metals investigated, characterized by a function independent of the nature of the metal and containing a single characteristic constant. (2) At moderate and high temperature, the functional dependence agreed well with Einstein's formula, but at the lowest temperatures of observation *the measured values fell systematically more slowly than predicted by the formula*, with even a hint of a tangent to the temperature axis (indicated by the dashed extensions to the measured curves). (3) Nernst's own heat theorem seemed confirmed, that is, $\lim dU/dT = 0$ at $T = 0$. (4) Fine differences in the curves for potassium chloride and chloride of mercury could be explained by the atomic constituents of the former substance having similar eigenfrequencies, which was not the case for the latter compound. (5) Collectively, the data constituted excellent confirmation of Planck and Einstein's quantum theory, the discrepancy at the

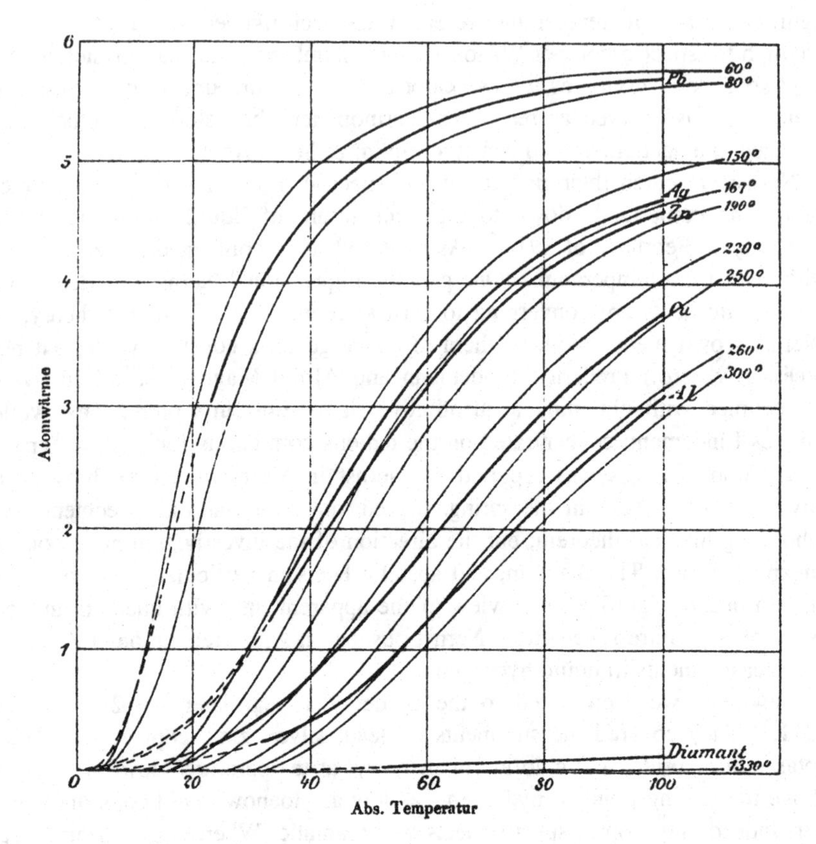

*Figure 2-4. Atomic heat versus temperature. Thin curves are calculated values for various eigenfrequencies; dashed curves are extrapolations of the measured curves. Nernst, Zeitschrift für Elektrochemie, **17** (1911), 274.*

lowest temperatures being interpreted simply to mean that the eigenfrequency is not actually sharp but that a spectrum is involved. (6) The similarity in the trend of metals and nonmetals cast doubt on the electron theory of galvanic resistance—a theory Nernst had reason to doubt on other grounds. (7) Nernst's eigenfrequency values deduced from fitting the observational data agreed quite well with values calculated from a formula for ν derived by Frederick Lindemann the previous summer.

The wealth of evidence for some unsuspected link with quantum theory animated Nernst as nothing could before. He promptly spent a weekend consulting with Einstein in Zürich on the matter, and had long discussions with Planck (who had been instrumental in securing Nernst's appointment in Berlin in the first place). Among his laboratory staff, Nernst was affectionately rumored to have "joined the Peripatetic school of philosophy."[30] As for his own theorem, he warned, "if anyone disputes the theorem I must ask him in the first place to fight it out with Messrs. Einstein and Planck. If he succeeds in defeating them, I will go into action."[31]

To focus on the emerging fine deviations from Einstein's formula required calibrating platinum and lead resistance thermometers against certified gas thermometers. In particular, it meant determining their temperature function over the rather awkward interval from 20 to 80 °K. Nernst soon rediscovered what had long been known in Leiden and London, namely, that the full resistance curve from the ice point to boiling hydrogen is much less linear for platinum than for lead. Examining the platinum curve more closely, he quickly retraced the behavior that had taken Dewar a decade to map out. The temperature coefficient was nearly constant down to about 65 °K, but, he found, it fell extremely rapidly at lower temperatures. Nernst quickly grasped the significance of this: The gross temperature behavior was remarkably similar to that predicted by Einstein's curves.

> The temperature coefficient of electrical resistance exhibits a certain parallel to atomic heats; furthermore, both quantities tend, at higher temperatures, toward a value independent of the nature of the substance. It suggests, in other words, a clear analogy between the law of Dulong-Petit and the rule of Clausius. ... It also appears that, at the lowest temperatures, the temperature coefficient falls off directly with the atomic heat, but noticeably slower than predicted by Einstein's formula.[32]

There seemed little doubt that Nernst was, unexpectedly, facing an apparent relationship between the two heretofore presumed disparate phenomena of specific heats and specific resistance—something the electron theory had missed and had even failed to impress Dewar. Nernst also saw a way of capitalizing on this purely empirical connection. Taking a cue from Einstein, he fitted the resistance curves with the expression

$$R = A(e^x - 1)^{-1} + B\,, \tag{2-6}$$

with a judicious choice of the eigenfrequency ν. A and B are constants characteristic of the metal; B, in particular, is a temperature-independent impurity contribution.

The expression agreed surprisingly well with Nernst's own measurements on platinum down to 20 °K, and also with Onnes and Clay's old data on lead to 14 °K. It even hinted of real deviations below 15 °K, which was slightly below Nernst's observational limit, suggestive of Einstein's "failure," and a point Nernst could easily account for. Still, for now Nernst cautiously preferred to view the expression simply as "a convenient interpolation formula for certain regions of electrical conductivity."[33] At the very least, it offered a promising test for the purity of a metal, and possibly for certain changes of structure. Whether the constants converge toward zero for pure metals, only further investigation would show. Eigenfrequencies determined from curve-fitting the resistance data agreed, at least in order of magnitude, with those obtained from the specific heat data or, where necessary, from Lindemann's formula.

Although Nernst was hardly conservative by nature, Frederick Lindemann was bolder—and a better theorist. The formula he deduced in 1910 provided an element missing from Einstein's pioneering specific heat analysis of 1907. Originally Einstein had suggested that ν could be obtained from optical absorption frequencies determined from "residual rays,"[34] but he had not offered a method for deriving it explicitly from other measurable properties of metals. Lindemann assumed that the melting point of a solid is the temperature at which the amplitude of vibration of the atoms is of the same magnitude as the atomic separation[35]—as did Riecke's calculational model the year before. In addition, he assumed the atomic volume to be cubical and retained the principle of equipartition; consequently, he was able to express the vibrational frequency in terms of the melting point, T_s, molecular weight M and volume V thusly:

$$\nu = (\text{constant})\left(\frac{T_s}{MV^{2/3}}\right)^{1/2} . \tag{2–7}$$

His eigenfrequencies agree rather well with those from optical wavelengths obtained by Rubens and Hollnagel[36] and with values he and Alfred Magnus obtained from empirically curve-fitting the specific heats in 1910.[37] As noted, the agreement with specific heats still held in 1911.

Encouraged, Lindemann turned to a derivation of Eq. (2-6) from first principles,[38] despite the evident failure of the electron theory to explain specific heats. Lindemann's starting point was again basically Riecke's model.[39] However, instead of assuming λ to be inversely proportional to the atomic radius, he assumed it to be inversely proportional to the amplitude of atomic vibrations, the energy of which he took from Planck's formula. Lindemann also, with Thomson, assumed a square-root dependence of n on temperature. Substituting λ, n and v (from equipartition) in Eq. (1-6) expressed in terms of resistance R gave:

$$R = A^2(e^x - 1)^{-1} + 2AB(e^x - 1)^{-1/2} + B^2 . \tag{2-8}$$

At high temperatures (small x), the amplitude of oscillation becomes large compared to the size of an atom. The first term then dominates, and

$$R = A^2(e^x - 1)^{-1} + B^2 ,$$

or, using the series expansion for e^x:

$$R = A^2X^{-1} + B^2 = \left(\frac{A^2k}{h\nu}\right) T + \text{const.} ,$$

in agreement with the resistance data in general. At low temperatures, Eq. (2-8) reduces to:

$$R = A^2e^{-x} + 2ABe^{-x/2} + B^2 ,$$

where now the second term is the principal term. Onnes's data for lead and silver[40] seemed to be well represented by Eq. (2-8) using ν-values by calculation or from the specific heats, judging by Lindemann's tabular comparison, and apparently also from the data for Pt, Cu, and Al. At the very lowest temperatures of observation, however, or for $T < 0.15\ \beta\nu$, definite disagreement was evident. Here, certainly for silver at 13.9 °K, the measured resistance was higher, again similar to the specific heat deviations. The fact that the constants *A*, *B* did not have quite the relationship indicated by Eq. (2-8) troubled Lindemann more. Perhaps, he suggested, it was a consequence of the simplistic assumption that the "sphere of action" (Wirkungssphäre) presented by the atoms to the moving electrons is rigorously proportional to the volume occupied by the atomic oscillations.

Still, in its broad features Lindemann's theory seemed to account for the temperature variation of resistance without resorting to "the usual complicated ad hoc assumptions" for the free path in spite of the "various doubtful basic assumptions of the free electron theory."[41] Lindemann shelved the problem, but would return to it with a more promising way of circumventing the doubtful assumptions three years later.

2.3 Onnes's Hypothesis and Prospects for a Definitive Test

Kamerlingh Onnes, too, was struck by the remarkable analogy between specific heats and specific resistance. Nor was the promise that Einstein's successful analysis held for resolving an experimental question he had doggedly pursued for years lost on him.

> The marked decrease in the resistance until it becomes practically zero at a temperature just above 4 °K and its remaining at this value as the temperature is lowered further as has been shown over a range of about two and a half degrees, so that, as far as resistance of these metals is concerned, the boiling point of helium is practically the absolute zero, points in another direction. It seems to me to be connected with the change with temperature of the heat energy of molecular motion of solid substances that has been deduced by Einstein in his theory of the specific heats, on the assumption that it is the energy of vibrators determined by radiation equilibrium.[42]

With the latest Leiden results, the theory "that [had] served for years as a guide in our Leiden researches"—the notion of electrons freezing to the atoms—was obviously no longer tenable. It is not the electrons that "freeze," Onnes decided, but rather the Planckian vibrators that under normal conditions limit the electron free path by thermal agitation.

> It seems that the free electrons in the main remain free, and it seems to be the movable parts of the vibrators that are now bound, their motion at ordinary temperature forming the obstacles to conduction; these disappear when the temperature is lowered sufficiently as the vibrators become then practically immovable.[43]

Apparently, Onnes was not yet aware of Nernst's instinctive and Lindemann's semitheoretical attempts to emulate Einstein's analysis. Onnes himself possessed considerable mathematical skills and respect for a theoretical point of view, despite his apparent casual attitude in this regard. Thus, his reputed remark to G. E. Uhlenbeck in 1924 to the effect that "these Maxwell equations, what good are they?" is easily dismissed as a personality trait.[44] His doctoral thesis, for instance, dealt with applications of Hamilton's and Jacobi's mathematical formulations; a much later (1914) example is a paper he co-authored with Paul Ehrenfest on Planck's combinatorial formula for the distribution of energy over oscillators. On the other hand, Einstein has described him as "extraordinarily accurate in his intuitive reasoning, [but] less able to express himself rigorously in abstract matters."[45] Onnes's "derivation" of a resistance formula, at this juncture, was hardly rigorous (as he subsequently admitted). Using a potpourri of arguments concocted from Riecke's latest electron theory, Drude's electro-optics, and quantum theory, he reasoned as follows. Expressing the conductivity as Riecke did in 1909, which was basically Drude's formula [Eq. (1-6)] except for a factor of two and a factor given by the velocity of light squared (since Riecke used electromagnetic units), or:

$$\sigma = \frac{1}{3}\,\frac{e^2 n\lambda v}{c^2 kT}$$

and using the relationship:

$$P = \frac{1}{3}\,\frac{e^2 v}{c^2 k\sqrt{T}}$$

Onnes wrote

$$\sigma = \frac{Pn\lambda}{\sqrt{T}}\,.$$

Here P is the number of conduction electrons per atom, following Drude's notation in his treatment of optical dispersion. (Onnes actually used Riecke's notation for most quantities, viz., L for the free path, g for the electron velocity, and α for $3k/2$.) Now, according to Riecke,

$$\lambda = \frac{a\sqrt{T_S}}{\pi(1+\beta S)^2}\,\frac{1}{\sqrt{T}}\,.$$

In this expression, a is the separation between cubically arranged atoms, T and S the melting temperature on the absolute and Celsius scales, and β the linear coefficient of thermal expansion. More simply:

$$\lambda = \frac{q}{\sqrt{T}}\,.$$

If the atomic impediments are Planckian vibrators, Onnes reasoned, then their influence on the free path should be proportional to the amplitude of vibration—a rote assumption that was by now becoming widespread. He therefore rewrote the expression for λ as follows:

$$\lambda = \frac{q'}{\sqrt{E_T}};$$

here $E_T = U$ of Eq. (2-4). Retaining, however, Riecke's quite unjustified assumption of n (and P) being independent of temperature, Onnes at last obtained the *ratio* of the conductivity σ_T at any absolute temperature T to the conductivity σ_0 at 0 °C (= T_0):

$$\frac{\sigma_T}{\sigma_0} = \left(\frac{T_0 E_0}{T E_T}\right)^{1/2}. \tag{2–9}$$

This handy formula, Onnes claimed, gives "good expression to the decrease with temperature of the resistance of pure monatomic metals,"[46] though curiously he did not (in *Communication 119* or anywhere else) offer an example in the form of a plot. First, for all practical purposes it predicts vanishing resistance well *above* the absolute zero (not simply at the absolute zero, the most Nernst claimed). For example, taking $\beta\nu = xT = 54$ (Onnes uses a for $\beta\nu$) for lead, the formula yields $R_T/R_0 \sim 1 \times 10^{-4}$ for $T = 4$ °K. In addition, since $\beta\nu/T$ is already small compared to unity at 0 °C, the exponential terms in Eq. (2-8), expanded in powers of x, can be approximated by retaining only the first two terms, giving:

$$\frac{R_T}{R_0} = \frac{T - \frac{1}{4}(h\nu/k)}{273.1 - \frac{1}{4}(h\nu/k)} \equiv \frac{T}{273.1}\left(\frac{1 - (h\nu/4kT)}{1 - \frac{1}{4}(h\nu/k \cdot 273.1)}\right) \tag{2–10}$$

or $R_T/R_0 \sim T/273.1 = 0.0037T$, which is the rule of Clausius.

Finally, the formula "also expresses the fact that the diminution of resistance diminishes in quantity at hydrogen temperatures, and that in greater degree for substances of high melting point than those of low melting point."[47]

Quantitatively, however, to fit the bend in the resistance curve at the lowest temperatures required a shift in frequency toward lower values; thus definite values could not be ascribed to ν. For instance, Einstein had effectively deduced (using wavelengths in cm) from elastic constants[48] a $\beta\nu$-value for silver of 200, compared with Nernst's value of 160 from specific heats—good agreement in Einstein's opinion. But for lead Nernst's value, $\beta\nu = 58$, was a factor of two below Einstein's value. Worse, compressibility arguments suggested an *increase* in frequency at lower temperatures, and specific heats suggested a decrease. With so much uncertainty, Onnes opted—purely empirically, it would seem—for half of Einstein's $\beta\nu$-values, "as it is more a question of showing that the introduction of vibrators leads to a qualitative explanation of the sense in which the observed change of resistance deviates from proportionality to temperature."[49] Adopting $\beta\nu = 111$ for Pt, 100 for Ag, 92 for Au, and 54 for

TABLE 2-2 W_T/W_0

T	T/273.1	Platinum C	Platinum O	Silver C	Silver O	Gold C	Gold O	Lead C	Lead O
379°86	1.365	1.405		1.401	1.411	1.397		0.384	
273.1	1.	0.579	1.	1.	1.	1.	1.	1.	1.
169.29	0.617	0.213	0.581	0.583	0.581	0.586	0.593	0.601	0.594
77.93	0.285	0.012	0.199	0.220	0.197	0.225	0.219	0.250	0.253
20.8	0.074	0.003	0.014	0.015	0.009	0.018	0.008	0.035	0.030
13.88	0.054	0.000	0.010	0.004	0.007	0.005	0.003	0.015	0.012
4.30	0.016		[0.009]	0.000		0.000	[0.002]	0.000	

Pb, Onnes compiled the following tabular comparison between calculated resistance ratios ("C") and observed ratios ("O," values from Onnes and Clay), showing a qualitative correspondence at least down to 4.30 °K. (The observed ratios for platinum are those for Pt_I in Figure 2-2; the gold values represent Au_V.)

In the printed version of his report in the Dutch proceedings, Onnes had more to say about the comparison in a long footnote, appended as the journal went to press. The measured ratios in Table 2-2 are for the purest available wire samples, omitting corrections for the influence of actual admixtures and of heat treatment during wire drawing. Taking $\beta v = 30$ for mercury, because of its lower melting point, he added a further tabular comparison as shown below in Table 2-3. Mercury was a good choice, not simply because of its low eigenfrequency. To dispose of the question whether the resistance of an absolutely pure metal actually vanishes above absolute zero, a metallic conductor of the highest possible purity was needed. Since mercury is a liquid at room temperature, it can easily be distilled repeatedly in vacuo to an even higher degree of purity than gold.

In a footnote to a much later Leiden *Communication*, Onnes explains that he derived his βv-value for mercury somewhat empirically by curve-fitting his formula to the older mercury data, but guided by van der Waals's law of corresponding states.[50] Only after the present round of experiments did he become aware of Lindemann's formula for determining frequencies. Onnes's arguments from corresponding states follow immediately from Lindemann's formula, since the "principle of similarity" applies equally well to Lindemann's reasoning.

TABLE 2-3

T =	77°.93 K	20°.18 K	13°.88 K
calculated	0.263	0.050	0.027
observed	0.279	0.056	0.033

While I wish to take this opportunity [he wrote]—as I should have done in my first paper on the subject—of referring to these important calculations of Lindemann it is perhaps fortunate that I had not seen them sooner, for they give $\beta\nu$ = 46 for mercury, while it was just from the estimate $\beta\nu = 30$ that I was able to forecast that at some of the helium temperatures the resistance of mercury could be measured and at lower temperatures it would disappear, and it was this forecast that made experiments with mercury so particularly inviting joined with the prospects of working with a pure metal.[51]

CHAPTER 3

Discovery and Its Aftermath

3.1 Superconductivity

Following the initial round of resistance measurements in liquid helium, Onnes and his team took time out to improve the experimental apparatus and procedures. The improvement largely consisted of modifying the helium liquefier in one essential respect: introducing the ability to transfer liquid helium from the liquefier to an adjacent but attached cryostat. This long-frustrated operation finally met with success, allowing resistance measurements on larger samples with greater ease and with improved accuracy. One major source of error had been the inability to stir the liquid bath, which was now possible as well. In the interim, it had also been established that liquid helium is an excellent electrical insulator, enabling resistances to be measured directly on bare wires; good thermal contact with the helium bath was thereby assured. Dorsman and Holst, as before, prepared and carried out the actual experiments. Holst, in particular, performed the resistance measurements.

The new experiments concentrated on mercury, the key to the question of sample purity. First, however, a few measurements on gold at helium temperatures were necessary, since the resistance of gold in liquid helium had thus far been inferred solely by extrapolation based on analogy with the platinum results. The terse announcement in late April of the preliminary results on mercury[1] had even less to say about gold. Onnes simply noted that the previous estimate for the resistance of the gold wire Au_{III} in liquid helium (dashed curve, Figure 2-2) was "supported by direct measurement. The conclusion that the resistance of pure gold within the limits of accuracy experimentally obtainable vanishes at helium temperatures is hereby greatly strengthened."[2]

The *experimental crucis* lay in mercury, however. A single experiment on the resistance of a frozen mercury thread was reported "with all reserve" to the Amsterdam Academy on 28 April 1911.[3] Onnes prefaced his report with his resistance formula and noted the excellent agreement it gave with experiments at hydrogen temperatures (Table 2-3). With a suitably chosen frequency, Onnes added, the formula had led them to expect:

1. That the resistance of pure mercury would be found to be much smaller at the boiling point of helium than at hydrogen temperatures, although its accurate quantitative determination would still be obtainable by experiment;
2. That the resistance at this stage would not yet be independent of the temperature, and;
3. That at very low temperatures such as could be obtained by helium evaporating under reduced pressure the resistance would, within the limits of experimental accuracy, become zero.

Indeed, the experiment appeared to confirm this forecast remarkably well. "While the resistance at 13.9 °K is still 0.034 times the resistance of solid mercury extrapolated at 0 °C, at 4.3 °K it is only 0.0013, while at 3 °K it falls to less than 0.0001." The experimental revelation that "a pure metal can be brought to such a condition that its electrical resistance becomes zero, or at least differs inappreciably from that value," Onnes observed, was "certainly of itself of the highest importance."[4] In addition, however, the confirmation of his forecast[5] of this behavior was important for lending support to his thesis "that the resistance of pure metals ... is a function of the Planck vibrators in a state of radiation equilibrium."[6]

Onnes's thesis was unexpectedly supported by an interesting piece of evidence that had just come to his attention, he added. The wavelength in vacuo corresponding to the eigenfrequency deduced for mercury, or $\beta\nu = 30$, is about 0.5 mm. Coincidentally, H. Rubens had found that a mercury lamp emits vibrations of very long wavelengths, in fact of about 0.3 mm. Onnes took this as evidence for some connection between the resistance behavior and optical phenomena. Presumably, it also pointed indirectly to a connection with specific heats and thence back to Planck's quantum.

The sparse preliminary report is devoid of experimental details concerning cryostat or mercury resistance (including degree of purity), or for that matter actual data. One senses from the guarded but casual tone of the *Communication* that the resistance behavior had, in fact, become a matter of some urgency at Leiden. One reason was the successful deployment of the modified cryostat. Only with its larger, more accessible sample chamber—the handiwork of Kesselring—had it been possible to accommodate the glass capillaries, special leads, and so forth, required with mercury. But another reason for the urgency is evident from a footnote inserted as the *Communication* went to press. Onnes had just then learned of the resumption of specific heat investigations in Berlin (i.e., those reported by Nernst on 23 February). Moreover, copies of two papers dispatched posthaste from Berlin informed him of Nernst being "independently led to assume a connection between the energy of vibrators and electrical resistance," and of Lindemann's "further development" of this hypothesis.[7]

No time was lost in repeating the experiment. True, Onnes's expectation as to the gross resistance behavior seemed to have been more or less confirmed. Certainly, the resistance of mercury sank to an extremely low value below the boiling point of helium. However, the fit of his formula to the observed resistance curve between the melting point of hydrogen and the boiling point of helium was perhaps less satisfactory for mercury than for gold. Quite another matter was the *rapidity* of the transition from a finite value below the boiling point in the case of mercury: a drop in resistance of at least a factor of ten per degree Kelvin. (The estimated resistance ratio at 3.0 °K, 10^{-4} times the value at 0 °C, was actually an upper limit.) To pin down a more definite resistance value, Holst turned to the differential galvanometer, using Kohlrausch's method of overlapping shunts. The new cryostat arrangement also made possible separate measurements of current strength and potential differences by appropriate current and potential taps.

Onnes communicated the new results to the academy in May.[8] Holst had checked and rechecked his measurements. There was no longer any doubt. The residual resistance at 3 °K, if not truly zero, had to be far lower than that indicated by the upper limit recorded in April. For all practical purposes, it had vanished. Onnes had encountered a new physical regime. His report was bluntly titled "The disappearance of the resistance of mercury."

Extrapolating the value of the mercury resistance in the solid state at 0 °C from its measured value in the liquid condition at the melting point (by the known temperature coefficient of solid mercury) gave a reference value $R_0 = 39.7\Omega$. At 4.3 °K the resistance had sunk to 0.084Ω, or $R/R_0 = 2.1 \times 10^{-3}$. At 3 °K, the upper limit on resistance was now $3 \times 10^{-6}\Omega$, corresponding to a ratio $R/R_0 = 0.75 \times 10^{-7}$; this value remained the upper limit at 1.5 °K as well.

In a second set of measurements, the resistance was monitored as the temperature was gradually raised again from below the boiling point. The temperature of the point at which the resistance first became detectable was found to be "slightly more than 4.2 °K."[9] Here, the resistance measured 230 micro-ohms, or $R/R_0 \sim 0.6 \times 10^{-5}$. As the temperature was raised to the boiling point (which Onnes gives as 4.3 °K), the resistance rose once more to its value of 0.084Ω. (At the time the most probable value for the boiling point of helium was actually 4.26 °K.) This was clearly a much faster rate of change of resistance than even the exponential term in Onnes's formula could handle; exactly how much faster could not yet be determined. Nor was all well *above* the boiling point. There were signs of a point of inflection somewhere between the boiling point of helium and the melting point of hydrogen—a matter of less significance but again something not predicted by the formula.

In any case, the shortcomings of the formula, which Onnes now shrugged off as "incomplete also on account of the method by which it was deduced," was suddenly of much less concern than the implication of vanishing resistance in liquid helium.

> The more the upper limit which can be ascribed to the resistance remaining as helium temperatures decrease, the more important becomes the observed phenomenon that the resistance becomes practically zero. When the specific resistance of a circuit becomes a million times smaller than that of the best conductors at ordinary temperatures it will, in the majority of cases, be just as if electrical resistance no longer existed under those conditions. If conductors could be obtained which could be regarded as being devoid of resistance as long as their cross section was not excessively small, or conductors of the smallest possible sections, either cylindrical with diameters of the order of the wave length of light, or films of molecular dimensions, whose resistance would be but small, if there had no more to be reckoned with the Joule development of heat in increasing the current in a bobbin to exceedingly high values, because the development of heat in a circuit of constant current strength could be made extremely small compared with the latent heat of vaporization of the liquid which can be used for cooling,—then further experiments in all possible directions would give the fullest promise, notwithstanding the great difficulties which are encountered when working with liquid helium. It is therefore all the more necessary to establish beyond all possibility of doubt the property of which advantage would be taken in such experiments. With this end in view modified measurements are being made.[10]

Details of the modified liquefier/cryostat were finally released in a *Communication* dated June 1911.[11] Perhaps the most important feature in introducing a separate but physically attached cryostat with independent experimental access from above was that it permitted stirring of the liquid helium bath. In this way the bath temperature could be held quite uniformly. The four current and potential leads were now of mercury as well. Moreover, the cryostat arrangement made it possible to attempt baths of temperatures between the boiling point of helium and the melting point of hydrogen—an awkward temperature regime important for resistance as well as specific heat experiments in general.

The apparatus, shown in the Leiden depiction of Figure 3-1, may be compared to the older arrangement of Figure 2-1. The essential detail of the new cryostat was the silvered siphon tube *Eah* that transferred the liquid helium from the vacuum glass *Ea″* (inner unit of third dewar from left) of the liquefier to the vacuum glass *S7* of the cryostat (innermost of the nested dewars, far right). At the extremity of the siphon opening into the cryostat glass was a valve *Eak* that was operated from above the cryostat cover by a handle communicating through a glass shaft with a worm gear. The lower portion of the vacuum glass of the liquefier as well as the rising limb of the siphon were surrounded with liquid air from which they were separated by a German-silver case *S*1. To this case a ring *S*4 supporting the cryostat glass *S*7 was rigidly fixed. When the cryostat glass was in position with its rubber ring, and the cover with its rubber ring was placed over the cryostat chamber, the German-silver case with the connecting tube and cryostat glass formed an air-tight enclosure. The stirring arrangement is shown separately beside the cryostat in Figure 3-1. It consisted of a German-silver pump *Sb*, with valved piston Sb_1 and outlet valves Sb_2; by means of a wire Sb_4 and a soft iron cylinder Sb_6 inside the glass tube Sb_7 the piston followed the course of a magnet *Sd* that was

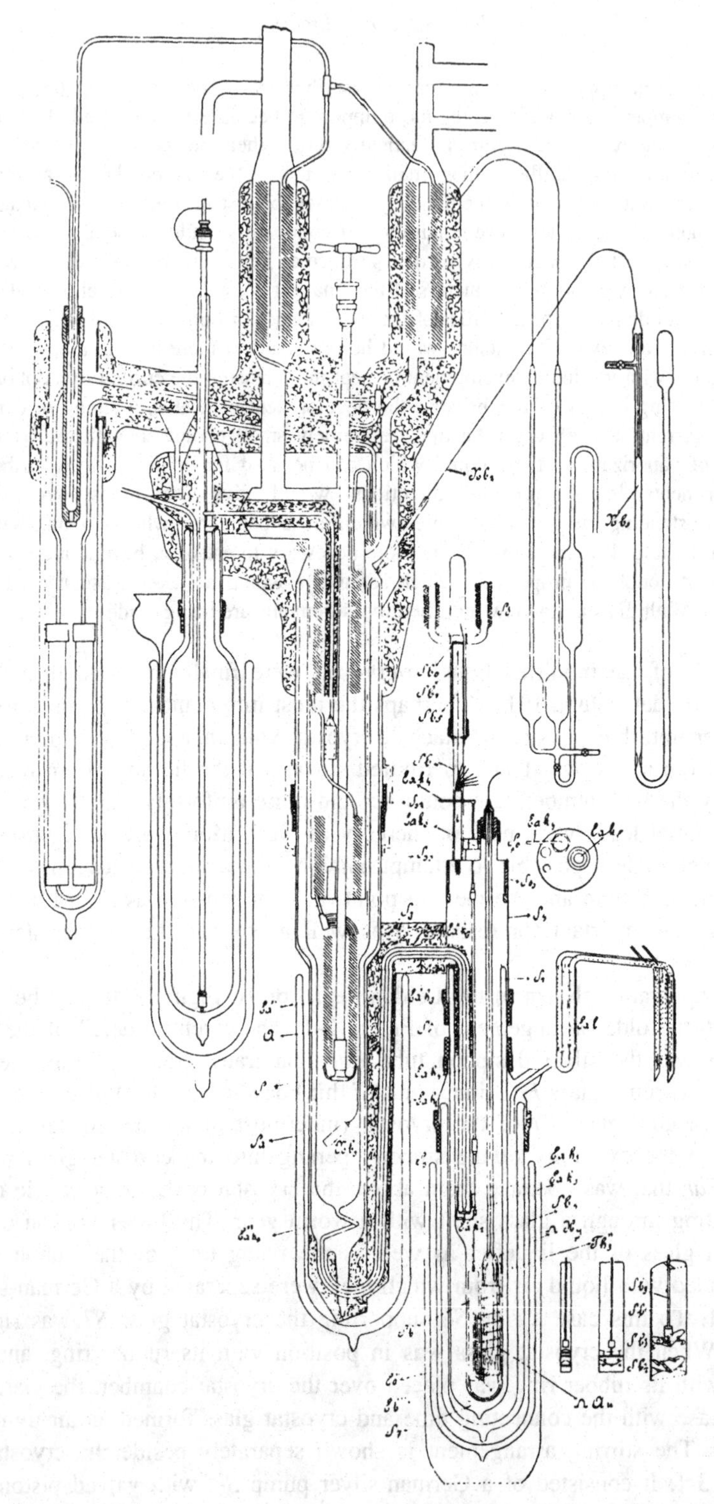

*Figure 3-1. Liquefier with separate but attached cryostat, 1911. Kamerlingh Onnes, RN, **14** (1912), 208.*

moved up and down by an electric motor. By this method vigorous circulation was possible. (Stirring was only necessary at liquid temperatures above 2.19 °K, because of the greatly enhanced thermal conductivity of helium at lower temperatures.)

The vacuum glass of the cryostat was, as earlier, surrounded by dewars containing liquid hydrogen and liquid air, successively. Portions of the glass surfaces were left unsilvered to facilitate viewing the liquid helium meniscus and to monitor the action of the valve. In the cryostat depicted in Figure 3-1 (the one in fact used in the April 1911 single experiment on mercury) the helium thermometer Th3″, the gold resistance thermometer ΩAu (an insulated gold wire wound on a glass cylinder), and the mercury resistance ΩHg may be discerned. To accommodate the freezing and thawing of the mercury thread without division of the thread or injury to the glass, the mercury was poured into a series of U-shaped capillaries with expansion reservoirs. This feature is not evident from the scale of the figure; for reasons of his own Onnes still put off publishing a detailed description of the all-important mercury resistance.

3.2 The Quantum of Low Temperature Revisited

In the spring of 1910 Walther Nernst, by happenstance, encountered the Belgian industrial chemist Ernest Solvay at the home of a chemistry colleague in Brussels. Solvay had invented a process for manufacturing sodium carbonate, which became the basis for a worldwide industrial enterprise. He used a substantial portion of his fortune to support various philanthropic causes. He was also an amateur dabbler in theoretical physics, and saw in Nernst a means of bringing some of his ideas on that subject to the attention of Planck, Einstein, and others. Nernst, for his part, saw Solvay's desire and enthusiastic philanthropy as the opportunity to organize an international conference to "review current questions in connection with the kinetic theory of matter and the quantum theory of radiation." Solvay concurred, enjoining Nernst to go ahead with the necessary preparations. Nernst lost no time, even though Planck cautioned that the time was premature for a concerted focus on quantum problems. "I believe," he had replied to Nernst's short memorandum, "that, other than ourselves, only Einstein, Lorentz, W. Wien and [Joseph] Larmor will be seriously interested in the matter."[12] Planck was, in fact, overly optimistic, since Larmor—a conservative classical theorist usually critical of new ideas—took scant interest in that particular "matter." In June Nernst secured Solvay's approval of a draft invitation for a suggested list of participants and a broad conference agenda. After the inevitable delays, the conference was scheduled for late October of 1911, and Nernst's revised invitation went out in June of that year over Solvay's signature.

The upshot was that some two dozen prominent physicists, among them Planck and Einstein, converged on the Hotel Metropole in Brussels (Figure 3-2) for a historical gathering from 30 October to 3 November.[13] Both Onnes and Lorentz represented Holland. Onnes had been invited to report on his ongoing

Figure 3-2. Solvay Congress, 1911. Standing, left to right: Goldschmidt, Planck, Rubens, Sommerfeld, Lindemann, de Broglie, Knudsen, Hasenöhrl, Hostelet, Herzen, Jeans, Rutherford, Kamerlingh Onnes, Einstein, Langevin. Steated left to right: Nernst, Brillouin, Solvay, Lorentz, Warburg, Perrin, Wien, Mme Curie, Poincaré. Courtesy International Institute of Physics and Chemistry, and AIP Niels Bohr Library.

investigations of electrical resistance. In addition to discussing the equipartition theorem, Lorentz served as session chairman. Frederick Lindemann, with Maurice de Broglie (older brother of Louis) and Robert Goldschmidt (who had put Nernst in touch with Solvay) acted as scientific secretaries to the congress.[14]

Since Nernst had drawn up the list of topics to be discussed, specific heats naturally had a prominent place on the agenda; the failure of Dulong and Petit's law, and its reconciliation with quantum theory, were even singled out for mention in the opening lines of Solvay's letter.[15] In his keynote talk to the congress, Nernst reviewed the latest specific heat measurements with his colleagues.[16] He pointed out that Einstein's treatment of the problem was clearly in the right direction, but it disagreed with the measurements at the lowest temperatures of observation. The failure no doubt lay in Einstein's simplifying assumption that a single eigenfrequency was at work. Nernst and Lindemann had in fact obtained better agreement with a specific heat formula of their own. This was a revised version of Einstein's formula, derived by trial and error and published in July of 1911:[17]

$$C_V = (3R/2)\,[x^2e^x/(e^x - 1)^2 + (x/2)^2\,e^{x/2}/(e^{x/2} - 1)^2] \tag{3-1}$$

Nernst and Lindemann associated the first term with the kinetic energy of the vibrating atoms, the second term containing "half quanta" with their potential energy. They even argued for the theoretical plausibility of their formula in terms of a model of heated solid bodies.[18]

Einstein had obtained advance proofs of the Nernst–Lindemann paper with the new formula from Nernst the previous summer, just as he too struggled—indeed, "tormented himself"[19]—with the shortcomings of his own formula. He remained unconvinced of its theoretical basis, however. Nevertheless, he concurred in the open discussion following Nernst's report that, "the formula ... is without doubt a great achievement, but I caution that it is an empirical formula. It is very evident a priori that the atoms of solid bodies cannot themselves behave exactly as infinitely small resonators; I regard the cause of the difference between experiment and theory as due to the fact that the atomic oscillations are not necessarily monochromatic."[20]

In his own report to the congress on specific heats,[21] Einstein emphasized that a more sophisticated treatment of the problem called for a true sum over a spectrum of eigenfrequencies as in Eq. (2-5). Such a model would have to involve transfer of vibrational energy between the interacting atoms of a solid. Personally, however, he could make little quantitative progress on the problem (that was not high on his list of priorities), aside from deriving a relationship between frequency and compressibility.[22] The specific heat problem was to be tackled more successfully before long by Peter Debye and, independently, by Max Born and Theodore von Karman in a joint collaboration. Debye's derivation was based on the classical theory of elasticity, but it left out considerations of the atomic and crystal lattice. His well-known formula for C_V, predicting a cubic dependence of C_V on absolute temperature, nevertheless falls much more slowly with temperature at low temperatures than the Einstein or Nernst–Lindemann exponential formulae. Born and von Karman's derivation took into account the actual atomic or ionic lattice, but the tedious calculation failed to yield definite values of specific heats for particular substances.

Nernst did not dwell on electrical resistance in his keynote report to the congress. His review of the Berlin program in this area occupied less than 1 of the 36 pages comprising his share of the printed proceedings. Essentially it was a repetition of his remarks on the subject before the 83rd *Naturforscher Versammlung* at Karlsruhe on 24–30 September,[23] or just prior to the Solvay Congress. He noted the analogous behavior of electrical resistance and atomic heats at moderately low temperatures and above, and credited Onnes simply with "the discovery of a region of temperature where the resistance of platinum becomes constant."[24] He then drew attention to ongoing measurements at both Leiden and Berlin showing that the onset of the bend presaging flattening of the resistance curve occurred higher in temperature, the higher the eigenfrequency of the metal. This experimental point seemed particularly well borne out in the case of aluminum. The ν-value of aluminum, according to Lindemann's model, is much higher than that of platinum. And, sure enough, its measured resistance, reported by Nernst at Karlsruhe, attained a constant value in boiling hydrogen (Table 3-1).

TABLE 3-1

T	W
273.1	1.000
79.0	0.256
66.0	0.222
21.5	0.166
20.5	0.165
17.4	0.165

In contrast, the resistance of lead, whose characteristic frequency is one-third that of aluminum, was by Nernst's measurements still falling in subcooled liquid hydrogen, as low as Nernst could go. Indeed, had he possessed a helium liquefier, Nernst might well have had the honor of discovering the second superconducting element.

One gains the impression from Nernst's reports at Karlsruhe and Brussels that, to him—a neophyte quantum aficionado—the value of resistance measurements at low temperatures, aside from their thermometric utility, was mainly as an empirical vehicle for inferring eigenfrequencies of solids. Nernst chose not to comment on the Leiden results in liquid helium, presumably because Onnes was to follow him in the agenda of congress speakers with an authentic summary, and because Nernst was understandably preoccupied with his heat theorem as it related to quantum theory. Certainly Einstein, and surely Nernst, knew of the latest Leiden measurements by the fall of 1911. The news had even caused a minor stir in the higher echelon of physics in Japan.[25] At Karlsruhe Einstein had made a point of asking Nernst whether his measurements on aluminum could corroborate the role of impurities in obscuring Onnes's momentous conclusion that the conductivity of pure metals becomes infinite on the approach to the absolute zero.[26] Nernst had replied that the high resistance of aluminum (not the early onset of its plateau) was undoubtedly associated with impurities, but he suggested that a "finite" resistance nevertheless characterizes each metal.

Onnes, in the long discussion following Nernst's report to the congress, had little to say about resistance per se. He stressed the difficulty of fitting the bend in the resistance curve with a single frequency, but offered no explanation for why half of Einstein's ν-values from elastic constants provided a better fit than Nernst's specific heat values. Onnes allowed that the Nernst–Lindemann formula, Eq. (3-1), was a step in the right direction to explain the rapid change in frequency needed to fit the observations.

Onnes read his own report to the congress on 2 November.[27] This was an important element in the superconductivity chronicle as it was the first and only major account in 1911 of the ongoing resistance measurements at Leiden. It also succinctly bridged the earlier measurements in liquid hydrogen with those in liquid helium, including mercury. His presentation centered on four figures, reproduced here as Figures 3-3, 3-4, 3-5, and 3-6, plotting as usual the ratio

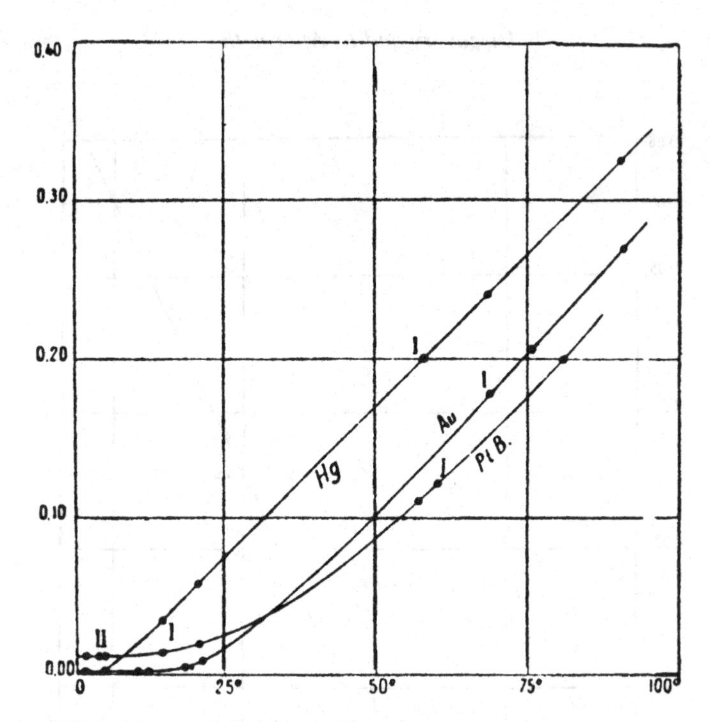

Figure 3-3. Resistance of platinum, gold, and mercury versus temperature, including Clay and Onnes's results in liquid hydrogen (I) and those of Onnes and Holst in liquid helium (II). Kamerlingh Onnes, Solvay 1911, 305.

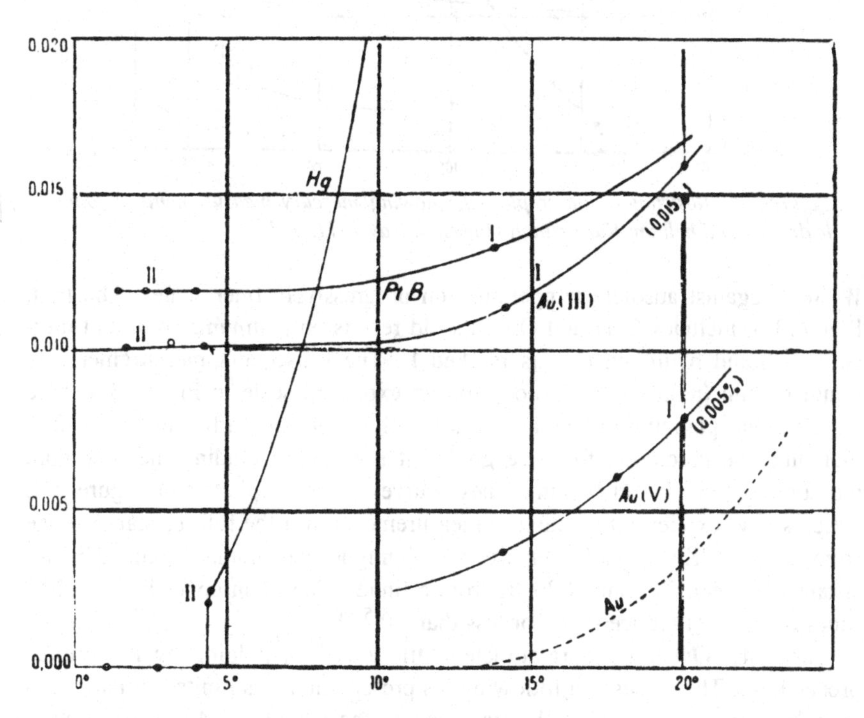

Figure 3-4. Resistance data replotted to a larger scale (compare with Figure 2-2). Kamerlingh Onnes, Solvay 1911, 306.

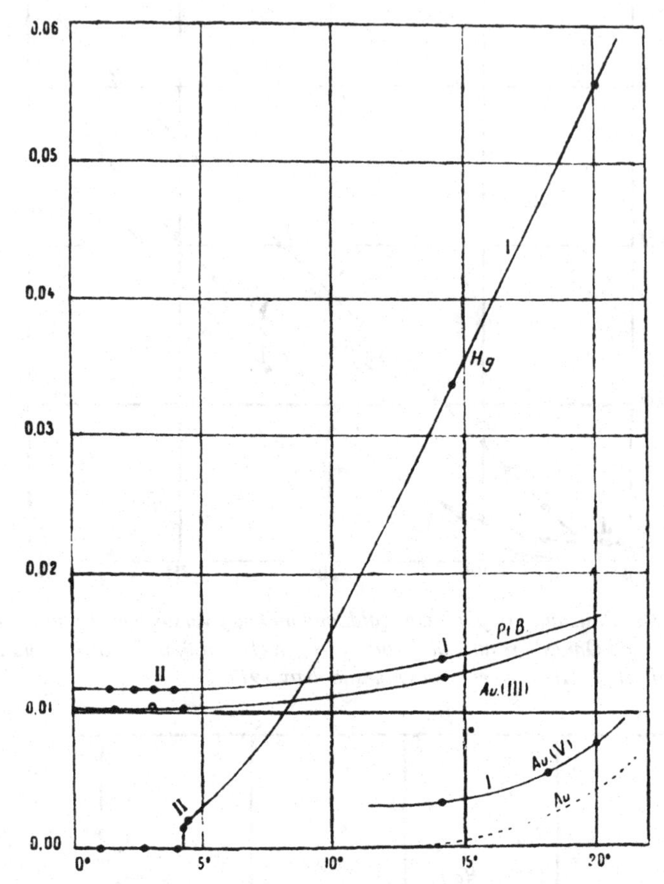

Figure 3-5. Resistance data replotted, showing mercury measurements in both hydrogen and helium. Kamerlingh Onnes, Solvay 1911, 307.

W_t/W_{273} against absolute temperature on progressively finer scales. The first, Figure 3-3, includes Clay and Onnes's old results on platinum, gold, and mercury in liquid hydrogen (points marked I). The subsequent measurements in liquid helium (points II) are shown on an expanded scale in Figure 3-4; here the gold and platinum curves are basically those plotted earlier in Figure 2-2, including the prediction for pure gold, but now also including the new gold measurements in liquid helium. These curves are all replotted in Figure 3-5, which shows explicitly the mercury measurements over the full resistance range from 2 to 20 °K.[28] Finally, Figure 3-6 highlights the dramatic plunge in resistance between 4.21 and 4.20 °K from a finite value to an upper limit of 10^{-6} times its value at the ice point, or less than 10^{-5} ohms.

Curiously, Onnes's report attracted little attention, judging by the printed proceedings. The discussion following his presentation was limited to Paul Langevin's query as to whether the transition in the vicinity of 4 °K was accompanied by any detectable structural alteration of the mercury thread, such as a

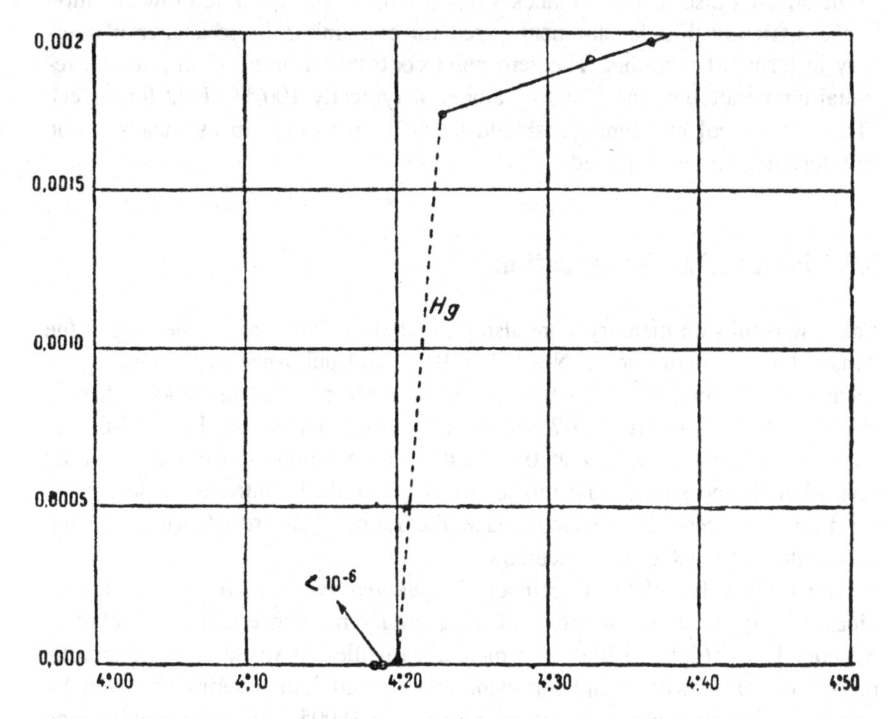

Figure 3-6. Mercury measurements near boiling point of helium. Kamerlingh Onnes, Solvay 1911, 309.

change in volume. The change in volume of substances seen during melting, Langevin pointed out, is accompanied as a rule by an enormous variation in electrical conductivity, probably because of a change in the number of free electrons. Perhaps, he suggested, the discontinuity in the variation of resistance can result from a discontinuity in the number of electrons liberated. Onnes replied that low-temperature measurements had been performed on the thermal conductivity, specific heats, density, expansion, and elastic properties of mercury. Although apparently nothing definitive came from this, he admitted that these measurements did not rule out the possibility of the increase in conductivity being a secondary result of a change in volume.[29] More promising, in his view, was the concept of a discontinuity in the period of Planckian vibrators. Simply doubling the frequency would be enough to account for the observed discontinuity in resistance, owing to Planck's exponential factor.

Planck himself did not help matters by dwelling on what became known as his "second quantum hypothesis," which embodied the concept of zero-point energy.[30] At Brussels, as a result of difficulties encountered with the assumption of discontinuous absorption of black body radiation, he gave for the mean energy of an oscillator the revised expression:

$$U = xkT\,[(e^x - 1)^{-1} + 1/2]\,. \qquad (3\text{-}2)$$

In the ensuing discussion of Planck's report, Onnes was quick to draw attention to the difficulty this development posed for explaining the resistance of mercury in terms of vibrators. The zero-point contribution implied an internal, residual temperature of the vibrating atoms of typically 100 °K ($h\nu/2$ for silver). "The [resistance] phenomenon should be difficult to explain by means of the new formula,"[31] he cautioned.

3.3 Hints of a Complication

The new results on mercury were also presented by Onnes at the meeting of the Amsterdam Academy on 25 November 1911, and published in the Dutch proceedings in December.[32] This time Onnes chose to plot the resistance values in absolute ohms, as in Figure 3-7, not as ratios of resistance as in Figure 3-6. The resistance of solid mercury at 0 °C had been extrapolated from the melting point, but the possibility that unknown effects could be introduced during the solidifying process, he decided, made the ratio of the resistance at helium temperatures to that at 0 °C uncertain.

The main value of the December *Communication*, however, is that it provides a long overdue description of the mercury resistance element, including mercury leads (Figure 3-8).[33] The mercury, distilled in vacuo at a temperature of 60° to 70 °C with liquid nitrogen, was poured into a series of seven U-shaped glass capillary tubes of approximately 0.005 mm cross section and frozen in them during the experiment. (The capillary had to be very fine, no larger than ~1/20 mm in diameter, to retain a resistance measurable with any precision.) The seven tubes were joined together at their upper ends by inverted *Y*-pieces that were sealed off above, and not completely filled with mercury; this gave the mercury an opportunity to contract or expand on freezing or liquefying without breaking the glass and without breaking the continuity of the mercury thread. To the *Y*-pieces b_0 and b_8 were attached two leading tubes Hg_1, Hg_2 and Hg_3, Hg_4 (whose lower portions are shown at Hg_{10}, Hg_{20}, Hg_{30}, Hg_{40}) filled with mercury which, on freezing, formed four leads of solid mercury. To the connector b_4 was attached a single tube Hg_5 whose lower part is shown at Hg_{50}. At b_0 and b_8 current entered and left through the tubes Hg_1 and Hg_4; the tubes Hg_2 and Hg_3 could be used for the same purpose or also for determining the potential difference between the ends of the mercury thread. The mercury-filled tube Hg_5 could be used for measuring the potential at the point b_4 in order to examine possible variation of resistance along the length of the thread.

Because of the cramped cryostat and in order to accommodate the bulky stirring pump *Sb* (Figure 3-1), the mercury tubes, shown as mounted in a plane in the cryostat drawing, were actually arranged in a cylindrical configuration as depicted in the lower middle insert to Figure 3-8. The leads projected above the cryostat cover Sb_1 as shown in perspective at the top. They too were provided with expansion spaces, while in the bent side pieces were fused platinum wires Hg_1', Hg_2', Hg_3', Hg_4', Hg_5'; these were connected to the measuring apparatus.

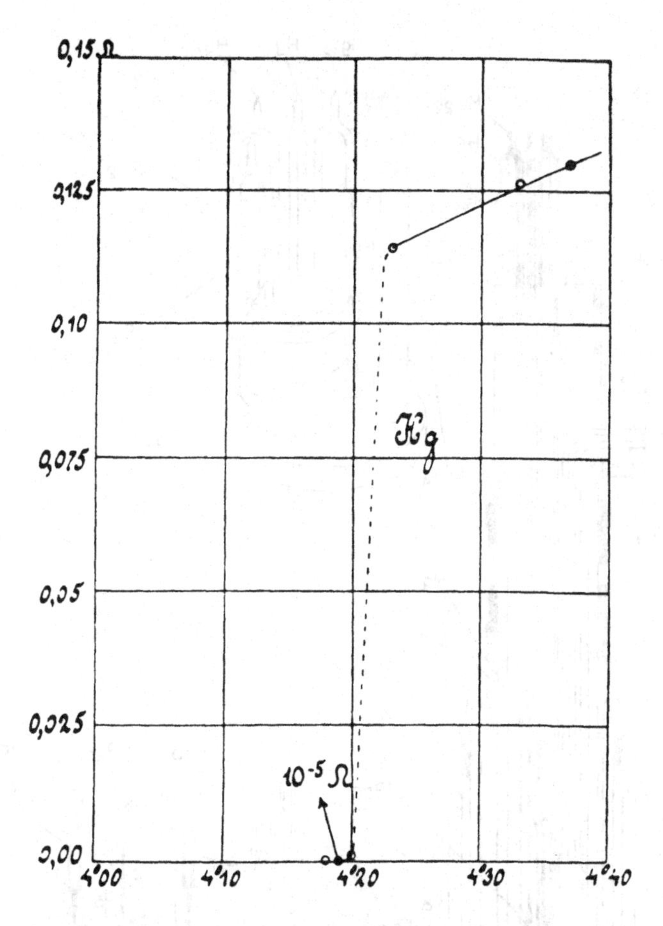

Figure 3-7. Superconducting transition plotted in absolute ohms. Kamerlingh Onnes, RN, ***14*** *(1912), 820.*

The junctions of the platinum wires with the copper leads of the measuring apparatus were shielded as effectively as possible from temperature variations. However, the mercury resistance itself, with the mercury leads, was found on immersion in liquid helium to harbor a considerable thermoelectric effect, in spite of the care taken to fill the circuit with pure mercury.

> A closer investigation of the true state of affairs was postponed for the meantime, and the thermoelectric force was directly annulled during the measurements by an opposed electromotive force taken from an auxiliary circuit. The magnitude of this thermoelectric force, which for one pair of the leads came to about half a millivolt, made it impracticable to reverse the auxiliary current as is usually done in the compensation method. The resistance of the mercury thread was then obtained from the differences between the deflections of the galvanometer placed in circuit with Hg_2 and Hg_3 and the compensating electromotive force, when the main current passing through the resistance was reversed. The galvanometer was calibrated for this purpose.[34]

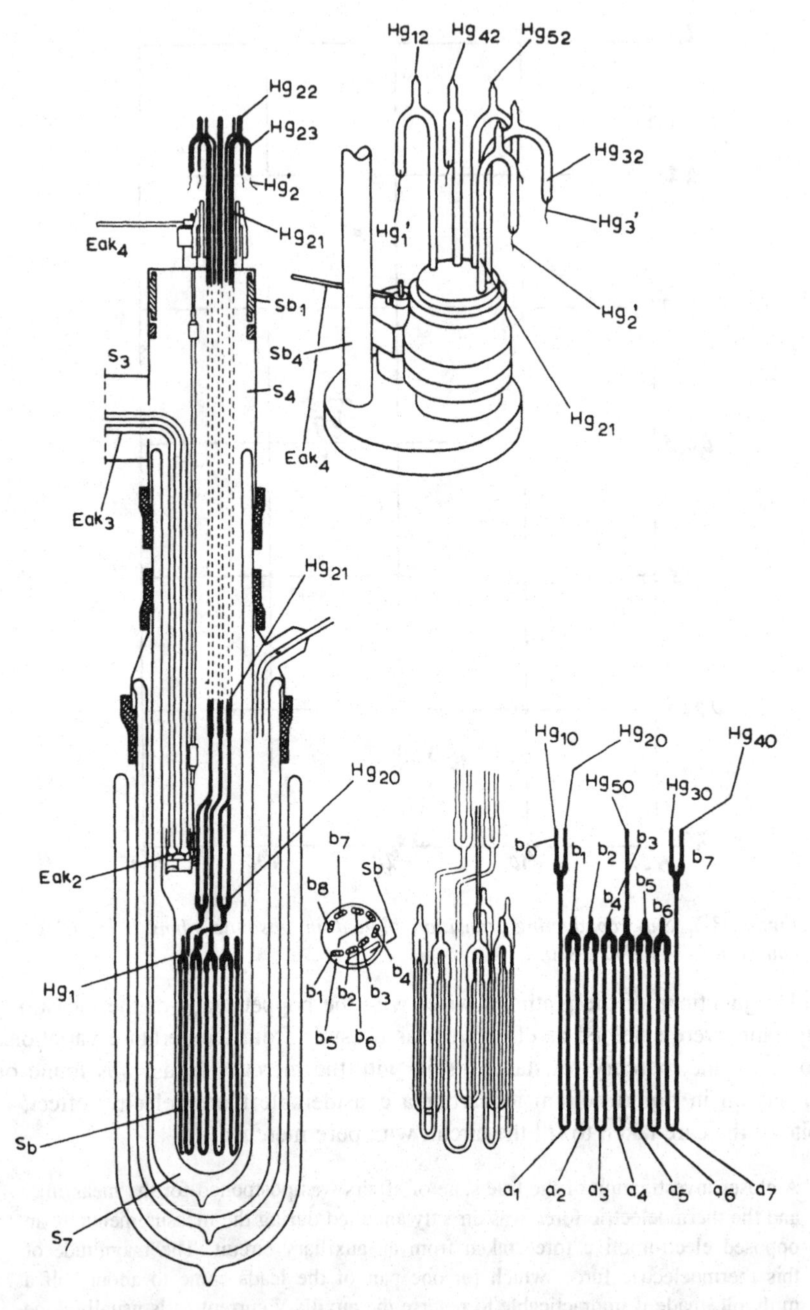

Figure 3-8. Helium cryostat, late 1911, including glass capillaries with mercury. Kamerlingh Onnes, RN, ***14*** *(1912), Plate I.*

Onnes concluded his report to the academy with a brief reference to an attempt, apparently repeated on several occasions, to pass a "comparatively" strong current through the resistance below the transition temperature. This seemed to be the only hope for improving the upper limit that must be ascribed to the "practically vanishing" resistance at, say, 3.5 °K. If the resistance was in fact finite, a stronger current should produce a larger potential difference. The attempt was thwarted, he noted curtly on an ominous note: "the peculiarities of the phenomena which then occur make it desirable to experiment first with a modified apparatus before proceeding further."[35]

In his holiday letter to Woldemar Vöigt at Göttingen that month Onnes noted his intention to continue to expound on the "singularly mysterious" behavior of mercury without delay.[36] Yet, 1912 came and went without additional word on the subject. Nevertheless his team was not idle. We know now that in the closing weeks of 1912 the resistance phenomenon was found not to be unique to mercury. As if this were not enough, nearly all of the year was devoted to clarifying the newfound "peculiarities," further distracting Onnes's team from the underlying phenomenon itself.

CHAPTER 4

Great Expectations

4.1 A Critical Current

The silence was broken in the spring of 1913 with the publication of a four-part definitive *Communication* covering a year of lively developments, as if to atone for the meager progress reports that had been issued until then. Onnes read the first installment at the meeting of the academy on 22 February.[1] Something new had cropped up, he could report. Already in the experiments of October of 1911, his team had indications of "special phenomena" setting in when an electric current of great density was passed through a mercury thread. They had invested more than a year in the matter, and the situation was still far from clear.

> Not until the experiments had been repeated many times with different mercury threads, which were provided with different leads chosen so as to exclude any possible disturbances, could we obtain a survey of these phenomena. They consist principally herein, that at every temperature below 4°.18 K for a mercury thread inclosed in a glass capillary tube a "threshold value", of the current density can be given, such that at the crossing of the "threshold value" the electricity goes through without any perceptible potential difference at the extremities of the thread being necessary. It appears therefore that the thread has no resistance, and for the residual resistance which it might possess, a higher limit can be given determined by the smallest potential difference which could be established in the experiments (here 0.03×10^{-6} V) and the "threshold value" of the current. At a lower temperature the threshold value becomes higher and thus the highest limit for the possible residual resistance can be pushed further back. As soon as the current density rises above the "threshold value", a potential difference appears which *increases more rapidly than the current*; this seemed at first to be about proportional to the square of the excess value of the current above the initial

value, but as a matter of fact at smaller excess values it increases less and at greater excess values much more rapidly.[2]

The "phenomena" were believed to be caused by a heating of the conductor, possibly a consequence of "peculiarities in the movement of the electrons"—or more likely simple Joule heating. Cooling the mercury thread by direct contact with helium might have been instructive, but the necessity of freezing the mercury in glass capillaries ruled that out. The awkward properties of mercury exasperated the investigation greatly. Following a day of laborious preparation for a run with liquid helium lasting a few hours at best, the delicate mercury thread was often found to have separated in the freezing process. If it did not, "very likely in consequence of some delay caused by the careful and difficult preparation of the resistance, the helium apparatus would have been taken into use for something else."[3] Or, in the odd likelihood of a successful run on the first attempt, the opportunity to check or repeat an observation, or follow up on a suggested experimental modification, was limited by the weekly quota of cryostats that could be filled. "Under these circumstances the detecting and elimination of the causes of unexpected and misleading disturbances took up a great deal of time."[4]

The new experiments seem to have been seriously under way in December of 1911. Being reasonably assured of the validity of Ohm's law for mercury above the "vanishing point" from experiments performed the previous summer (3 and 6 milliamperes gave about the same resistance at 4.23 °K, 84 mΩ), Onnes now had the temperature reduced to 4.20 °K. The resistance here was essentially zero, in agreement with an earlier run (Figure 3-7). Raising the current slowly with the temperature held constant, Holst reached a point where considerable time was required before the potential readings stabilized, at a current slightly in excess of 7 mA. Once conditions did stabilize, the resistance was estimated to be 0.000746Ω, calculated by Ohm's law "without wishing to give it beforehand any other meaning."[5] Cooling the thread to 4.19 °K with the same current gave $1.4 \times 10^{-5}\Omega$, in agreement with the measurements made in 1911. At this temperature, the current could be raised to 14 mA before a potential difference became perceptible across the ends of the resistance (Holst could measure 0.03 microvolts). Above 20 mA, the potential difference rose rapidly; the sample was no longer stable. The trick was to obtain an even lower temperature. Pumping on the helium reduced the bath to 3.65 °K. Now it required nearly an ampere to produce an observable potential difference.

The temperature trend of the threshold value restricted attempts to reduce further the resistance limit by the artifice of stronger currents to the lowest attainable temperature. The attempt at 3.65 °K gave a value of 10^{-9} of the resistance at 0 °C. Although it was two orders of magnitude below the limit established in May of 1911,[6] Onnes suspected that the new limit might be even further reduced if only the disturbances—possibly operative even *below* the threshold—could be fully eliminated. The obvious approach to try first was eliminating the possibility of any Joule and thermoelectric heat reaching the resistance by conduction from the external circuit. For this purpose, both the mercury current leads and potential wires were diverted en route to the measuring apparatus, first passing downward through the helium bath. In addition, considerable heating blamed on the Peltier effect had initially plagued the circuit, which was attributed to Peltier heating from impurities in the mercury legs. Although great care was bestowed on the distillation procedure, the troublesome E. M. F. was not fully eliminated and required compensation during the measurements. Despite these precautions, "the results ... obtained .... correspond pretty well with the previous ones (Dec. 1911)."[7] Nor was the sharpness of the transition affected; the plunge in resistance began at 4.21 °K and extended over an interval no more than 0.02 °K. (Much later measurements at Leiden would, in fact, reveal a dependence on crystalline purity and the strength of the measuring current.) As a result of the various experimental improvements, the resistance limit was nevertheless pushed to less than 4.1×10^{-10} in January of 1912, and to about half of that value in February.

In view of the "close correspondence of the phenomena"[8] (even replacing the original, U-shaped resistance tubes with a single tube of greater cross section did not qualitatively affect the results), Onnes was becoming convinced that the cause of the potential phenomena resided in the mercury resistance itself, not in some outside disturbance. For one thing, the possibility of a breakdown of Ohm's law for mercury *below* 4.19 °K could not be discounted. The electron theory, Onnes suggested, supplemented by his vibrator hypothesis and the additional stipulation that the electrons move freely through the lattice between collisions with the vibrators, offered the possibility of mean-free path lengths below the vanishing point comparable to the dimensions of the conductor—100 cm or longer. Under these conditions, the electron drift velocity under the applied field might no longer be negligible compared with the velocity of thermal agitation—a pillar of the classic electron theory. For a certain current density, dependent on temperature, the drift velocity "might be just sufficient to bring the vibrators into motion, which otherwise below 4.19 °K are stationary," rather like radiation stimulation of Planckian vibrators.[9]

More plausible, considering the high current densities used in some of the experiments (up to 1000 A/mm^2), was some kind of enhanced magnetoresistance of the conductor exposed to its own self-field. Substantial magnetoresistance had recently been measured by Onnes and Bengt Beckman (a visitor in residence from Uppsala) in solid mercury at hydrogen temperatures,[10] and a strong temperature dependence for the effect was known for other substances. Other nonexotic explanations seemed possible, even if they were farfetched. A very small residual resistance, either intrinsic to the pure metal or as a result of crystal admixtures, might be evenly distributed throughout the thread. About as plausible and equally improbable was highly localized heating by a great current density in an otherwise resistance-free conductor.

Experiments in January of 1912 were designed to determine more quantitatively the influence of the thread thickness—that is, current density—on the transition temperature and on the breadth of the transition. For this purpose two mercury threads were prepared, one with a resistance of about 50Ω and one of about 130Ω. Both were mounted in a circuit with a milliammeter, and across

TABLE 4-1

Current (A)	W_{130}	W_{50}
0.006	0.0545	0.0252
0.010		0.0250
0.016		0.0249
0.030	0.0549	0.0260

each of them one of the coils of a differential galvanometer was connected as a shunt. Using one coil at a time, the resistance of each of the mercury threads could be measured separately; by connecting the two coils in the opposite direction the changes in the resistance ratio with temperature could be investigated as long as the difference was small. A change in the resistance ratio W_{130}/W_{50} between the ice point and the boiling point of helium was observed (2.55 vs. 2.18, respectively). Such an effect had been noticed before, and could be readily explained by random variations in the freezing of the two mercury threads. Once at 4.25 °K, however, changing the current strength had a negligible effect (Table 4-1).

Ignoring the slight deviation at 30 mA, the constancy of these data still gave Onnes no reason to doubt Ohm's law *above* the vanishing point. Next, the resistances were monitored as the temperature was lowered through the transition in steps of a few millidegrees (Table 4-2).

"On lowering the temperature from the boiling point to where the appearance of the resistance begins, this ratio remained unchanged according to the observations with the differential galvanometer," wrote Onnes; "from that point

TABLE 4-2

T	3.7 amp/mm^2 W_{130}	16 amp/mm^2 W_{50}
4°.24 K	0.0532	0.0244
4.22	459	182
216	314	0.0069
214	264	34
213	190	13
210	128	0.0003
207	0.0087	1
205	50	1
201	46	0.0000
196	21	0.0000
190	0.0005	0.0000
180	0.0000	0.0000

downwards the resistance in which the current density was smaller, disappeared more quickly."[11] Although the resistance still disappeared over a finite temperature interval, Onnes suspected that the transition was actually instantaneous; he attributed the observed breadth in the transition curve to gradual cooling of the mercury thread over its length.

The experiments in January revealed a new problem that had to be guarded against. With W_{50} maintained at some temperature below the vanishing point, the current density threshold was not yet reached with a current of 1 ampere. On raising the current to 1.5 amperes, however, enough Joule heat was generated by the current in the platinum wires joining the mercury leg to bring quickly the thin mercury thread to a temperature above the vanishing point. "All this was accompanied by a rapid boiling of the helium, while the ammeter showed a strong falling off of the main current corresponding to a decided rise in the resistance. From the readings, it could be seen that the resistance of W_{50} had risen to that which it has at hydrogen temperature."[12]

This incident plainly showed that the effort to thwart an influx of heat by conduction from the outside had failed. The experiment was therefore modified once more. The mercury thread was extended at each end with an auxiliary mercury guard thread of a larger cross section. Any potential difference seen across the extremities of both auxiliary threads could not, by their larger cross section, be a consequence of heat entering the extremities of the thinner mercury thread. It was expected that the central thread would show a potential difference before the guard threads; any heating in the main thread would have to be developed within the thread itself. The experiment failed at first, however, confirming merely that accidental circumstances in the freezing of the mercury, in this case the guard threads, could indeed play a role in the threshold effect. Better results were obtained in February with more carefully frozen guard threads. Current densities as high as 1000 A/mm^2 (2.5 amperes in a thread 0.028 mm in diameter) could be employed at 2.45 °K without developing any heat in the wire. This threshold value produced a new record for the resistance limit, which was not to be broken for over a year:

$$\frac{W_{2.45}}{W_{273}} < 2.10 \times 10^{-10} .$$

The experiment also proved fairly conclusively that the development of heat at still higher densities *originated in the thread itself.*

A final set of definitive measurements, made in June of 1912, concluded this series of experiments. Curves of resistance versus temperature for the same wire carrying 1.3, 13, or 130 A/mm^2 established today's well-known rule according to which the threshold current density is fairly represented as being inversely proportional to the helium bath temperature, increasing approximately linearly with the difference between the bath temperature and the vanishing point, at least over a limited temperature range.

Although experiments in February left little doubt that the disturbing potential effects were due to heating, Onnes was no closer to the *cause* of the heat.

He still regarded impurities as the most likely candidate, in spite of the labor expended on the purification of mercury.

> Now the experiments had realized the expectations [Onnes reported to the Academy on March 22], that mercury could be so far freed from impurities, as to make the resistance practically nothing. But if one may judge by the additive resistance which even very pure gold exhibits [e.g., Au_V in Figure 2-2], then with the residual resistance of mercury which is only perceptible at the threshold value of current density for the lowest temperatures, it would be a question of an impurity of the order of a millionth of the trace that could possibly be present in the most carefully purified gold. And it was *a priori* doubtful if the mercury could be procured in so much greater a state of purity than gold.[13]

The experiments of February were therefore repeated with trace impurities added to the solid mercury.[14] First distilled in a vacuum by means of liquid air, the mercury was brought into contact with gold, and then it was mixed with a larger quantity of pure gold. In a second attempt, the mercury was similarly treated with cadmium. To Onnes's chagrin, "the resistance disappeared in the same way as with pure mercury; much of the time spent on the preparation of pure mercury by distillation with liquid air, might therefore have been saved."[15] Not only that, but even common amalgamated tin of the kind used for backing mirrors was found to have zero resistance at helium temperatures. (It almost seemed to be mocking them when its vanishing point was found, in December of 1912, to lie somewhat *higher* than for pure mercury, 4.29 °K.)

The new experiments eliminated Onnes's strongest impurity candidate, mixed crystals in the solid mercury. "Less conductive particles" separated out during the freezing seemed unlikely, assuming the laws of current division between two conductors in contact still held when one is pure mercury at a temperature below 4.19 °K. To be sure on this point, "experiments on the possible influence of contact with an ordinary conductor upon the superconductivity of mercury"[16] were tried, and in the process a new and lasting term in the vocabulary of physics was coined. (Henceforth, Onnes and his followers would use *super*conductivity and *supra*conductivity interchangeably, an inconsistent practice occasionally plaguing the literature even today.) This time the test seemed simple enough: replacing the glass capillary tube with one of steel. By Kirchhoff's rule, the superconductor should short-circuit the steel. On the other hand, Onnes argued, "if the electrons can hit against the vibrators of an ordinary conductor, they [ought to] give off work in this collision. Thus a thread of superconducting mercury, if an ordinary conducting particle were present anywhere in the current path, could show resistance at that spot, even although the particle did not entirely bar the section which was otherwise free from resistance."[17]

In fact, the resistance did not disappear, and the experiment proved nothing. The resistance could equally well have been brought on by the smallest gap between the mercury and steel. Had the resistance disappeared, nothing more would have been learned, Onnes decided, since the contact between mercury and steel would have been in doubt.

Communication 133b ended with the not very enlightening conclusion that the heating was associated with an as yet unexplained local "microresidual" resistance, in contradistinction to an "additive mixture" resistance such as that which would have been caused by impurities.

The third of the spring reports of 1913[18] (the last Leiden *Communication* on the superconducting state of mercury) was no more conclusive. Inhomogeneities in the form of different states of crystallization along the length of the mercury threads were explored. Joule heating might occur at the dividing surface between such regions, and a source of Peltier heat at the interface could still not be ruled out. But Onnes appears to have exhausted the store of convincing possibilities, referring vaguely to "one or two things that seem to confirm [these suppositions]" which, however, were "too indefinite" for publication. "Taking all this together," he concluded simply, "we are brought back to the idea that the potential phenomena must be ascribed to 'bad places' ... "; yet, "the regularity of the phenomena" remained "a weighty objection to this hypothesis."[19]

4.2 Superconducting Magnets

December of 1912 brought new surprises. Judging by the experience with gold and platinum, Onnes had not been very optimistic about prospects for additional superconductors, he confessed in his report to the Amsterdam Academy the following May.[20] Both elements, the traditional candidates in the quest for zero resistance, had failed to deliver on their promise; mercury was an anomaly. Besides, preoccupation with the critical current phenomena had left no time for a systematic search. The biggest drawback of mercury, being a liquid at room temperature, was that it could not be easily wound into a coil. The solution proved surprisingly simple when, on 3 December, a wire of especially pure Kahlbaum tin was tested for superconductivity. It had been melted in a vacuum and poured into a glass capillary U-tube; its resistance at the ice point was 0.27Ω. At the boiling point of helium, its resistance was low but still finite: $1.3 \times 10^{-4}\Omega$. At 3 °K, however, it had vanished or was below $10^{-6}\Omega$. Tracing the temperature interval from 3 °K back to 4.25 °K in finer increments, the disappearance was found to occur suddenly at 3.78 °K.[21]

The new superconductor was well-timed for the present round of studies, which had gone about as far as it could with mercury. Two small solenoids were prepared without delay in order to increase the sample length and measuring sensitivity. These modest coils, constructed by the resourceful Flim, share the distinction of constituting the first authentic superconducting solenoids. A steel core was covered with a substantial layer of pure tin, turned down in a lathe, and a thin spiral shaving was cut off with a sharp chisel. This method avoided the work-hardening invariably associated with drawing a wire.[22] Several such spirals, rolled to a final strand thickness of 0.01 mm, were joined into a longer wire by melting them on to each other. The resulting wires, each about 2 m long, were wound on glass cylinders between a spiral of silk thread.

Current leads of tin were fastened to the wire ends; they led downward through the helium bath, and attached to copper wires.

The first measurements in liquid helium, with Holst at the galvanometer as usual, confirmed a transition as sharp as that observed in mercury but lower in temperature (3.785 °K for either coil). They revealed a microresidual resistance less than 10^{-7} times the resistance at 0 °C, and a threshold current density accompanied by potential "phenomena" that "increased rapidly with the increase of the excess of the current above the threshold value. In a word [Onnes added], the tin wire behaves below the vanishing temperature of the tin, 3.8 °K, qualitatively precisely the same as a thread of mercury below the vanishing point of that metal."[23]

The next superconductor came swiftly: The superconducting state of lead lagged behind that of tin by only a few days. One reason for focusing attention on these two metals just then may have been their simultaneous application in Bengt Beckman's ongoing, parallel program on Hall-effects at low temperatures. For one thing, the method of preparing long tin wires without drawing had been devised by Beckman. A lead wire prepared this way was found to be superconducting on immersion in liquid helium without even pumping on the helium bath. It remained superconducting when the temperature was raised as high as the cryostat permitted, 4.29 °K. Based on the trend in the current threshold with temperature up to this point, the vanishing point for lead was estimated to lie near 6.0 °K. (The temperature interval between 4.3 and 14 °K, the lowest temperature within reach in liquid hydrogen, is awkward to bridge with helium and hydrogen refrigerators, and remained uncrossed at Leiden until 1923.)

The experiments continued, concentrating both on reducing the ceiling on any remaining microresidual resistance and on illuminating the stubborn threshold behavior. At best, marginal progress was made. First, a "branching" tin wire, like the mercury array of U-shaped threads, was wound around a glass tube. As with mercury, it incorporated a pair of hefty guard wires since Onnes had been unduly optimistic about the maximum usable current strength, and he wanted to be sure no Joule heat penetrated to the wire. The critical current was disappointing: 1.6 amperes (150 A/mm^2) at the lowest temperature (1.6 °K). Just below the vanishing point it fell to 6 mA. A new, larger coil of tin insulated with picein did relatively worse. The disappearance of the resistance extended over a much broader temperature range, which Onnes blamed on reduced cooling due to the picein. More frustrating yet, the critical current at the lowest temperature was now only a third of its former value. Experiments with lead were no more promising, other than serving to reduce the resistance limit by a factor of four, or from 2.1×10^{-10} to 0.5×10^{-10}. But Onnes remained guardedly optimistic, in spite of the lackluster performance.

> Our results with tin and lead make it seem probable that all metals, or at least a class of them, if they can be procured sufficiently pure, pass into the super-conducting state when reduced to a low enough temperature. Perhaps in all it would also be suddenly. But the additive admixture-resistance which can be caused by mere traces of admixtures, will in general make the detection of the phenomena a difficult one.

> A number of experiments with resistance-free conductors of which several suggest themselves at once, now that we can use the easily workable super-conductors tin and lead, can be undertaken with good prospects of success.
>
> In this way the preparing of nonresistive coils of wire, with a great number of windings in a small space, changes from a theoretical possibility into a practical one. We come to new difficulties [however] when we want not only to make a nonresisting coil, but to supply it as a magnetic coil with a strong current.
>
> I have been engaged for some time making a preliminary estimate of these difficulties.[24]

The new difficulties alluded to here held the key to those plaguing Onnes all along. They refer specifically to experiments undertaken in late February with a pair of small coils. One was the aforementioned tin-picein coil and one coil was wound with lead insulated by silk soaked in liquid helium. Both coils, prepared by Flim, are still extant.[25] Before winding the first one, a sample of the tin wire was wisely tested in liquid helium; 8 amperes could be passed through the wire without exceeding the threshold current density (560 A/mm^2 for a wire cross section of 1/70 A/mm^2). However, *after* the wire was wound into a coil, the threshold was reached with only 1.0 amperes. The lead coil, of similar geometry and size,[26] exhibited essentially the same threshold current, 0.8 amperes. Onnes neglected to give its "short sample" threshold current (to use a modern term) but noted that a coil current of 0.8 amperes corresponded to about 800 ampere-turns per cm^2. He had expected to pass 9000 ampere-turns per cm^2, which would imply a short sample current of 9.0 amperes.

Onnes attributed the disparity in performance between wire and coil to a lack of adequate cooling in the insulated coils compared to the bare, stretched wire (short sample) in liquid helium. (He had originally expected that the helium permeating the coil through the silk texture would promote uniform heat transfer over the coil.) It is tempting but not convincing to conclude from the evidence that Onnes anticipated at this time quantitative experiments on the effects of a magnetic field—experiments announced only 8 months later. As for the threshold effects in general, he concluded that the presence of "bad places" in the wire was still the most plausible explanation—that is, not phenomena intrinsic to the superconducting state but extrinsic effects: inhomogeneities in the form of, for example, different states of crystallization or mechanical tension along the wire or thread.

> If we may therefore be confident that they can be removed (for instance by fractionising [sic] the wire) and if moreover the magnetic field of the coil itself does not produce any disturbance ... then this miniature coil may be the prototype of magnetic coils without iron, by which in future much stronger magnetic fields may be realised than are at present reached in the interferrum of the strongest electromagnets.[27]

4.3 A Vision of 100 Kilogauss

In the fall of 1913 Onnes attended the Third International Congress of Refrigeration, which was held in Washington and Chicago from 15 September to

1 October.[28] He took the occasion to thank personally Director H. S. Miner of the Welsbach Light Company in Gloucester City, New Jersey for the precious gift of a quantity of helium gas, supplanting the dwindling stock from monazite; the new helium gas was derived from thorianite. The importance of this contribution is clear from the tone of a letter Onnes wrote to Miner the previous spring, which acknowledges the receipt "in good order 4 drums with helium gas," and continues:

> I thank you very much indeed for the precious and liberal gift which you have made once more to my laboratory. Until about a year ago our stock of helium was quite sufficient for our experiments; but then we had a serious loss so that from that time our stock was only just sufficient to continue this work. From this you will see how absolutely necessary it was to have a new supply. We could have made the helium ourselves so as we have done before, but at the cost of an enormous amount of work and time.[29]

At Chicago, Onnes delivered a major review paper on cryogenic research at Leiden since the last congress on refrigeration (held in 1910), including the new phenomenon of superconductivity.[30] (Onnes came straight from the second Solvay Congress, held again in Brussels immediately before the congress in America and devoted to the structure of matter. Electrical conduction was not on the formal agenda this time.) He still seems to have regarded superconductivity as an extreme case of the normal mechanism of electrical conduction in metals. Although he described mercury as having "passed into a new state, which on account of its extraordinary electrical properties may be called the superconductive state," he added confidently that "there is left little doubt, that, if gold and platinum could be obtained absolutely pure, they would also pass into the superconducting state at helium temperatures."[31] He said this in spite of the strong evidence that superconductivity is not simply a matter of purity. (Much later, in Onnes's last effort to reach lower temperatures in 1922, he still had not given up on gold: "A simple example [of problems awaiting resolution beyond the low-temperature barrier] is the question whether a metal such as gold can be made super-conducting by cooling it more than we have been able to do.")[32]

In his report to the congress, Onnes dwelt mainly on the exciting possibilities for high-field, air-core magnets wound from the new superconductors. The field intensity from iron-core magnets, the standard route to high magnetic fields, was normally limited by saturation of the core to approximately 20 kilogauss; higher fields could be reached, but only with heroic efforts. Jean Perrin's old proposal (we met Perrin before in connection with cathode rays) to generate 100-kilogauss fields with air-core copper magnets cooled with liquid air (for exploiting the Zeeman effect) had been squashed by Charles Fabry's closer analysis in 1910 of the technical requirements for such an ambitious project.[33] Fabry had found that the amount of power required to generate a given magnetic field in a cylindrical coil, expressed in watts, could be represented by the formula:

$$W = \rho\eta a \frac{H^2}{K^2}. \quad (4\text{–}1)$$

Here, a may be taken as the inner radius of the coil (in cm), ρ is the resistivity of the copper in ohm-cm, η the packing fraction of the winding (ratio of copper area to total winding cross section), H is the field in gauss, and K is a dimensionless constant depending on the winding geometry, but approximately 0.18 for a cylindrical coil of uniform current density. (In the literature K is commonly denoted by G for "geometry factor," and sometimes known as the "Fabry factor.") Assuming a copper resistivity in liquid air of 0.27×10^{-6} ohm-cm (1/6 its ordinary value) and that the Joule heat could be withdrawn at the same rate it is generated, Fabry estimated that to produce 100 kilogauss in a 1-cm coil would require 100 kilowatts. The electrical power for charging the winding would pose no real difficulty; rather, the problem was the Joule heat generated in the small coil volume (25 kilocalories per second). For it to be removed by evaporation of liquid air would require approximately 0.4 liters of liquid per second, or 1500 liters per hour. About seven times more power would be needed for the refrigeration than for powering the coil.

Even more problematic, Fabry concluded, was ensuring adequate heat transfer between the supposedly compact coil and the coolant. This would, in fact, require a coil of considerably greater volume than originally envisaged, increasing the electric power demand (and amount of liquid air) in the same proportion. The cost of implementing Perrin's scheme "might be about comparable to that of building a cruiser!"[34]

Onnes had reexamined Perrin's proposition in light of the cryogenic realities viewed from Leiden. Substituting silver in Eq. (4-1), still with $a = 1$ cm, and assuming liquid hydrogen instead of liquid air, did not alter the prospects significantly. The required hourly rate of liquid was halved, but the refrigeration watts per watt of electric power became even more unfavorable. Allowing, moreover, for a realistic coil size, Onnes concluded that a "fantastic extension" of the Leiden plant would still be required. Even liquid helium did not seem practical, as long as the coil was not superconducting.

But with superconducting wires of tin or lead, the prospects changed drastically. Joule heat would no longer be a problem, even at the highest current densities contemplated, in Onnes's view. Due to the high current density, the coil could be of relatively modest dimensions. There remained the unresolved possibility of a resistance developing in the superconductor by the magnetic field. But here Onnes felt the coil experiments at Leiden offered some hope. As we noted earlier, he attributed the poor coil performance relative to that of the stretched wire sample to poor heat conduction from "bad places" (mechanical defects) in the wire, due to the interposed insulation of silk. He did not believe magnetoresistive heating played a role, he remarked at the congress in Chicago, because no appreciable resistance was apparent below the threshold value of current where magnetoresistance usually became measurable. (Nowhere does Onnes give the field in the small lead coil explicitly, but it is readily calculated from the dimensions of the coil and from the number of ampere-turns; the geometric factor involved is related to the Fabry factor. A current of 0.8 amperes—the critical current—in 1000 turns wound in a cylindrical coil of 1-cm cross section with an inner diameter of 1 cm, would produce a field on the coil axis of 400 gauss. The peak field in the winding, i.e., that seen by the conductor, would be approximately 450 gauss.)

Improved coil performance could be expected by interposing metal foil between the windings, instead of silk, since an ordinary metal is an insulator compared to a superconductor.

> In a coil of bare lead wire wound on a copper tube the current will take its way, when the whole is cooled to 1.5 °K, practically exclusively through the windings of the superconductor. If the projected contrivance succeeds and the current through the coil can be brought to 8 amperes as for the stretched wire, we shall approach to a field of 1000 gauss. The solution of the problem of obtaining a field of 100000 gauss could then be obtained by a coil of say 30 centimeters in diameter and the cooling with helium would require a plant which could be realised at Leiden with a relatively modest financial support. If we cannot wind the wires so closely as was done in the experiment which I have described, the dimensions of the coil of which I have spoken will have to be taken greater and of course the difficulties and cost will increase proportionally. When all outstanding questions will have been studied and all difficulties overcome, the miniature coil referred to may prove to be the prototype of magnetic coils without iron, by which in future much stronger and at the same time much more extensive fields may be realised then [sic] are at present reached in the interferrum of the strongest electromagnets. As we may trust in an accelerated development of experimental science this future ought not to be far away.[35]

Apparently Onnes entertained real hope for the "modest support" with the *Association Internationale du Froid* acting as intermediary. According to his colleague J. P. Kuenen he had secured promise of support for such a venture from, among others, Arséne D'Arsonval (introducer of, among other things, the first moving-coil galvanometer) and Georges Claude[36] (co-inventor, with Linde, of the process for producing liquid air in quantity), both prominent members of the association. Nevertheless, he did not discount the seriousness of the potential phenomena above the threshold current. As a lesson of what to expect from the enormous increase in resistance, Onnes quoted an experiment on a bare lead wire immersed in liquid helium. When the critical current was exceeded, the Joule heat generated could not be removed fast enough because of the formation of gaseous helium bubbles; the temperature rose rapidly to the melting point and the wire parted.

> If matters are scrutinized more closely it seems that generally, when the current surpasses the threshold value, a local heating of a very small portion of the superconductive wire takes place, so that the step-up point of resistance is reached; at greater densities this temperature spreads over the wire, till there is equilibrium between the heat given off and the Joule-heat generated. I cannot go into all the details of the phenomena, but the bearing upon the question of experiments with superconductors as soon as we go to greater current densities is evident. The great question is, whether we have to do with phenomena that are to be attributed to

> local disturbances by bad places, places at which Joule-heat or Peltier-heat is developed, or that they will occur at a given current density in the metal itself even if it is homogeneous, unstrained, everywhere crystallized in the same manner.[37]

If the potential disturbances were indeed associated with bad places, so could the limiting microresidual resistance. Assuming means could be found to prepare wires free of bad places, "there would then be no limit to the current density at all and the magneto Joule-heat would become the determining factor for the dimensions and the intensity of the field."[38] But should the premature onset of resistance prove to be intrinsic to pure metals, this would certainly place a severe limit on the realizable current density, and on the minimum practical coil size. A consolation, in this case, would be the apparent realization of a domain of electrical conductivity impervious to Ohm's law.

Returning to the "simpler" case of superconductivity below the threshold, Onnes concluded his presentation in Chicago by raising several intriguing experimental possibilities. One he had posed in his report to the academy the previous May. If the superconducting state is a quirk of extremely long mean-free paths, it suggested the possibility of devising a "Lenard window" of superconducting foil. Had not amalgamated, though not ordinary commercial, tin foil been found to be superconducting? In May, Onnes had stated that he had interested Lenard himself in collaborating on such an experiment;[39] in Chicago he hinted that experiments in this direction were forthcoming, and in Stockholm, subsequently, he noted that "relevant preliminary experiments" were actually in progress. Apparently the matter was not followed up at Leiden, however, or perhaps it was simply dropped, although somewhat related experiments were conducted much later at Toronto and Berlin.[40] Another question raised earlier, whether a superconductor remains so in contact with an ordinary conductor, had already been resolved in the affirmative. A thin layer of tin, produced by tinning over a constantine wire, became superconducting despite the normal substrate. (This experiment was to influence Einstein's views on the superconducting mechanism some years later.)

Here perhaps we would not be amiss to chide or defend Onnes for attaching curiously little importance to "explanations" of "phenomena." He appears to have harbored a characteristic reluctance to draw premature conclusions without exhausting the experimental possibilities. In the opinion of some, Kamerlingh Onnes was "never really interested in the explanatory schema as such, but only in so far as these schemata can facilitate the further investigations of the different facets of what is being studied."[41]

On 13 November, Onnes was informed by telegram from the Royal Swedish Academy of Sciences that he was the recipient of the Nobel Prize in Physics for the year 1913 for "his investigations on the properties of matter at low temperatures which led, *inter alia*, to the production of liquid helium."[42] News of the bestowal brought the normal laboratory routine at Leiden to a halt, with 14 November declared a special "helium day" in Onnes's honor by the laboratory staff.[43] (The term derived from the fact that, because of its scarcity, liquid helium was only available on select days, usually just once a week.)

In his presentation speech in Stockholm on 10 December, Th. Nordenström, President of the Swedish Academy, drew attention not only to superconductivity but also gave brief credit to the specific heat determinations of Nernst and his students.[44] (In fact, Planck had unsuccessfully nominated Nernst for the Nobel Prize in 1911.) Onnes's Nobel Laureate Lecture given the next day[45] is mainly of interest now as an authentic summary of the "Leiden und Freuden," a wonderful phrase borrowed from Ernest Cohen,[46] that eventually led to the podium of the Great Hall in Stockholm. He also used the opportunity to express his appreciation of support from colleagues and collaborators—above all van der Waals.[47] It was an uplifting address, replete with Onnes's characteristic poetic language. He ended on a note of optimism for the Leiden program's contributions "toward lifting the veil which thermal motion at normal temperature spreads over the inner world of atoms and electrons."[48]

From a footnote to the printed version of Onnes's Nobel Lecture we know that until shortly before the lecture he was still confident about high-field superconducting coils. The small coil experiments carried out the previous spring had given no cause for serious concern about magnetoresistive effects, he wrote. "From analogy it was to be assumed that [magnetic resistance] rises regularly with the field and even when it is proportional to the square of the field it would not be of very great importance even in fields of 100,000 gauss, as the resistance produced in the coil by its own field was anyway still unnoticeable [unnoticeable, that is, at a critical current of 0.8 amperes, or in a field of a few hundred gauss!]. Only experiments, however, could make certain of this."[49] Alas, the experiments were not long in coming.

CHAPTER 5

A Technical Snag; The War Intervenes

5.1 Another Threshold, and Dashed Expectations

The decisive experiment was delayed, needlessly as it turned out, by Onnes's dogged optimism. Ironically, the change of resistance produced by a magnetic field had been the subject of a minor research program in its own right off and on for 20 years at Leiden, primarily as an adjunct to studies on the Hall-effect. It had been pursued to liquid air temperatures by van Everdingen and colleagues, but continuation to lower temperatures "had been forced aside from other researches which could not be delayed, until the study of [magnetoresistance] and of allied problems for various metals at the lowest possible temperatures [had] been rendered essential to the further development of the theory of electrons by the discovery of the fact that the resistance of pure mercury disappears at liquid helium temperatures."[1] The investigation was taken up again in 1912, first by Beckman assisted by his wife Anna, and then by K. Hof, under Onnes's tutelage as always. Although magnetoresistive studies on lead and tin in liquid hydrogen were barely under way with Hof in early 1913,[2] the general trend in the change of resistance with field for normal metals was by then fairly well established by Beckman: increasing as H^2 in weak fields and proportional to H or slower in stronger fields.

The enormous rate of increase of the potential effects in tin and lead above the threshold current could not be reconciled with either a linear, or a quadratic field dependence. Moreover, because the magnitude of the observed magnetoresistance was extremely modest even at 10 kilogauss (except for bismuth, long known for its anomalous behavior in this regard), typically showing a rise of

0.01% at the melting point of hydrogen in the case of mercury,[3] Onnes firmly believed that the effect of a magnetic field on superconductors was insignificant—certainly below 10 kilogauss. Therefore the relevant experiments were put off, since a 2-kilogauss magnet was the only one then available that could accommodate cryostats with superconducting coils.

Fortunately, the delay did not last long. A new, powerful electromagnet of the Weiss-type had been under contract with Machienfabrik Oerlikon for several years; Onnes had personally negotiated the specifications[4] with the helpful advice of Pierre Weiss himself. The magnet arrived in Leiden in early January of 1914. It was normally capable of 20 kilogauss; with an iron-cobalt pole-tip insert (and 500 kilowatts of cooling) it could reach 50 kilogauss. One kilogauss was, in fact, to prove quite adequate.

On 17 January, a small noninductively wound lead coil was exposed to a field of 10 kilogauss, where it exhibited "considerable" resistance. Onnes's *Communication* on the experiment, reported at the meeting of the academy on 28 February, is obscure on a vital point. He wrote that "we had not been so successful in the construction of this coil as in the previous one, as it did not become supra-conducting. It was, therefore, possible that not much value could be attached to this experiment."[5] Thus it is unclear whether the coil had been tested first in zero background field, and hence whether it was, in fact, known to be a dud even in the absence of a field. A noninductively wound tin coil also showed considerable resistance at 10 kilogauss when cooled to 2 °K; at 5 kilogauss the resistance decreased "more slowly than in proportion to the field." This experiment was considered no more instructive, as it "did not seem to be reconcilable with the ... observations in which the magnetic field generated no resistance in supra-conductors."[6]

With the superconducting properties of both coils in doubt, the experiments were repeated with a pair of lead and tin coils known to be superconducting in earlier tests, "notwithstanding that the windings were in a magnetic field." These were the two coils that were utilized the previous May.[7] Being wound induction-free "was of no consequence, now that it was a question of such comparatively large resistance."[8]

The old lead coil was tested first. It was ascertained, initially, that the coil was still superconducting in zero applied field at the boiling point of helium, and that as before, it remained superconducting to 0.4 amperes (or a maximum self-field of 200 gauss). To be quite sure, they checked that the current actually passed through the coil by observing a small compass needle suspended near the cryostat. Next, a field of 10 kilogauss was applied. A "considerable" resistance developed, which decreased "somewhat" at 5 kilogauss. Basically, the matter was settled then and there. The experiment had "made it fairly certain that the magnetic field created resistance in supraconductors at larger intensities, and not at smaller ones. The apparent contradiction that so far had existed between the different experiments, was hereby solved."[9] That is, a *threshold effect* was once more at work. Further measurements narrowed the magnetic threshold value to between a paltry 500 and 700 gauss. Soon curves of resistance as a function of field were obtained for bath temperatures of 4.24 and

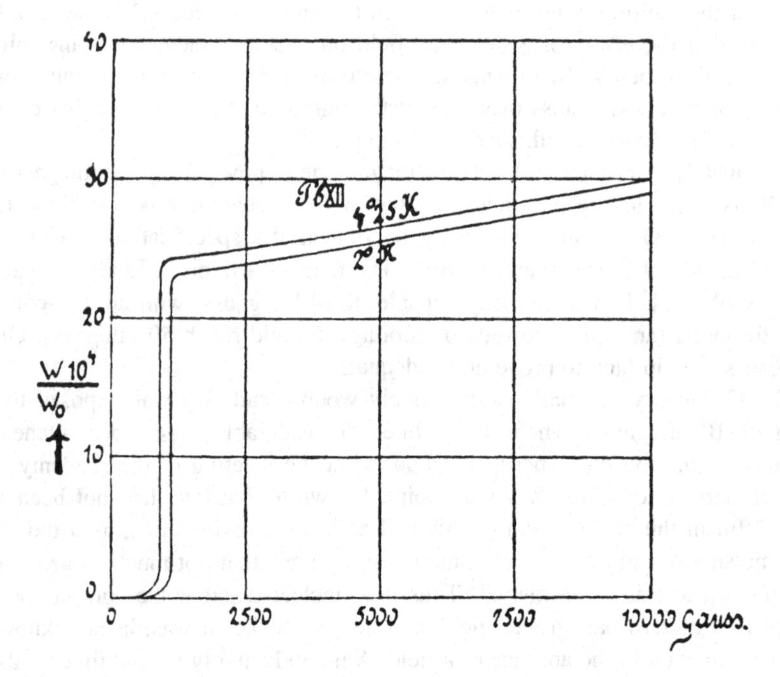

Figure 5-1. Resistance of a lead coil as a function of magnetic field for two different helium bath temperatures. Kamerlingh Onnes, RN, ***16*** *(1914), 989.*

2 °K (both well below the transition temperature) and a current of 0.004 amperes (below the threshold current at either temperature); both curves[10] are reproduced in Figure 5-1.

> It will be seen that the transition from the supra-conducting condition to the ordinary conducting condition through the magnetic field takes place fairly suddenly. The curve, which represents the change of the resistance with the field is closely analogous to that which represents the change of the resistance with the temperature. ... The sudden change in the resistance moves at low temperatures towards higher fields; beyond this point the resistance increases at lower temperatures (2°K) almost in the same way as at higher ones, it seems as if *the introduction of the magnetic field has the same effect as heating the conductor.*[11]

The original tin coil showed essentially the same behavior: a sudden but even lower threshold field, estimated to lie near 200 gauss.

Both coils had been tested with the plane of the winding parallel to the external field—that is, with the field lines partly transverse to the conductor, partly parallel. Since considerable differences were known, from Hof's studies, between the ordinary transverse and longitudinal magnetoresistive effects, a new lead coil was constructed by a procedure devised earlier by Beckman and Onnes. Pressed lead wire was wound on a plate of mica, covering it with a few flat layers of insulated windings; this geometry enabled the coil and its cryostat to be mounted such that the plane of the winding could be unequivocally oriented transversely or longitudinally to the field. The results presented no great surprises. "The sudden change in both effects takes place almost at the same threshold value of the field. The longitudinal effect is weaker than the transverse effect [a feature of the normal magnetoresistive effect]."[12]

"There is no doubt," Onnes concluded, "that the phenomenon discovered here is connected with the sudden appearance of ordinary resistance in the supra-conductors at a certain temperature. The analogy between the influence of heating upon the resistance and that of the introduction of the magnetic field, is so far complete."[13] Interestingly, Onnes did not infer any relationship between threshold current and threshold field—the Silsbee hypothesis of 1916. Perhaps he was still too preoccupied with the more basic question. He saw no reason to doubt that "vibrators" were still at work. If the effect of a threshold current was the manifestation of a critical electron drift velocity stimulating the Planckian vibrators into motion, the magnetic threshold might in a similar way reflect stimulation of the atoms by the onset of a critical energy of rotation imparted to the electrons by the magnetic field.[14]

There was, however, another side to the magnetic behavior, as Onnes evidently realized. It was no more than a hunch, and he failed to follow it up. The February report ended on a prophetic note: " ... it is certain that the phenomenon described here is connected with the laws of magnetisation [sic] of supra-conductors which are as yet unknown."[15]

5.2 Persistent Currents

The onset of the war in 1914 essentially forced a halt to all work with liquid helium. Onnes had time for only one last set of superconducting experiments in the late spring of 1914, albeit experiments he had contemplated "from the moment [he] had found in mercury a supra-conductor."[16] Unraveling the various troublesome "phenomena" had subsequently demanded his full attention and caused repeated postponements of these particular experiments. They were, in fact, predicated on cryogenic facilities not in place in 1911. Although the new experiments shed no further light on the nature of the superconducting state per se, they offered (at least in principle) a marvelously effective method for pinning down a much better value for the microresidual resistance than was possible with any of the potentiometric methods. They also provided a marvelous demonstration of the phenomenon of superconductivity, and of Onnes's experimental flair.

The basic idea was simple. A current set up in a closed electrical circuit will decay with a time constant (the time for the current to fall to $1/e$ times its initial value) given by $\tau = L/R$, where L is the self-inductance of the circuit and R the resistance. The time constant for normal conductors is very short; thus a lead coil resurrected from the earlier experiments for the present purpose had a room temperature resistance of $R = 734\Omega$ and an inductance of 10 mH, giving $\tau = 1.4 \times 10^{-5}$ sec—a time too short for Onnes to measure. But a microresidual resistance lower by a factor of 2×10^{10} at 1.8 °K implied a time constant of at least the order of a day. Onnes proposed to start the current in a superconducting ring by induction: switching on a field (below H_c) with the ring in the normal state, cooling it below T_c, and removing the field, thereby inducing a current whose disappearance could actually be observed. In preparing for a first proof-in-principle experiment in mid-April,[17] a sample of the conductor was tested to make sure its superconducting properties were intact. Then the current leads of a coil wound from the conductor were fused together, forming a closed superconducting loop. (Onnes had previously satisfied himself that the joint resistance was negligible.) Meanwhile, rather elaborate calculations had given assurance of the feasibility of the essential idea: that a field strength well below the threshold field at 1.8 °K (1000 gauss) was adequate to generate by induction a current in the conductive loop less than the threshold current but potent enough to produce an easily detectable field.

The coil was mounted in the same cryostat used in the threshold field experiments conducted earlier in the spring, with the plane of the winding oriented vertically. Suspended from a vertical shaft leading down into the cryostat, it could be raised and lowered or turned about a vertical axis. The large Weiss electromagnet could be rolled on casters to and from the cryostat; its magnetic axis was oriented horizontally with the pole-tip gap wide enough to accommodate the multiple-walled glass dewar of the cryostat. By twisting the vertical shaft, the plane of the winding could be oriented perpendicular or parallel to the field.

The experiment began with the coil in the "interferrum" oriented in the perpendicular mode. A field (well below the threshold value) was switched on with the coil still at room temperature; the current generated was immediately dissipated by the ordinary resistance of the lead winding. The coil was then cooled by siphoning liquid helium into the cryostat with the field held constant. Obtaining a strong current in the superconducting coil by induction required cooling down the coil as far as possible, thereby maximizing the threshold value of both the field to be used for the induction and of the current. The first attempt was made at 1.8 °K, "the lowest temperature which [could] be reached comparatively easily and maintained for a long time." The presence of a current in the coil could be inferred from a magnetic needle placed on one side of the cryostat on a level with the coil—the same ploy devised during the earlier threshold field experiments. The magnitude of the current was determined by a second coil, geometrically similar to the superconducting one, placed equidistant to the compass needle; its current was adjusted to neutralize the action of the superconducting coil on the needle. When the lead coil was superconducting, the applied field was reduced from 400 to 200 gauss, and immediately afterward the electromagnet was rolled away. The measured compensating current was approximately 0.6 A (compared to the critical current at 1.8 °K of 0.8A).

> During an hour the current was observed not to decrease perceptibly (as far as could be judged by the deviation of the needle with an accuracy of 10%). During the last half hour the coil was no longer at 1.8 °K but at 4.25 °K, the temperature

> of helium boiling under normal atmospheric pressure. Undoubtedly even at this temperature the observation might have been continued much longer without much diminution of the current. A coil cooled in liquid helium and provided with current at Leiden, might, if kept immersed in liquid helium, be conveyed to a considerable distance and there be used to demonstrate the permanent magnetic action of a supra-conductor carrying a current. I should have liked to show the phenomenon in this meeting (Kon. Acad. Amsterdam), in the same way as I brought liquid hydrogen here in 1906, but the appliances at my disposal do not yet allow the transportation of liquid helium.[18]

Well over a decade later a coil of this type was indeed conveyed a considerable distance from Leiden, but not in Onnes's hands.

Various control experiments were deemed necessary. Thus, the induced current was immediately quenched when an ordinary resistance was generated in the circuit or the coil was lifted out of the helium bath; reimmersion produced no further action. Essentially no magnetic effect was observed with the coil winding parallel to the field, as was expected. Rising fields were as effective as falling fields, and could be made to compensate for each other. "Taken together [these checks] may be said to confirm the main experiment which shows that it is possible in a conductor without electromotive force or leads from outside to maintain a current permanently and thus approximately to imitate a permanent magnet or better a molecular current as imagined by Ampere."

> The electrons once set in motion in the conductor continue their course practically undisturbed. The electrokinetic energy, represented by Maxwell by the mechanism of the rotating masses coupled to the current, retains its value, the rotating fly-wheels go on with their velocities unchanged, as long as no other than supraconductors come into play: the application of a small ordinary resistance however stops the mechanism instantaneously. Although the experiment mainly confirmed my deductions as to what had to be expected, a deep impression is made by the striking realization which it gives of the mechanism imagined by Maxwell completed by the conception of electrons.[19]

A deep impression was apparently made on Paul Ehrenfest as well. After visiting Onnes's laboratory that month, Ehrenfest conveyed his excitement to Lorentz, whose theoretical chair he had assumed two years earlier (in place of Einstein, who had declined the position). "It is uncanny to see the influence of these 'permanent' currents on a magnetic needle. You can feel almost tangibly how the ring of electrons in the wire turns around, around, around—slowly, and almost without friction."[20]

Aside from impressing Ehrenfest, the first persistent current experiment had, in fact, not succeeded in quantitatively reducing the microresidual resistance limit. About all that could be concluded was that the current decay was very slow—too slow to detect with the rather crude technique of compensating a compass needle. The experiment was nevertheless repeated in May,[21] now with the compensating coil suspended in a liquid air dewar. Each time the action of the superconducting coil on the needle was nulled, the primary coil was turned 180° about the vertical axis and again compensated with the current in the second coil reversed. The magnetic moment of the primary coil was deduced by averaging the two readings.

Assuming the validity of calculating the diminution in current from the best available estimate for the microresidual resistance, $R_{1.8}/R_{290} = 0.5 \times 10^{-10}$ as deduced in February of 1913, a time constant of about 2.7×10^5 sec (75 hours) could be expected. This implied a 4% fall in current over 3 hours, a change that might be barely detectable with the improvements in the apparatus. The new experiment was made at 1.7 °K with a field of about 200 gauss employed to induce a current of 0.4 A. In spite of the need to replenish the liquid helium level in about 2 1/2 hours, an operation that unavoidably raised the bath temperature for a short period to the normal boiling point (4.25 °K), the experiment "rendered it probable" that the change in current was less than 4% in 3 hours. Still, the accuracy was not much better than before. Onnes could think of several possible contributing errors, but he was unable to do much about them. Changes in the shape of the helium liquefier and cryostat during liquefaction and liquid transfer was one possible error; shifts in zero of the compass needle from distortions in the earth's field by nearby iron machinery, and possible effects of magnetized coil materials, mainly brass and lead, were also possible. In repeating the experiment a second time, two liquid air-cooled compensating coils were used, enabling compensations from either side of the cryostat. The induced current was halved to minimize the disturbing effects of replenishing the helium. Nevertheless, the accuracy was judged to be not greater than about 2% of the measured moment; since it was not possible to continue the experiment beyond 3 hours,[22] "again only an upper limit for the change could be established, to be put at about 2/3% for current and induced magnetization combined. Taking all the experiments together," Onnes concluded, "it may be considered as probable, that the change of the current is less than 1% per hour which raises the time of relaxation to above 4 days."[23] Thence, the upper limit for the resistance ratio of lead was moved back to about $0.3 - 0.2 \times 10^{-10}$.

In a variation on these experiments, devised to establish conclusively that the magnetic moment of the superconducting coil was indeed produced by a current, Onnes demonstrated that a persistent current initiated by induction (or a battery) could be extinguished by breaking the circuit.[24] In one version, a current was induced in the usual way by removing the field; having established a magnetic moment corresponding to an essentially steady circulating current, the coil was severed under liquid by pulling on a bronze loop provided with a knife edge. The compass needle fell back to a residual deviation that was perhaps 10% of the induced current; the latter was verified by the impulse in a ballistic galvanometer now also in the circuit.

A modification of the experiment employed a clever superconducting key, or switch, first suggested by Jan Peter Kuenen, a younger colleague (and former student) of Onnes. At first Onnes had doubted that a contact resistance "manipulated under helium" could be made small enough for this purpose, since "the transitional resistance of a stop-contact treated with all due care at ordinary temperature is not likely to be less than 0.0001Ω, which is still 100,000 C. G. S. while the micro-resistance of the coil itself is only 37."[25] In fact, a quite modest pressure sufficed with a switch consisting of two lead blocks, one of which had three small conical points on its surface. By pressing the blocks together in liquid helium, a contact of negligible resistance could be readily opened or closed, allowing the superconducting coil to be short-circuited at will. The basic experiment now went as follows: With the key and a galvanometer both in the circuit and open, a current was sent through the coil while it was superconducting. The circuit was then closed, which produced no change in the deviation of the compass needle. Nor did closing the galvanometer affect the current. Thereupon if the switch was opened, however, and accompanied by a throw of the ballistic galvanometer, the current was instantly extinguished.

The last of the persistent current experiments was prompted by the desire to repeat the demonstration with mercury. This would necessitate a single loop of larger cross section, produced by freezing mercury in a ring-shaped capillary with an expansion head, not unlike the U-shaped resistances. It also raised the question of whether the effectiveness in maintaining a persistent current might be influenced by the current density, in view of the fact that the critical current itself was actually a current density effect. In the event, time ran out before the experiment could be tried, but Ehrenfest's interest in this led to a simpler substitute test with a lead ring. Ehrenfest suggested that the experiment could be performed equally well with the windings "parallel" as it had been done with the windings in "series," in effect replacing the multiturn lead coil by a single ring of lead. In doing so, a substantial current was stored without difficulty, corresponding to a current density "not much smaller" than in the coil experiments. "This," Onnes ended his final report of 1914 on the superconducting program, "may for the present be regarded as a confirmation of the supposition that the threshold value of current strength of a conductor is mainly a threshold value of current density for the material of the conductor."[26]

5.3 The Dilemma of Superconductivity

Although the superconducting experiments came to a halt in 1914, speculations on the underlying mechanism did not. The hiatus in measurements was perhaps for the best. Since the time of Lindemann's first theoretical attack on the problem, renewed attempts to explain the phenomenon had been made, and Lindemann himself had not given up. Nor had Onnes ceased to speculate. In spite of his preoccupation with the experimental distractions that were constantly cropping up, his progress reports to the academy in this period were sprinkled with much intuitive guesswork on the significance of the superconducting state for atomic interactions and quantum states, most recently in the third installment of the bountiful *Communication 133* of 1913.[27] Perhaps the most convincing clue that superconductivity involved a breakdown of classical electron behavior had come from another branch of Onnes's liquid helium program: measurements made with Holst on the thermal conductivity of mercury in the liquid helium temperature range. Preliminary measurements in June of 1912 had indicated no distinct discontinuity in conductivity (or specific heat) at the transition point. Surprised, and dissatisfied with the accuracy of the measurements (which are considerably more difficult to perform than those of electrical resistance), Onnes had intended to repeat the measurements at the first opportunity. In this effort he was frustrated time and again by the various "circumstances"; when the war halted work with liquid helium altogether, he went ahead and published the results anyway.[28] This time he shared the authorship with Holst, who had by then finished the work for his dissertation[29] and already left for Eindhoven.

Onnes had not overlooked the importance of this particular finding when he reviewed the experiments performed on mercury to date in 1913:

> The experiments ... leave no doubt that for mercury below 4.19 °K there is no question of an approximate validity even as regards the order of magnitude of the relations established by Wiedemann and Franz and by Lorentz. The failure of this relation between λ, k and T indicates a difference between the super-conducting and the ordinary conducting state which may be regarded as a *characteristic difference* of both.[30]

With the law of Wiedemann and Franz (i.e., that the thermal conductivity divided by the electrical conductivity is proportional to the absolute temperature) in jeopardy, Onnes grew extremely weary of classical mechanisms. By the late spring of 1913 he questioned whether "the whole hypothesis developed [by himself in 1911 and reiterated in 1913] concerning the movement of free electrons through the metal ... must not be replaced by an essentially different one for the superconducting condition, according to which the movement of the electrons is carried by the current for considerable distances, but each separate electron which takes part in the progress, only moves one molecular distance."[31]

Just such a collective ("billiard balls") mechanism had been proposed by Johannes Stark the year before.[32] Stark probably inherited his peripheral interest in metallic conduction while he served as an assistant to Riecke at Göttingen. He had hoped to succeed Riecke in his chair at Göttingen in 1914, but his analytical competence was questioned by the faculty tenure committee; the ensuing bitter dispute was not resolved. Stark, in fact, quarreled with nearly everybody (except with Lenard, a man of the same philosophical and political ilk who, it happened, was also working on a resistance theory) and all accepted points of view.[33] Perhaps one can credit his contrary nature for his model of electrical conduction, which was equally contrary to accepted views on conduction—a harbinger of a whole new class of ("lattice") theories. Stark's model assumed that each atom in a metal releases a valence electron, and that these electrons form a regular lattice maintaining the atoms in position. An electron in a lattice can be displaced only on certain shearing surfaces (*Schussflächen*) of the metal crystal and only in unison with the simultaneous movement of other electrons.

Combining Stark's representation with his own Planckian vibrators, Onnes interpreted the transition point as the temperature where the distance between

Schussflächen, determined as usual by the action of the vibrators, becomes too large for the electrons to "jump from one atom to another without doing work."[34]

Ironically, Stark's political leanings would, some day, prove to be more influential for superconductivity than his physics. But for now, although an "interesting hint"[35] of a mechanism for circumventing the contrivance of unrealistic electron-free paths, Stark's model did not lead to any quantitative results; it was not even able to account for thermal conductivity. Not only that, it was "to a similar degree open to the objections raised against the classic electron gas theory."[36]

Considerably more promising, also in Onnes's opinion, in pointing away from the simple gas theory, was Wilhelm ("Willy") Wien's attempt in 1913 to reconcile the free electron concept with the older quantum theory.[37] The enormously gifted Wien had participated in the Solvay Congress, and he was familiar with the early Leiden measurements on mercury as well as with the ongoing specific heat investigations in Berlin. He was, in fact, not a neophyte to the field; he had collaborated with Ludvig Holborn on resistance thermometry at the Physikalisch-Technische Reichsanstalt in the mid-1890s. He was an equally gifted theorist. In spite of expressing serious doubts, off and on, about Planck's radiation theory,[38] Wien developed the first theory of electrical conduction in which the atomic vibrations are treated quantitatively. His approach to the problem, developed at Würzburg (where he was eventually succeeded by Stark), embodied modifications of the electron theory and mathematical features that were to remain an integral part of later theoretical attacks on both the problem of electrical conductivity *and* specific heats.

Wien's derivation was based on Drude's expression for Ohm's law, Eq. (1-7). But, from the evidence furnished by specific heats of solids that electrons do not appear to participate appreciably in thermal energy, he was forced (much to Lorentz's consternation) to discard the very foundation of the free electron theory—that the kinetic energy of the electrons is proportional to the absolute temperature.[39] In so doing, Wien purposely ignored thermal conductivity and forfeited the most successful result of the classical theory, the Wiedemann-Franz law. He also assumed, along with Riecke and Lorentz (and Onnes, for that matter) that the electron density is independent of temperature, which follows more or less automatically from the constancy of velocity. Thus he left the mean-free path as the only variable. One further quite general assumption was necessary even "without a deeper understanding of the collision mechanism," namely, the functional dependence of the number of collisions between the electrons and atoms on the amplitude of the atomic vibrations. Wien's particular contribution was a formal justification for a dependence proportional to the square of the amplitude—by 1913 a generally accepted relationship—by demonstrating that this is the only relationship that makes the number of collisions per unit time independent of the exact distribution of energy among the atoms. ("There are ... just as many collisions, whether $nh\nu$ is equally distributed among n atoms or whether one atom possesses all n quanta.")[40] Finally, Wien's derivation of a formula for the resistance as a function of temperature involved integrating Planck's first expression for oscillator energy, Eq. (2-4), over a spectrum of frequencies. The derivation is similar to Peter Debye's treatment of specific heats in this period (ca. 1912), although there ν has a somewhat different meaning; Debye's ν is not a frequency of atomic vibrations, but rather a frequency of elastic waves in a solid. (Debye's Habilitation thesis in 1909 had also been on the electrical conductivity in metals, but, as he would comment in retrospect, his thesis "was no good ... just a chewing up of the old business.")[41]

At high temperatures Wien's formula, apart from a numerical constant, is:

$$R \propto \frac{k\nu_M}{h} T , \tag{5–1}$$

where ν_M is an upper limit in the frequency spectrum. At low temperatures, Wien predicted a quadratic dependence on T:

$$R \propto \left(\frac{kT}{h}\right)^2 . \tag{5–2}$$

If expressed as a ratio of resistances, Eq. (5-2) reduces to Onnes's Eq. (2-10) as long as $h\nu_M/kT$ is small. The expression reproduces the gross dependence of resistance on temperature moderately well (e.g., predicting vanishing temperature coefficient for $T \rightarrow 0$, and a higher coefficient the higher the characteristic frequency ν_M). However, the quadratic dependence, despite a prematurely optimistic survey of a few published resistance measurements, proved to be insufficient to account for the precipitous drop in resistance of most pure metals at the lowest temperatures—certainly not for mercury.

As for the implications of Planck's new hypothesis involving zero-point energy, Wien, like Onnes, could only speculate.[42] In the end he too was forced to ignore it. Nor did Wilhelmus Keesom, one-time assistant to Onnes and now a close colleague at Leiden, have more success with it. Like Wien, Keesom was both a highly competent theorist and experimentalist. At that time, he was working on the application of the quantum theory to the equation of state of an ideal monatomic gas.[43] With Wien, he attempted to extend the formalism to the theory of free electrons in metals, but only for thermoelectric effects.[44] Both topics were discussed by him at the Wolfskehl Congress on the kinetic theory of matter at Göttingen in April of 1913, and communicated on his behalf by Onnes to the Amsterdam Academy at its meeting the following month. Keesom came to the same conclusion as Wien on the necessity of removing the temperature dependence of the electron velocity at low temperatures, a domain he courteously referred to as the "Wien field." At high temperatures (the "Richardson field"), applications of the quantum theory to free electrons were less sensitive to "Planck I or II"[45] and led to results in fair agreement with equipartition theory. Keesom discussed these matters in a letter to Wien early in May. In his reply, posted 9 May immediately after returning from a lecture tour in

America, Wien expressed agreement on the difficulties of reconciling zero-point energy with metallic conduction.[46]

In the summer of 1914 Lindemann returned to Britain, joining the Royal Aircraft Establishment at Farnborough. His departure, hastily forced by the outbreak of the war, coincided with Holst's departure for the Philips Company in Eindhoven; in this way both Nernst and Onnes lost valued students and collaborators. Amidst the challenging problems of combat aviation, Lindemann found time to take another crack at the metallic state. In reviewing, in late 1914, the various phenomena of electrical resistance, heat conduction, and secondary (thermoelectric) effects and theories offered to account for them, he observed that they were all "in absolute contradiction with one another or with the facts."[47] Thus, specific heats had shown that there can be at most one free electron per hundred atoms if the electrons obey the laws of equipartition (i.e., that the energy of monatomic solids is equally distributed over all degrees of freedom), whereas conductivity demanded more free electrons than atoms. Giving up equipartition would forsake heat conduction and the Wiedemann–Franz relation. Assumption of a long free path to compensate for a small number of free electrons led to contradictions with the optical properties of metals.

The expression "free electron," Lindemann concluded, "might almost be called a contradiction in terms." If the electrons are not attracted by the ions (which evidently they are not, or they would combine with them), the mutual forces between the electrons should be sufficient to obviate any similarity to a perfect gas. Far from it, he argued, for the electrons in a metal may be viewed as a "perfect solid." This conception,[48] a radical departure from the free electron viewpoint, embodied the idea of a crystal lattice in a much more satisfactory way than Stark's cumbersome construction. Lindemann associated electrons, in the absence of a current, with a space lattice interpenetrating the atomic lattice. Electrical conduction was treated as a drift of the electron lattice as a whole through the atomic lattice, with its motion hindered by atomic vibrations. The theory thus bore a certain similarity to Wien's, except that the electrons were no longer visualized as moving independently through localized channels, but instead they were coupled in a lattice of considerable rigidity. Below a certain critical temperature, the amplitude of vibrations was too small to interfere, giving rise to superconductivity. Since the energy of lattice vibrations was quantized, as usual, the specific heat problem was avoided, because the small mass of the electrons ensured that their lattice vibrations would be of a much higher frequency than the vibrations of the atomic lattice.

Lindemann did not attempt to calculate an explicit dependence of resistance on temperature, in view of the complex spectrum of vibrational amplitudes below the Debye–Wien limit, and the difficulties posed by averaging all the probable forces acting on the electron space lattice. However, he postulated a repulsive force of the general type $kf(r)$. (Such a force is, in fact, implicitly involved in almost all of the older electron theories as well.) The unknown force function was assumed to entail a resistance proportional to the square of the amplitude, hence to the energy (as in Wien's theory), since $E = \alpha A^2$. Since the quasi-elastic force holding the atoms in position, α, could be assumed to be roughly proportional to the number of electrons per cm^3 (n), the resistance would be a function of n and k, say $\phi(n, k)E$. General dimensional arguments suggested a formula of the type

$$R = \frac{\rho^{1/2}}{n^{2/3}k^{1/2}} E, \qquad (5\text{–}3)$$

where ρ is the density of the electron space lattice. As ρ, n and k are independent of temperature, the resistance is proportional to E, in accord with the experimental facts.

Lindemann's theory, therefore, offered a qualitative explanation for electrical conductivity, even if it failed to provide quantitative predictions. It was modified somewhat by J. J. Thomson and others during and after the war, but these attempts led no further. Thomson's curious "dipole" theory was actually put forth in 1907,[49] but it was reworked in 1915 in a form that sought to cope with the various problems inherent in the older electron theory, particularly superconductivity.[50] Atoms were regarded as electrical doublets, normally aligned at random. In the presence of an electrical field, they experience a moment tending to align them in chains parallel to the field, bound together by both the field and forces between the dipoles. Superconductivity occurs when the forces between the dipoles exceed the lattice forces. Thermal vibrations and the lattice forces eventually combine to exceed the dipole restoring forces above a certain temperature.

The theory was clever, imaginative, and implausible, typical of J. J. Thomson's penchant for developing ingenious mechanical models in the tradition of Maxwell and Kelvin.[51]

5.4 Holst vis-à-vis Kamerlingh Onnes

Superconductivity appears to have been discovered partly as a result of Onnes's seemingly narrow inclination for precision measurements. Actually his style was deceptive and certainly was not a legacy of his predecessor, Rijke, who in the opinion of one whose opinion counts, "had not had such higher ambitions than to keep the collection of instruments in good order."[52] And if Onnes was perhaps not in the front rank of deep conceptual thinkers, he had a remarkable intuition for basic principles and the experimental opportunities they offered—most specifically the law of corresponding states, but also quantum theory—and a strong sense for the generally promising directions in which to seek "new physics." Reaching low temperatures, or low resistances, was to Onnes not an end in itself but the key to new physics, as exemplified in the Nernst heat theorem, Planck's quantum theory, and the concept of zero-point energy.

> It is true [Onnes stated in his Nobel address] that Faraday's problem as to whether all gases can be liquefied has been solved step by step in the sense of Van der

> Waals' "matter will always show attraction," and thus a fundamental problem has been removed. At the same time, however, the question asked by Planck introduces a problem which is probably no less fundamental, to the solution of which investigations into the properties of substances at low temperatures can contribute.

In other words, *door meten tot weten* was Onnes's motto. Above all, Onnes was an excellent organizer and administrator who relied on a hierarchy of students, assistants, and younger colleagues in the pursuit of his ambitious low-temperature program. Unfortunately his autocratic style left little room for independent initiative among those who worked under him; all contributions had to be in the mainstream of Onnes's personal program. Kuenen, one of Onnes's first students, needed a decade abroad to find himself before returning to play a major—albeit short-lived—role in the laboratory. Another frustrated student was Keesom. Recalls Burgers, who also worked briefly under Onnes, Keesom had to postpone his work on crystals until he obtained his own professorship relatively late in his career.[53] One of the few who did not linger on was Gilles Holst.

It is interesting, but not very surprising, that the actual detection of superconductivity was made by an assistant, Gilles Holst, somewhat by chance. Nor is it surprising that Onnes's recognition of Holst's role in the discovery was limited to acknowledgments accompanying the various *Communications*. Because he was of the old school and the one responsible for determining the research program, Onnes considered it only proper that all results should appear in print under his own name only; coauthorships were reserved for more senior staff members. The year of the discovery, 1911, was only Holst's second year at Leiden. After completing his preliminary schooling in Holland, Holst had gone to Zürich in 1904 to study electrotechnical engineering at the prestigious Eidgenössische Technische Hochschule (ETH). Two and a half years later, with the intermediate examination behind him, he switched to physics and mathematics, obtaining the degree of Geprüfter Fachlehrer in 1908. He stayed on for another year at ETH as assistant to H. F. Weber, the Weber whose specific heat measurements had goaded Einstein and Nernst. (Weber had, in fact, turned down Einstein's application for an assistantship under him.) Holst then returned to Holland and became assistant to Onnes—a position, curiously, which Einstein again had unsuccessfully applied for in 1901![54]

In 1914, when the time was right, Holst did in fact coauthor several papers with Onnes. One was the aforementioned paper on thermal measurements in superconducting mercury; another dealt with resistance measurements on conventional, pure metals.[55] That year Holst obtained his doctorate at Zürich, on the basis of other, independent thermodynamic work done late in his Leiden tenure. An abstract of his thesis appears in a 1914 Leiden *Communication*,[56] at the end of which he acknowledges the opportunity of working under Onnes:

> At the end of this work it is my pleasant duty to thank Professor H. Kamerlingh Onnes for his kindness to put all the equipment necessary for this research at my disposal. The time, during which I had the honour to be his assistant and more in particular the years during which I had the pleasure to assist him with his own

> research, will always remain a period of my life of which I will only be able to think with a feeling of deep gratitude.

In January of the same year, Holst joined the Philips Company where he soon became the driving force in establishing the Philips Natuurkundig Laboratorium.[57] That challenge had been declined by, among others, Onnes's other chief assistant in the early superconducting experiments, Cornelis Dorsman (who had been offered a research position at Philips as early as the memorable year of 1911). Finally, in Onnes's confidential recommendation for Holst's membership in the Royal Dutch Academy, he emphasized Holst's collaboration in the discovery of superconductivity.[58]

More interesting, and even surprising, is Holst's contrasting scientific temperament, which is hardly the type one identifies with galvanometers and cathetometers. In view of what we know of his style in his later years, it could be construed as odd that Onnes singled him out for such a mundane chore. It was more likely, though, that Onnes thought it promoted good scientific discipline. Moreover, the measurements had high priority by then, and Holst was entrusted with them because of his considerable background. Hendrick Casimir, who knew Holst well in later years at the Philips Laboratory, stresses that Holst "had no special liking for precision measurements and even less for least-square fitting of semi-empirical formulae to empirical data."[59] This attitude, in fact, became a guiding principle of the research program Holst directed at Eindhoven. "I soon discovered," recalls Casimir (who succeeded Holst in the directorate at Philips much later), "that in the Philips Laboratory precision measurements were the exception rather than the rule, and what has been called the physics of the next decimal place was not held in high esteem."[60]

Holst's Leiden years cannot have been very easy for him. Whether it was to his credit or not, he inherited few of Onnes's principles or traits, which are usually a lasting impression of a professor's influence on his student. As a research director at Philips, according to Casimir, Holst would never insist on being coauthor, let alone sole author, of papers based essentially on the work of others.[61] Possibly, though, notes Casimir, "he may have at times been disappointed by the fact that few of his co-workers proposed themselves that he should be coauthor. I know he was grateful to the few who did. His attitude was a noble reaction to what must have been a great disappointment of his younger years."[62] Holst not only encouraged his staff to publish, but he believed in the importance of leaving the staff members by and large free to pursue projects on their own—naturally within the broad guidelines and research policies set forth by laboratory management. To the extent that tasks had to be apportioned, Holst tried to assign them in accordance with the personal characteristics of the individual researchers.

On a purely personal level, Casimir speaks of Holst's "lack of formal oratorical skills," his "quick and witty repartee" in approaching people—"essentially kind but wavering between shyness and bluntness."[63] These traits are somewhat surprising for someone in Holst's position. Dual traits of vaguely this sort have been attributed to Onnes as well, but he was much more complicated.

Burgers describes Onnes as having "had a certain roundness and also a certain abruptness." He also "had a nobility in his heart," he quickly adds. At the same time "Kamerlingh Onnes pounced upon you," something that "gave a certain attraction in the long run."[64] Marian von Smoluchowski, writing to Ehrenfest, found Onnes "thoroughly pleasant and interesting in many respects, but his is another kind of talent."[65] Ehrenfest himself, according to Martin Klein, felt some of Onnes's strangeness but found a way of dealing with it. He considered Onnes "a friendly gentleman," but thought he might be a little like Chwolson of Petersburg which, adds Klein, "was not a recommendation."[66]

There was, at any rate, an undeniable element of charm in Onnes's old-fashioned manner, and it must be remembered that at the time he was carrying out the superconducting experiments with Holst, Onnes was 60 years old; Holst was in his mid-20s. Thus, Onnes was rather forgetful, recalls Burgers.

> He worked in the afternoon, and then [Flim] sat until late in the evening. And Mrs. Onnes from time to time sent—not a taxi cab—a horsedrawn cab to wait for him. And the cab was waiting before the laboratory. Onnes quite forgot and walked home without observing it.

Or, he had very particular working principles:

> His rule for his assistants was that you should work all during the day in the laboratory, and in the evening you should write up your data. You had a little pocketbook, and you should write them neatly into a report in the evening. On the weekends you should make your weekly reports. And you could get ... a day off for love; I mean when you were engaged. But you could not get a day off really to listen to a theoretical lecture because experimental physics requires the whole man. [This was so in spite of Onnes's professed conviction of the importance of theoretical knowledge for successful experimentalists!][67]

CHAPTER 6

A Fresh Start

6.1 The War Years; Resumption of Experiments

World War I erupted in July of 1914. It did not directly affect the Netherlands, whose neutrality was preserved during the war's four long years, albeit precariously in view of the horrors in adjacent Belgium. Thus, life at the University of Leiden went on with some semblance of normalcy. Although the Dutch borders were officially sealed, a modicum of outside scientific contact remained possible. Those discussions that did take place were usually dominated by the war. For instance, in 1915 Lorentz met in Berlin with Planck, a spokesman for German science, and Planck, in turn, visited Leiden; the two discussed the attitude of scientists, manifestos, and countercharges.[1] Einstein, also living in Berlin, was able to pay a visit to Ehrenfest and Lorentz in September of 1916—to be sure, only after coping with much red tape between the Dutch authorities, Leiden University, and the Ministry of Foreign Affairs in Berlin.[2] And Onnes maintained limited correspondence with colleagues abroad, including Woldemar Vöigt in Göttingen and Arnold Sommerfeld in Munich.[3]

Nevertheless, meaningful contacts became more and more difficult, and the war was increasingly felt in Leiden. A number of the University faculty drilled regularly in a local Home Guard unit, at Katwijk, a nearby fishing village. The war forced postponement of a planned and rather desperately needed reconstruction and expansion of the cryogenic laboratory. And Onnes's experimental program, particularly the superconducting investigations, coasted to a halt. The main reason was a dwindling supply of helium gas, with no immediate prospects for replenishing it. The original stock from monazite sand had long since been exhausted. In 1910, the stock had been augmented by a gift of gas from Georges Claude at Boulogne-sur-Seine and, as noted earlier, from Director Miner of the Welsbach Light Company in America. But after 1914 further supplies, gratis or imported, were no longer possible, and by 1918 only 300 liters of gas remained. Oddly, thanks to the war the helium program would be off to a good start once hostilities ended.

Meanwhile, the war was less kind to—in fact, decimated—the principal low-temperature programs abroad. In London, Dewar's research, no longer in the forefront in any case, was stopped altogether. Unable to recover from the defeat in 1908, for Dewar the war was the final blow. Dewar now turned to thin films and bubbles, pathetic subjects by his earlier research standards, but subjects ideally suited to his passion for public discourse and oral demonstrations.

In Berlin, by the same token, Nernst's work was terminated, and Nernst threw himself into war work. Both he and his star pupil, Frederick Lindemann, wound up devoting all of their energies to the service of their respective native lands and causes. Nernst's commitment was more spirited and idealistic, in keeping with his strong personality. In the early phase of the war, when he was 50 years old and an academician of great stature, he volunteered his services to the driver's corps of the German army in its dash from Belgium to France.[4] Nernst then turned to advising the German High Command on chemical weapons—the program that Fritz Haber subsequently inherited and spearheaded on a large scale—and the development of explosives, all the while convinced of Germany's ultimate defeat. He lost both of his sons to the war. Near the end of the war he found solace in returning, one more time, to something closer to his heart. His 1918 monograph, offering the theoretical and experimental basis for his heat theorem,[5] was to be Nernst's swan song to low-temperature research. Fortunately, another student of his, Francis E. Simon, would carry on.

Across the English Channel, Lindemann was rejected for active military duty, and faced some difficulties because of his perceived German background and associations. In the event, he spent the war at Farnborough absorbed in hazardous, experimental tests on aeronautical problems for the Royal Flying Corps (such as how to extricate a flying machine from a tailspin), at considerable personal risk. He also designed and stabilized gun sights. Nevertheless he did find some time for superconductivity, as we know. When the war ended, Lindemann was appointed to a chair in physics at Oxford, succeeding Ralph Clifton. In so doing, his scientific productivity came to a somewhat premature end; perhaps he "fell a victim to the blandishments of the English aristocracy."[6] Henceforth, his forte would lie in the administration of science, initially in strengthening the physics program at Oxford—in effect restoring the languishing Clarendon Laboratory. He did not, however, entirely abandon personal research, at least for a period, particularly research on specific heats. In this connection, and partly for nostalgic reasons, he paid a visit to Berlin in the early 1920s. Nernst was gone, but Lindemann obtained a copy of Nernst's original liquefier from Alfred Hoenow, who faithfully hung on. "Needless to say, it never worked."[7]

Of the low-temperature laboratories established before the war, other than Leiden, only the Physikalisch-Technische Reichsanstalt (PTR) in Charlottenburg survived. In due time, it would rival Onnes's laboratory in the postwar unraveling of superconductivity, although not under Holborn, who initiated the low-temperature program at the PTR. And Lindemann's revived Clarendon Laboratory would eventually take center stage.

With the Leiden cryogenic program essentially stopped, little further progress was made on superconductivity there in these years. Unexpectedly, an important advance was nevertheless made in 1916, but at the Bureau of Standards in America. Because experimental investigations had essentially ground to a halt everywhere, the war forced scientists to take a closer look at existing data. In particular, Francis Briggs Silsbee, who had joined the Bureau staff in 1911, the first year of superconductivity, had gone over Onnes's superconductivity data with considerable care. In late 1916 he published a "note" on metallic conduction at low temperatures in the *Journal of the Washington Academy.*[8] While by no means shedding any real light on the mechanism of superconductivity, the article nevertheless constituted a milestone by providing the needed correlation between the critical current density and the critical field that had eluded Onnes. "Though the relationship seems quite obvious," Silsbee modestly wrote, "I have come across no mention of it in the literature on the subject, and think it worthy of notice as furnishing a possible clue to further theories of metallic conduction."

Silsbee summarized the experimental situation regarding the superconductivity of mercury, tin, and lead, including what was known concerning the threshold current and its temperature variation, and the threshold field. Then he got right to the essence of his argument.

> The particular point which is the subject of this note is that *the "threshold" value of the current is that at which the magnetic field due to the current itself is equal to the critical magnetic field.* In other words the phenomenon of threshold current need not be regarded as a distinct phenomenon, to be explained by heating, or otherwise, but is a direct result of the existence of the phenomenon of threshold magnetic field.[9]

In a coil, he continued, "owing to the cumulative effect of the successive turns, the field produced by a given current is much greater in the coil than in the same wire 'when straight,' and, consequently, the current required to give the critical field strength will be much less."[10] He supported this point by Onnes's observation of a drastically degraded critical current in the coil, compared to its value for the same wire when straight.[11] In the case of a straight wire of circular section, the effect to be expected was more complicated, he warned. The field intensity at any radius r inside a wire of radius r_0 carrying a current I distributed uniformly over the cross section is given by

$$H = 2\,Ir/r_0^2\,, \tag{6-1}$$

and the field at the surface, consequently, is

$$H = 2\,I/r_0\,. \tag{6-2}$$

As soon as the current exceeds the value $H_c r_0/2$, where H_c is the critical field for the material, the outermost surface layer of the wire becomes resistive, shunted by the still superconducting core, Silsbee argued. This, however, means by Eq. (6-2) that the field at the surface of the core is even greater, due to the inverse dependence of field on radius. The system is therefore unstable and the current will shift suddenly to a new distribution. But the precise form of the current distribution would depend on the generally unknown relationship between resistivity and field. About all one could say was that the resistance changes virtually discontinously at a certain field.

Silsbee had reviewed Onnes's data on mercury, tin, and lead (*Communications* 133a, b, c, and d). His note tabulated for each temperature of observation, and for different wire sizes, the threshold current, the corresponding threshold current density, and the self-field due to the threshold current, calculated from Eq. (6-2) with $I = I_c$. For wires of different diameters, but at the same temperature, he found that the latter ratio, I/r_0, seemed a better constant than either the current or the current density. Moreover, where measured threshold fields were available, for lead and tin but not mercury, they agreed more or less with the values computed from the threshold current. (The discrepancies, consistently in the direction of the measured threshold fields exceeding the calculated fields, could be explained by necked-down, thin spots in the wire—the "bad places"—where the field intensity would be correspondingly greater.) All of this gave strong support to the hypothesis "that the magnetic effect is the more fundamental."[12]

While Silsbee was pondering the significance of the critical field, Onnes's superconducting program was not entirely idle. In spite of the dry cryostats, preparations were made for a renewed search for additional superconductors, even though the actual tests would have to wait. Thallium, lying between mercury and lead in the periodic table, seemed a good starting point. In December of 1916 the first bifilar samples of thallium were prepared.[13] They were extruded from rods into wire, wound bifilarly on threaded porcelain tubes, provided with electrical leads, and sealed in a glass tube with helium gas. Without much helium, the only measurements possible were zero-point calibrations. For these, the resistances were sealed in an additional tube filled with liquid paraffin or distilled benzene, and placed in crushed ice. The calibration measurements were made, in part, by Burgers during January–February of 1917 with the aid of overlapping shunts or by potential compensation.

While Onnes's stock of helium was dwindling fast, helium suddenly acquired significance as a strategic war material. In 1915 the British Admiralty instituted a search for helium-bearing gas for replacing hydrogen in naval and military balloons and airships. John C. McLennan of the University of Toronto was put in charge of the project, which moved slowly, however. A small pilot plant for extracting helium from natural gas was eventually constructed near Hamilton, Canada, and later moved to Calgary where trial runs were finally under way in late 1919. Although McLennan's project bore fruit too late to aid the war effort, it led to the creation of a fledgling cryogenic laboratory at Toronto in the postwar years—a laboratory that would play a role in future developments in superconductivity. Meanwhile, when America entered the war the U.S. Bureau of Mines was given the same task as McLennan's, and a helium search on a much larger scale was mounted with characteristic American energy. Three experimental extraction plants were constructed in Texas. The Linde plant at Fort Worth proved to be the most successful, and became the first helium production plant in the United States. It was constructed under the auspices of the U.S. Navy during 1919–1921. Thus it came about that the war, having stopped Onnes's experiments, was indirectly responsible for their resumption as well. In 1919 Admiral Griffin, on behalf of the U.S. Government, presented Onnes with a gift of 30 m^3 of helium gas. (The gift was augmented by Canada in 1921, when McLennan personally carried a cylinder of 2 m^3 of helium gas to Leiden.)

The fourth superconductor was verified rather casually, not long after the war ended. The formal occasion was the 17th Dutch Scientific and Medical Congress, held in Leiden during 24–26 August 1919. The agenda included public demonstration of liquid helium, highlighting as well some of the low-temperature discoveries in the last years before the war. Onnes, with his new student Willem Tuyn, availed himself of the opportunity to make a preliminary examination, shortly before the congress opened, of the change of resistance with temperature of one of the thallium samples, mounted and waiting in a cryostat.[14] Indeed, the galvanometer reading went off scale; thallium was clearly a superconductor. More accurate measurements in June showed its vanishing point to be 2.32 °K.[15]

While occupied with thallium, Onnes and Tuyn also found time to prepare additional resistances, some of lead and some of its isotope uranium lead (Radium G). Modifications of the cryostat, then in progress, promised a more accurate value of T_c for lead (previously estimated to be about 6 °K by simple extrapolation from the boiling point of helium), and would allow tracing the full resistance curve into the hydrogen temperature region. Onnes seems not to have had doubts about uranium lead also being a superconductor. He wished, rather, to check on any influence of the *mass* of the atomic nucleus on the transition temperature[16]—an idea suggested quite naturally by a mechanism involving Planckian vibrators. In 1916, while procuring fresh samples of pure lead (and thallium) from the firm of Kahlbaum, Onnes additionally obtained 16.5 g of the lead isotope from Professor Otto Hönigschmid, the well-known expert on atomic weight determinations at the Radium Institute in Vienna. (At that time, comparative determinations of the atomic weights of lead and its isotopes were playing an important role in confirming the Rutherford disintegration theory and the displacement law of Soddy and Fajans.) The sample, one personally used by Hönigschmid in weight determinations, was delivered by "friendly mediators"[17] via a circuitous route including the Dutch legation in Vienna and the Technischen Hochschule in Prague. Perhaps because of the ballyhoo about security, and with the war on, the latter transaction caused some misapprehension among the institute staff. In a letter from the radiochemist Fritz Paneth to Otto Stern, whose Army regiment was then fighting near Lomscha on the Eastern Front, Paneth discussed the problem of isotopes generally. Something new had cropped up, he added on a nervous note.

Already now I can tell you what it concerns, asking you to exercise caution, since there is much anxiousness in this area and it concerns a foreign investigation. The point is that Kamelingh Onnes has received from Hönigschmid a sample of his uranium lead (atomic weight 206.0) and is presently preparing to make comparative investigations on the supraconductivity of it and ordinary lead at very low temperatures. I believe ... that the results will be of importance to you, and we shall write to you about it as soon as we ourselves are informed.[18]

Not until 1919 could the new resistances be made ready. Wires of lead and RaG were drawn and prepared in much the same way as the thallium resistances, although the chemical properties of lead obviated the need for sealing the samples in an inert atmosphere. The essential feature of the new cryostat, constructed under Flim's supervision, was that test samples could be immersed in helium vapor or gas. One or more samples, with a helium gas thermometer, were mounted in a silvered vacuum glass bell suspended within a larger, silvered glass. A regulating tap and electric heater determined whether the bell-shaped experimental chamber was filled with liquid or vapor of a desired temperature. Temperatures were determined by a constant volume gas thermometer with an open manometer; the relationship between vapor pressure and temperature of the helium bath was derived graphically from experiments conducted before the war.[19] The resistances were determined by Kohlrausch's standard procedure of comparing the deflection of a galvanometer when connected across the unknown and a known, standard resistance; the resistance values were taken as proportional to the mean of the deflections for both directions of current.

In two sets of measurements in May of 1920 both the lead and uranium lead resistances proved superconducting in liquid helium. The temperature of the cryostat was slowly raised by means of the heater. "At a certain moment the galvanometer moved quickly over 35 cm on the scale and the vanishing point was apparently reached," wrote Onnes; "the suddenness of the deflection speaks well for the usefulness of the cryostat if not too high demands are put upon it." The critical temperature of lead was judged to be 7.2 °K. Moreover, it was concluded, " 'Kahlbaum' lead, atomic weight 207.20, and uranium lead (RaG), atomic weight 206.06, have the same vanishing point temperature within the accuracy of 1/40 degree [determined by the accuracy in the cathetometer reading of the meniscus level]."[20]

The question of an isotope effect rested there, but would resurface as a prominent issue after World War II. It is now known that T_c varies inversely with the square root of the isotopic mass.[21] This is quite consistent with Onnes's negative result, since the respective atomic masses imply a difference in T_c of only 0.02 °K—which by chance was Onnes's measurement limit.

6.2 Brussels, 1921: Ignored Clues

Onnes was responsible, with Ernest Rutherford, Mme Curie and Lorentz, for getting the activities of the Solvay Foundation going again after the war. In due time the third Solvay Conference was scheduled for Brussels during the first week of April 1921. The conference theme this time was to be atoms and electrons. Onnes had insisted that relativity was too abstract a subject, but he got superconductivity included during final amendments to the program.

Actually, Onnes read two separate reports at Brussels. One covered paramagnetism, because illness had prevented Paul Langevin from preparing an invited presentation on the topic (though Langevin recovered in time to take part in the formal discussion at Brussels), the other superconductivity. Onnes's printed report on superconductivity,[22] including the open discussion, ran well over 30 pages. Indeed, there was much to cover since his first appearance at Brussels 10 years earlier. In addition to mercury, the inventory of superconductors now included lead, uranium lead, tin, and thallium. A search down to 1.5 °K—the lowest temperature within reach in 1921—had revealed no more superconductors, although further candidates were by no means exhausted. Attempts to link superconductivity with certain material properties had been no more successful; at most, Onnes saw a possible significance in the fact that Hg, Tl, and Pb form a group of elements with consecutive atomic numbers, analogous to the ferromagnetic triad Fe, Co, and Ni. As to whether the resistance in the superconducting state is truly zero, Onnes thought not. In his opinion the microresidual resistance was real.

Onnes reviewed the critical current and field, unsuspected complications at the time of Solvay I, and the elegant persistent current experiments of 1914. Silsbee's hypothesis appeared to have neatly reconciled the two thresholds, although further experiments were needed to settle the matter unequivocally. On the other hand, the persistent currents, he noted, were in accord with what became known as "the rule of Lippmann." The rule, published by Lippmann in 1919,[23] was actually stimulated by the 1914 experiments and is simply the consequence of Maxwell's electrodynamics for perfect conductors. It states that the magnetic flux linking any closed circuit within a body of zero resistance—a superconducting lead ring, for example—cannot change; circulating currents are induced on the surface of the conductor so as to create a magnetic flux density in the interior that cancels the flux density due to the applied field. This is perfectly true for Onnes's particular experimental sequence: cooling a ring *in* a field, *then* removing the field. Onnes himself was fond of distinguishing between two points of view by calling the persistent current behavior the "antilogon" of the reverse experiment, which is more analogous to Wilhelm Weber's diamagnetism: exposing an already superconducting specimen to a field (less than the critical field). In this case, again by virtue of Lippmann's rule, circulating currents are induced. However, if now the field is reduced to zero, the specimen returns to its initial current-free state. It seems this indeterminacy in the final state had never been seriously questioned. Since 1911, moreover, it had been tacitly—and erroneously—assumed by Onnes, and by everybody else, that the same was true for a (superconducting) *solid*, or any "simply connected" body of infinite conductivity, such as a sphere. This supposition, and the failure to follow it up, except for a single—albeit inconclusive—experiment, retarded a significant advance in superconductivity for more than a decade. Lippmann, who knew Onnes during his Wanderjahr at Heidelberg, by invoking the authority of Maxwell in the case of superconductivity, was perhaps partly to blame for thwarting a major twentieth-century development in electrodynamics!

At Brussels, Onnes was inclined to view the abrupt *discontinuity* of the transition as the fundamental characteristic of superconductivity, not the fact that the conductivity attains extremely high values. Vanishing resistance could be accommodated, at least qualitatively, by virtually any of the many theories making the rounds, but the discontinuous transition defied them all, however implausible they may have been. The question whether the transition was accompanied by other changes in structural or physical properties, raised by Langevin at Brussels in 1911, was still not resolved. Onnes suggested that the thermal transition is an allotropic transformation similar to a ferromagnetic transition at the Curie point, somewhat in the same vein as J. J. Thomson's dubious dipole theory.

Onnes's views on this differed considerably from those of Percy Bridgman of Harvard who, by coincidence, was also pondering superconductivity at about the time of the Solvay Conference. Bridgman's views on the nature of metallic conduction arose from his attempts to reconcile the effects of pressure (as well as temperature) on the electrical resistance of metals with the prevailing electron theory of metals,[24] and the phenomenon of superconductivity seemingly went hand-in-hand with his early work on polymorphic transformations. "The thesis which I wish to support," wrote Bridgman in a 5-page note to the *Journal of the Washington Academy*, "is that the discontinuous change in resistance on entering the supraconducting state is also a mark of a polymorphic change, and that a discontinuity and supraconductivity are not as intimately related as sometimes supposed, but are due to quite distinct mechanisms."[25] In Bridgman's view, the normal condition of *any* metal at very low temperature is superconducting, in the sense that the electrons may pass freely from atom to atom, providing the atoms of the metal are in contact at rest as in a solid near absolute zero. However, as the temperature is raised, the atomic centers become separated, and if the separation exceeds a certain critical value, electrical resistance sets in. This viewpoint sidestepped the need to invoke a new mechanism to explain the discontinuity; to clinch his argument, he pointed to Onnes's initial belief that the phenomenon is somehow tied to metallic purity—a belief undermined when it was found that amalgams show the effect as well. The appearance of superconductivity, he argued, is "to a certain extent capricious": The fact that sometimes samples do not become superconducting is due to an accident of structure or handling. Emphasizing Silsbee's hypothesis, Bridgman noted that the temperature of a polymorphic transition is altered by a magnetic field, so here too it was unnecessary to invoke some new connections between the conduction mechanism at low temperature and the magnetic field.

Bridgman's essential argument, that the clue to the superconducting transition should be sought in thermodynamics, was a bold suggestion, since there was no evidence for a heat of transformation at the vanishing point. (In fact, the first experimental test for this was yet to come.) Nor was there, apparently, a change in volume—another thermodynamic signpost. These circumstances were to be responsible for perpetuating what H.B.G. Casimir has called the "first superstition" of superconductivity, that the phenomenon is a peculiar trait of the electron free path, not a special state of the electron gas.[26] (Casimir's second superstition relates to the uncritical acceptance of the rule of Lippmann.) Bridgman would have more to say on the matter at the next Solvay Conference, where he participated actively. But for now, the electron free path held sway, despite growing doubts and arguments to the contrary, for example those of Lindemann and Keesom, long before Bridgman. Onnes wrapped up his presentation by expressing confidence that the Rutherford–Bohr atom, with quantum theory, would eventually deal successfully with superconductivity. He phrased the problem in the form of a series of unresolved questions:

1. When the atoms of the Rutherford–Bohr model combine to form a metal, what becomes of their peripheral electrons? Do they lose all or part of their kinetic energy?
2. How many types of electrons ("free," or fixed more or less strongly) are involved and what statistics do their movements obey according to the quantum theory?
3. Is the movement of conduction electrons in adjacent atoms coherent?
4. What is the mechanism by which the conduction electrons in ordinary conductors transmit the energy acquired under the action of external electromagnetic forces while subject to thermal agitation?
5. Do the normal velocity electrons play a role?
6. How can the atoms have a superconducting contact; how can they form superconducting filaments that provide macroscopical paths along which the electrons slide without transmitting energy consistent with the freedom of motion of the superconducting atoms; and how are the conduction electrons guided in paths in the midst of atoms in thermal motion?
7. Under what circumstances does the superconducting state cease to exist?
8. What is the reason for the equivalent role of a magnetic field and temperature in the destruction of a stable adiabatic flux of electrons amidst an agglomeration of atoms in thermal motion?[27]

The discussion following Onnes's report was dominated by Langevin and his former student Leon Brillouin who, with his father Marcel, also participated in the Solvay Conference of 1921. Langevin opened the exchange by returning to Silsbee's hypothesis. Assuming that the resistance of a wire of circular section increases discontinuously by a large factor k at a well-defined critical field (he began), Silsbee had concluded that, for a current slightly in excess of the critical value, there will be a superconducting core of radius r_0/k in which the current density is k times the average value, or $1/k$ of the total current flows there. Outside the core, the material is in a field greater or equal to H_c and, by hypothesis, the material has an increased and uniform resistance. Since k is very large and r_0 is very small, the effect of the core should, in fact, be negligible. Langevin had examined the current distribution more closely, assuming Silsbee's relation between resistivity and field. He found that there are actually three distinct regions in the wire cross section: Silsbee's inner region where the magnetic field due to the current is less than H_c; an intermediate region where the current density varies such as to maintain a constant field equal to H_c; and an outer layer where the field exceeds H_c. Langevin remarked that he had communicated his results to Silsbee, who found them sufficiently interesting to warrant a follow-up note on the subject in 1917.[28] Langevin had also calculated the current relaxation time. For Onnes's lead coil, he estimated a time constant longer than 10^5 sec, in good agreement with the actual measurements.

The discussion continued, with Ehrenfest, Rutherford, de Haas, and de Broglie joining in, but it was mainly dominated by Langevin who stressed the utility of Lindemann's dual-lattice model, and Bragg who, naturally, dwelt on the relationship between superconductivity and the crystal structure. (Germans were excluded from the 1921 conference, except for Einstein whose nationality was in any case a matter of dispute.) Superconducting metals, Bragg observed, occur in a region of the periodic table characterized by relatively close atomic spacing, with the peripheral electrons interpenetrating adjacent atoms to some extent, lending credibility to Onnes's picture of electrons sliding adiabatically along interatomic filaments. The discussion session concluded with an impromptu, ponderous lecture by Leon Brillouin on his own recent work on electrical and thermal conductivity of metals,[29] utilizing the formalism devised by Einstein and L. Hopf for treating resonances in a radiation field.[30] Brillouin's treatment of the problem accomplished little more than rederiving, by a slightly different route, Wien's results and Eduard Grüneisen's contrary and purely empirical rule of 1913[31] that, at low or moderate temperatures, the resistance of a metal varies approximately as the product of the atomic heat and the absolute temperature, or

$$R \propto C_p T . \qquad (6\text{-}3)$$

Einstein himself was hardly optimistic in his own review of superconductivity and the stubborn problem of metallic conduction in general the following year. His views on the subject were not voiced at the Solvay Conference. (Though invited because of his international reputation and accepting with pleasure, he failed to show up; he was in fact in the United States, just then, espousing the cause of Zionism.) Rather, they appear in the volume issued on the occasion of the celebration of the fortieth anniversary of Onnes's professorship.[32] Recapitulating the difficulties of the electron theory (e.g., the electron density and path length) and Onnes's quasi-successful but ad hoc circumvention of them, he admitted that even normal conduction—certainly superconductivity—still defied explanation. Superconductivity could not be a matter of *free* electrons moving under thermal agitation, in light of Onnes's demonstration that normal threads coated with a superconducting layer are also superconducting. The electrons of the coating would sooner or later penetrate into the normal conductor, and the mean, preferred motion constituting an electric current would be destroyed. "It appears as though, based on our present state of knowledge, free electrons simply do not exist in a metal," he concluded.[33] Rather, metallic conduction would have to be explained by transfer of peripheral atomic electrons along closed molecular conducting chains in the manner of Onnes's Amperian currents. A similar mechanism had been suggested by Fritz Haber in 1919. Its credibility was, however, undermined by Onnes's further discovery—acknowledged in a note appended to Einstein's musings—that a junction between two different superconductors (lead and tin) is also superconducting: It seemed most implausible that two different atoms could form conducting chains with each other.

"With our wide-ranging ignorance of the quantum-mechanics of composite systems," Einstein admitted, "we are far from able to compose a theory out of these vague ideas. We can only rely on experiment."[34]

6.3 Brussels, 1924: A Question of Reversibility

The fifth and last element added to Leiden's postwar inventory of superconductors, indium, was announced in June of 1923. Its location in the periodic table near the known superconductors—above thallium and beside tin—made it a likely candidate. Tuyn and Onnes did not state explicitly a value for T_c in their June report to the academy,[35] but their tabulations from December of 1922 show unmistakably a critical temperature of about 3.41 °K. Its value, the authors claimed, reinforced the weak evidence for some regular periodicity among the known vanishing points: rising slightly from In to Sn and strongly from Tl to Pb. In addition, because T_c rises toward the left, from Tl to Hg, and assuming the trend continues, T_c for gold ought to lie higher than T_c for mercury. In fact, Au showed no signs of becoming superconducting down to 1.5 °K; "the conclusion might be drawn," notes Tuyn and Onnes, "that Au—perhaps with other metals—can never become so."[36] And yet, a vanishing point below 1.5 °K could not be ruled out, as it was still the lowest temperature attainable.

Metallic conduction held center stage at the fourth Solvay Conference, held in 1924, which was specifically devoted to the electrical conductivity of metals and related problems. This time the opening address was given by Lorentz.[37] He reviewed attempts to understand metallic phenomena in terms of microscopic principles couched in various ways in the classical or quantum-theoretical framework. As for superconductivity, he had little to offer that was new beyond referring to Lindemann's belated attempt in 1915 to explain perfect (R = 0) conductivity and a renewed attempt by J. J. Thomson in 1922 to treat the problem as greatly enhanced *normal* conductivity in terms of ordered electrons.[38] Both thereby evaded the dilemma dogging the classic theory: specific heats. About all that Lindemann and Thomson accomplished was suggesting new approaches to the behavior of electrons, even in normal metals: Approaches avoiding the need for invoking free electrons (a view championed by Lindemann as well as Einstein), and relying instead on an electron space lattice (Lindemann) or electron chains (Thomson, Onnes)—either approach foreshadowing the need for collective ordering of the electrons in superconductors.[39] The hint was ignored by Owen Williams Richardson who, in his Solvay presentation following Lorentz's review, dwelt on a theory of metallic conduction of his own involving tangential Bohr orbits in solids.[40]

Onnes's report to the conference[41] was actually delivered by Keesom, because Onnes's health was declining at this time. (Keesom also fielded the discussion following the formal report.) The report began with a description of two new and important experiments at Leiden. Both were refined versions of the persistent current experiments in 1914, and were cleverly devised by Onnes and Tuyn to reduce even further the upper limit on any microresidual resistance by establishing more accurately the "degree of invariability" of the supercurrent. Their success depended critically on an important advance in the cryogenic resources at Leiden. Hitherto, or since 1912, the test dewar had been cryogenically separated from the helium liquefier, but still attached to it physically (Figure 3-1). But now, thanks largely to Flim, the dewar was *transportable*. Liquid helium could be transferred by siphoning it from the liquefier, and then the light dewar could be carried to an experimental cellar where there was much less vibration than in the cryogenic hall, rather constant ambient temperature, and the experiments could be carried out more leisurely without the disturbing effects of replenishing the liquid.

The first of the new experiments was based on the electrodynamic action between a pair of persistent current rings of lead. The rings were mounted concentrically in a helium dewar (Figure 6-1), between the poles of a Weiss electromagnet. The inner ring *A* was suspended from a torsion head by a spring and a glass rod, interrupted by a shock absorber in the form of a vane submerged in oil; it was free to turn within the outer, fixed ring *B*. The field was turned on, and the rings were made superconducting by filling the dewar from the liquefier with the aid of a siphon and a sliding seal. With the plane of the rings coinciding and perpendicular to the external field, persistent currents were induced, as before, by switching off the external field (simply by rolling the magnet away on casters). The plane of the inner ring was then rotated through a certain angle, for example 30 degrees, with respect to the outer ring by turning the torsion head. This turning caused new induced currents, which reached a steady state after about 20 minutes, whereupon the observations commenced. The electrodynamic forces between the currents, striving to keep the rings coplanar, produced a torque on the inner ring, counteracted by the torsion of the suspension. Any decrease in the magnitude of the currents could be observed as a change in the twist of the torsion fiber, or angle between the rings, by a mirror and a scale across the room. Observations continued as long as the rings remained immersed in liquid helium, or for about 6 hours. During this time, the currents remained constant within a limit of precision judged to be one part in 1750. Onnes estimated the variation in the relative current strengths to be less than 1/2100 per hour. This corresponded to $R_{4.2^\circ\mathrm{K}}/R_{0^\circ\mathrm{C}} \leq 10^{-12}$, reducing the previous microresidual resistance limit by another factor of ten.

The experiment was then repeated, this time with a hollow sphere of lead replacing the inner lead ring (Figure 6-2), for the purpose of establishing the "invariability of the *distribution* [emphasis added] of currents in a superconductor subject to pondermotive forces."[42] That is, Onnes sought to demonstrate

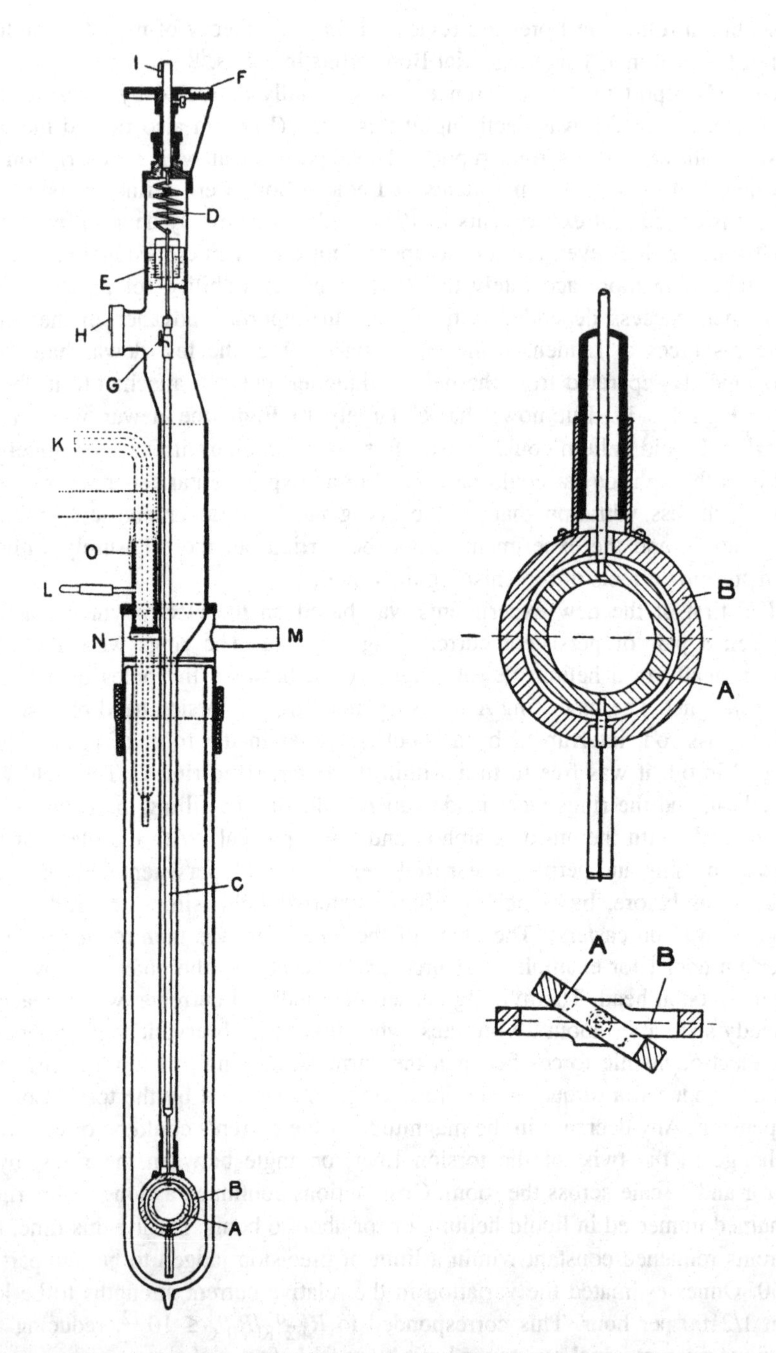

Figure 6-1. Cryostat assembly for experiment with concentric lead rings. Letters indicate rotating ring (A), fixed ring (B), glass support rod (C), mirror (G), damping vane in oil bath (E), spring (D), graduated disk (F), torsion head (I), details of siphoning mechanism (K, L, M, O, N). Kamerlingh Onnes, Solvay 1924, 253.

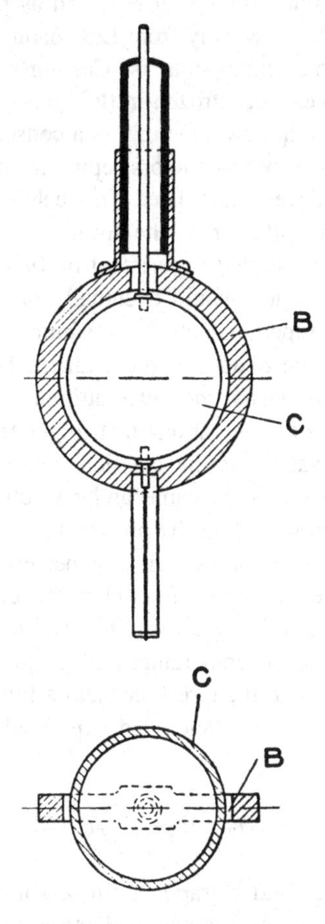

Figure 6-2. New experiment, with hollow lead sphere (C) replacing inner ring. Kamerlingh Onnes, Solvay 1924, 253.

explicitly that the paths of persistent currents in a superconducting body, his "tubular filaments," are rigidly fixed in the body. The hollow sphere (*C*) was simply a thin layer of lead evaporated on a glass ball, mounted with the vertical axis perpendicular to the plane of the ring encircling it on the equator. On removing the field, the sphere exhibited a magnetic moment corresponding to a distribution of persistent currents on the surface of the sphere. The torque on the sphere was measured as a function of angular position. When rotated, the moment appeared to rotate with it, "prov[ing] plainly the constancy of the magnetic moment," or "that the free electrons in a superconductor ... are guided in their trajectories ... in tubular filaments embedded in the superconductor."[43] Measurements analogous to the earlier experiment on the constancy of the torque between ring and sphere indicated a stability in current better than one part in 20,000 per hour, reducing the limit on microresidual resistance by another order of magnitude.

The hollow ball experiment was long regarded as prima facie evidence for the frozen-in flux expected classically from Lippmann's perfect conductor subject to a changing external field—that is, Casimir's second superstition. In hindsight, one can indeed expect a frozen-in flux in a multiply connected superconducting body, such as a hollow sphere; it is a consequence of the sensitivity to geometry exhibited by superconductors, epitomized by the concept of the "intermediate state" introduced much later. This behavior, unfortunately, is not representative of a superconductor in the form of a singly connected, solid sphere and therefore the celebrated experiment of 1924, repeated 5 years later, proved nothing. The distinction would eventually prove crucial, invoking the essential characteristic of superconductivity. Casimir has blamed the prevalence of the superstition and failure to address the magnetic behavior of superconductors at Leiden on the *Door meten tot weten* attitude—favoring precision measurements at the expense of observations that might lead to something new.[44] Perhaps Lorentz's rederivation of the classical frozen-in flux at that time[45] fueled the dogma. The matter only came under scrutiny 10 years later, when serious doubts were raised regarding the prevailing conception of the magnetic properties of superconductors through an unexpected chain of developments. But in 1924, the resistance measurements occupied everyone's attention.

Of more pressing concern that year was the need to elucidate the pervasive, inexplicably analogous role of temperature and magnetic field in suddenly restoring the resistive state, with the threshold field a function of temperature. At the Solvay Conference in 1921, Onnes had expressed the equivalence as follows:

$$H_T = H_0 - C_{HT}T \,, \tag{6-4}$$

where H_0 is the threshold field extrapolated to absolute zero, and C_{HT} a constant. H_0 is characteristic of each superconducting substance, and C_{HT} is the same for all bodies, presumably derivable from quantum theory. In the interim, the approximate validity of Eq. (6-4) had been qualitatively verified by Tuyn's measurements on lead, indium, and tin.[46] The measurements had, however, also shown that the "magnetic" transition from the superconducting state is not perfectly discontinuous at a well-defined critical field, but that the transition curve extends over a narrow but finite field interval. A good example is Tuyn's measurements on tin at various bath temperatures, as seen in Figure 6-3. Because of the nondiscontinuous transition, Onnes and Tuyn adopted as their definition of the critical field the field strength that raises the resistance to one-half its normal value, which they denoted by $H_{1/2}$. The measurements had shown that, to first approximation, $H_{1/2}$ is a linear function of temperature. Closer inspection showed (Figure 6-4) that curves of $H_{1/2}$ plotted against temperature are actually slightly concave toward the origin or have a negative slope (actually more nearly parabolic, as we now know). This fine detail was merely noted in passing by Onnes and Tuyn, but it would prove to be important in understanding the superconducting transition from a thermodynamic viewpoint. (The shape of these curves is, in fact, dictated by Nernst's heat theorem.)

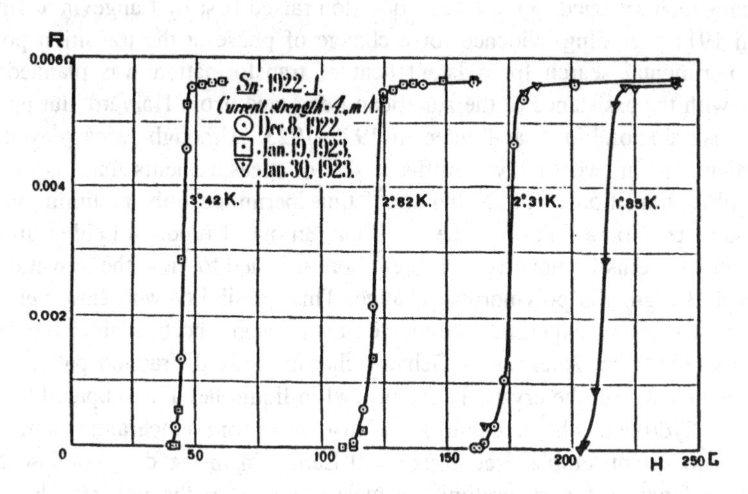

Figure 6-3. Magnetic transition curves for tin at various bath temperatures. Tuyn and Kamerlingh Onnes, Franklin Institute, Journal, **201** *(1926), 393.*

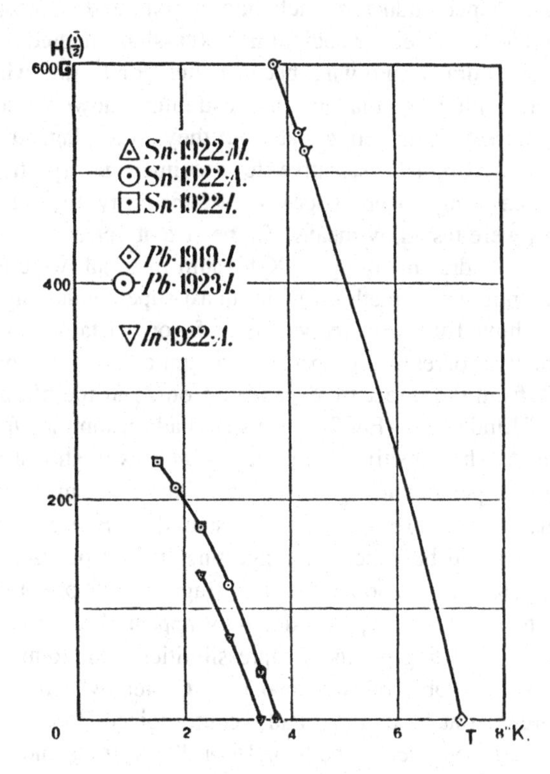

Figure 6-4. Threshold field (or $H_{1/2}$) versus temperature for indium, lead, and tin. The value of T for H = 0 is the "transition temperature." Tuyn and Kamerlingh Onnes, Franklin Institute, Journal, **201** *(1926), 395.*

Onnes then returned to the crucial question raised first by Langevin at Brussels in 1911 regarding evidence for a change of phase at the transition point. An experimental search for a latent heat of transformation was planned by Onnes with the assistance of the American Leo Dana, from Harvard, during the latter's postdoctoral stint at Leiden in 1922–1923. Although Dana played an important role in two highly significant sets of measurements that uncovered the lambda transition in liquid helium,[47] time permitted only preliminary, inconclusive tests on a tin sample before Dana ran out of funds in Leiden. In any case, Onnes seems by now to have been more inclined to view the transition in terms of Bridgman's polymorphic change. This possibility was, however, negated by a decisive experiment conducted by Keesom shortly before the 1924 Solvay Conference. Analysis of Debye–Scherrer x-ray diffraction patterns revealed no change in the crystal lattice of lead in liquid helium compared to that in liquid hydrogen which, in turn, was also the room-temperature pattern.[48] Equally important, emphasized Bragg and Langevin in the discussion section following Onnes's report, preliminary measurements of the *intensity* distribution in the pattern above and below the transition point of lead again showed no change, indicating little or no change in the *electronic* structure as well.

Onnes ended his Solvay report with a review of the known regularities among the various superconductors, including indium, and of prospects for additional superconductors. Neither sodium nor potassium, melted in glass capillaries, nor aluminum drawn into wire, became superconducting when cooled to 1.5 °K. However, gallium, germanium, and cadmium showed tantalizing hints of approaching the superconducting state, but they were plagued with a small residual resistance extremely sensitive to specimen purity, treatment, and strength of the measuring current. (Because of their rarity, the gallium and germanium samples were tested by means of a persistent current method; the cadmium samples, wires drawn from pure Kahlbaum material, were tested by one of the usual potentiometric methods.) The quasi-superconducting behavior of these substances hinted that superconductivity favors metals with low melting points (low attractive forces)—a property of indium and tin and an old idea that sprang naturally from the direct $\beta\nu$-dependence on T_s in the old prewar calculational model of Lindemann (Eq. 2-7). Einstein had reexamined this possibility two years earlier.[49] The sensitivity to purity as well as treatment was taken as evidence that the superconducting state is a mixture of superconducting and ordinary crystals or complexes—a view also shared by Einstein. If correct, anisotropic behavior could be expected, suggesting fruitful persistent current experiments on "aggregates of monocrystalline fragments"[50] oriented in different directions. The tendency for superconductivity appeared to be correlated with the reciprocal of the melting points, compressibilities, and atomic volumes. Inspection of the atomic table of Bohr and Dirk Coster (who received his doctorate under Ehrenfest at Leiden and had recently joined Bohr in Copenhagen) suggested spherically symmetric shells of 18 or 32 electrons and a few valence electrons as preconditions for superconductivity. (To Onnes's considerable credit, these rules are not so different from modern, also largely empirical rules—i.e., those of Matthias.)

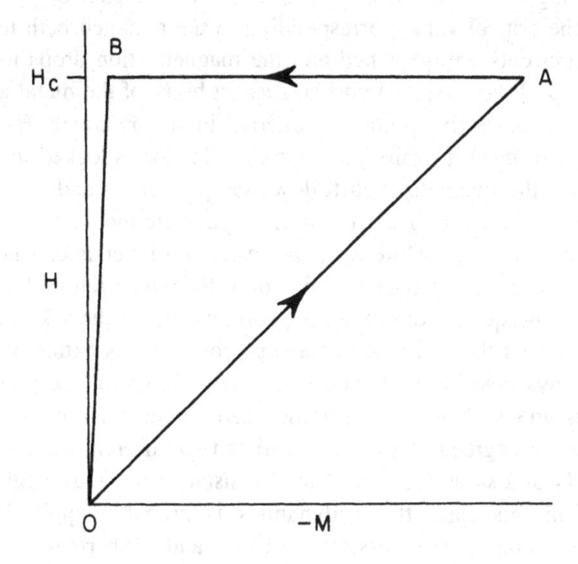

Figure 6-5. Behavior of the magnetization M of a superconductor when cycled in an external field, according to Keesom. Adapted from Discussion du Rapport de M. Kamerlingh Onnes, Solvay 1924, 287.

From the vantage of superconductivity, the Solvay Conference of 1924 is best remembered, not for Onnes's report, but for the discussion following it, and particularly for the exchange between Bridgman and Keesom on a certain thermodynamic point. Admitting that Keesom's own crystallographic evidence rendered an allotropic modification in the superconducting state unlikely, Bridgman argued that the effect of a magnetic field could be explained more simply by postulating a discontinuous change in the magnetic permeability between the normal and superconducting state. All that was necessary was a thermodynamic cycle, analogous to that used in deriving Clapeyron's equation for the usual change of state. He illustrated this by superimposing such a cycle on the curve of critical field versus temperature, with temperature and field replacing the usual variables of temperature and pressure.

Keesom (and Lorentz) objected that the transition in a magnetic field is irreversible, as evidenced by the irreversible magnetization behavior expected for ideal (infinite conductivity) conductors in an applied field—the reason again being basically Lippmann's currents. Consider a plot of field versus magnetization, Keesom argued, where the applied field H and magnetization M are related to the magnetic induction B of a sample by the usual expression:

$$B = H + 4\pi M . \qquad (6\text{-}5)$$

We start with a conductor made superconducting by cooling it below T_c in zero-applied field (the origin, Figure 6-5). If an external field is then applied, the sample behaves as if it were diamagnetically magnetized because of induced, circulating screening currents, and it follows the path *O–A*. If, at *A*, the field reaches the critical value corresponding to the assumed bath temperature, the screening currents are quenched and the magnetization drops to some low value (at *B*) given by the (small) normal susceptibility of the metal at the given temperature. (Hence, at this point, the internal induction equals H_c.) Upon reducing the field from H_c to zero, the internal induction is locked in by induced surface currents, the magnetization follows the path *B–O*, and in zero applied field the specimen is left with a net frozen-in magnetic moment—a state different from the original one. Moreover, the cycle is irreversible, since the path *A–B* cannot be traced in the opposite direction. Bridgman retorted that the irreversibility is a consequence of circulating currents. If, for the sake of argument, we posit instead that the resistance of a superconductor is extremely small but finite, it is always possible to traverse the cycle slowly in the absence of induced currents and with negligible Joule heating—in other words, reversible conditions. Keesom agreed that if we postulate two different states of magnetic susceptibility k_1 and k_2 as well as a small resistance, we may neglect induced currents and, in this case, thermodynamics is probably applicable. Taking Clapeyron's equation for a reversible, isobaric, and isothermal transition of a substance between two phases,

$$L = (v_2 - v_1)T\frac{dp}{dT},$$

but replacing p by the magnetic field and v by the magnetic moment (i.e., replacing the work done by the external pressure by that of the magnetic forces), Keesom wrote:

$$\lambda = (k_2 - k_1)TH\frac{dH}{dT}. \qquad (6\text{–}6)$$

This important relationship between dH/dT along the critical H_c–T curve and the heat of transformation λ for the magnetic transition had also been derived independently by Bridgman in his journal article of 1921.[51]

The argument, briefly, runs as follows.[52] The cycle consists of a transition from phase (1) to (2) at a temperature T and magnetic field H, transfer of the substance to temperature $T + dT$ and field $H + dH$, transition in the reverse direction from (2) to (1), and transfer of the substance back to the initial temperature and field. In Clapeyron's original equation, above, it is merely necessary to replace $(v_2 - v_1)dp$, the work done by the external forces, by the work done during the cycle by the magnetic forces. Now the work done by the magnetic forces during a change of magnetization is HdM. During the first transition at a constant field, the magnetization changes from k_1H to k_2H; hence, the total work is $H^2(k_2 - k_1)$. During the second transition, the susceptibility remains constant but H changes by dH; now the work is k_2HdH. Similar expressions give the work during the remaining two stages of the cycle, yielding the total work during the cycle of $HdH(k_2 - k_1)$ and, hence, Eq. (6-6).

Onnes's own reluctance to pay serious attention to the growing thermodynamic arguments, indeed his strong predisposition toward superconductivity all along, must be attributed to his own conservative personality, singleness of purpose, and his dogged adherence to the aforesaid prevailing attitudes regarding superconductivity: not as a change of state but as a change of free path, and as a phenomenon in accord with the irreversible behavior associated with the rule of Lippmann—neither of which encouraged a thermodynamic viewpoint.

The rest of the discussion, mainly between Keesom, Langevin, Lorentz, Bridgman, Richardson, Rosenhain, and Broniewski, drifted inconclusively over most of the other topics covered in Onnes's Solvay report. Perhaps Mme Curie struck the most perceptive chord, noting the most compelling evidence that superconductivity involves a fundamentally different mechanism from ordinary conductivity: namely, that good conventional conductors do *not* become superconductors at low temperatures—an anticorrelation that would stand the test of time.[53]

One topic mentioned only fleetingly by Onnes and quite overlooked in the overall discussion contained, unbeknownst to the group, the seed for eventual progress. Onnes referred, almost in passing, to certain measurements started by his student Gerardus Johannes Sizoo shortly before the Solvay Conference got under way. Although a compelling theoretical explanation of the superconducting state that might guide further investigations was still conspicuously lacking, Onnes thought one of the numerous hypotheses circulating was more promising than most. A test for the influence of an *elastic deformation* on the superconducting transition should confirm or put to rest a pervasive hypothesis "that a relatively large space between the atoms is favorable for the appearing of supraconductive state, when the further special conditions, shown only by the class of supraconductors, are present."[54] This hypothesis was a logical consequence of the melting point (βv) argument, and seemed well-nigh demanded by H. A. Kramer's "electron orbitals" that figured so prominently in the presentations at Brussels. The seed for this work may as well have been planted by the young visitor from Harvard, Leo Dana, who, on Ehrenfest's request, gave one or more talks at Leiden on Bridgman's experimental and theoretical high-pressure work at Harvard. But Onnes also hinted in the eventual report on the work with Sizoo that his hunch was just as much born of frustration in uncovering any other changes in metallic properties accompanying the transition. The alternative was to test whether the transition can be affected by physical persuasion—mechanical deformation, for instance. In pursuing this, Sizoo's measurements would somewhat unexpectedly point a way out of the theoretical stalemate, but not by the route visualized by Onnes.

The new experiments therefore involved the effects of mechanical tension and compression on the transition temperature of tin and indium samples. A report was issued in June of 1925.[55] In the first measurements, a tin wire was stretched vertically between two small tin blocks, the bottom one fixed and the upper block communicating with a spiral whose length was a measure of the applied tension. As usual, resistance measurements relied on a potentiometric method with Diesselhorst compensation for thermoeffects. In March of 1924,

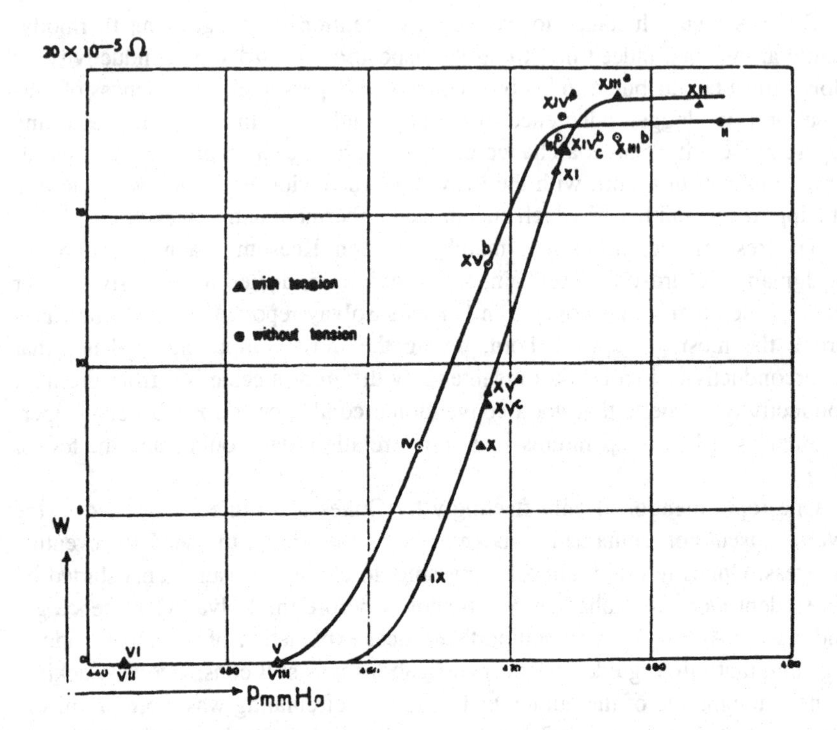

Figure 6-6. Effect of mechanical tension on transition curve for tin. Sizoo and Kamerlingh Onnes, RN, **28** *(1925), 658.*

preliminary results already indicated that "stretching of the tinwire [sic] was conducive to the appearing of the supraconductive state."[56] Results from April, Figure 6-6 (which appeared forthwith in Onnes's Solvay report), are more informative, showing clearly that the transition curve is displaced slightly toward higher temperatures under tension (by about 70 mK under 3 atmospheres). The converse effect—that of elastic compression—was less easily measured. For one thing, about 200 atmospheres were judged necessary to produce a discernible effect. Six months elapsed before measurements could resume. This time the resistance was wound on a small glass tube, around which a second glass tube was slipped, and the whole assembly was placed in a copper cylinder closed firmly at the top and bottom and pierced with capillaries for admitting helium gas and current and potential leads. With the cylinder under liquid helium, the leads were first filled with compressed helium; next, the cylinder was filled with liquid helium and then the pressure was increased in steps between zero and 1500 atmospheres with the aid of a hydraulic press. As expected, measurements with tin showed that "the 'vanishing-curve' is shifted by the application of the pressure to the side of the lower temperatures."[57] To make sure that these effects are "general properties of supraconductors," not intrinsic to tin alone, the pressure measurements were repeated with indium, giving about the same results.

These are very difficult measurements, because of the small effect involved. Nevertheless, the order of magnitude results (e.g., -5×10^{-5} °K/atmosphere for tin under pressure) agree rather well with modern measurements, which are—ironically, in view of the Keesom–Bridgman debate in 1924—explained in terms of the thermodynamics of the superconducting transition.[58] To Onnes, they seemed to corroborate his hypothesis nicely, assuming "the extension and the compression of the supraconductive metal may be considered as equivalent to an increase and decrease respectively, of the distance between the atoms."[59] Broadly speaking, this hypothesis also agrees with the modern interpretation that these phenomena are, indeed, partly associated with volumetric changes in the superconducting transition.

6.4 A Legacy of Nernst

The question of a pressure dependence of the superconducting transition temperature had, as one might expect, caught the attention of Bridgman—the doyen of high-pressure physicists who had strong opinions on superconductivity as well, although he was not a low-temperature physicist per se, as noted earlier. We know this question was also raised by Franz (Francis) E. Simon in a letter to Onnes dated 3 June 1924,[60] or about when Sizoo must have started preparations for his second round of measurements intended to address this particular effect. Simon might have contributed much to the discussion at large in Brussels, but once again German nationals were excluded. (As in 1921, Einstein was invited to the 4th Solvay Conference, but this time declined due to the continuing exclusion of his German colleagues and his displeasure with the French invasion of the Ruhr in 1923 to extract German reparation payments.) Simon was then a *Privatdozent* at Nernst's Physikalisch-Chemisches Institut in Berlin.[61] Recently recuperated from a second, serious war injury sustained on the eve of the Armistice, he had resumed his interrupted studies of physics at the Friedrich Wilhelm University of Berlin in 1919 (now known as Berlin's Humboldt University). Here he was exposed to Einstein, Haber, von Laue, Planck, and especially Planck's colleague Nernst, under whom he passed his doctorate in 1921 on, naturally, specific heats at low (i.e., hydrogen) temperatures. It had been left to Simon to prove the validity of Nernst's heat theorem in which Nernst himself was losing interest. Simon had stayed on at the Institut, for the time being as Nernst's assistant in a post of "Assistantship Extraordinary," which the influential Nernst had persuaded the Ministry of Education to create.

As the next step in his development, the most sensible course of action might have been a stint with Onnes at Leiden. Indeed, his letters to Onnes in 1924 on influencing the transition temperature struck a receptive chord, and Onnes invited him to spend a year in his laboratory. Unfortunately, a requirement for becoming a *Privatdozent*—the first rung on the academic ladder in Germany—was to "habilitate" or give an inaugural lecture. This granted the *Privatdozent* the right to give university lectures on a permanent basis and—of

Figure 6-7. Francis Simon. Courtesy Francis Simon Collection, AIP Niels Bohr Library.

more pressing importance in the chaotic economic times of Germany in 1925—to collect a fee for them. Fear of jeopardizing his Habilitation (which was duly given early in 1925 on "the size of atoms and molecules")[62] prevented Simon from accepting Onnes's tempting invitation. While Onnes's guidance might have helped,[63] even without it a new school of low-temperature physics was nevertheless forming without fanfare under his unassuming leadership in Berlin.

In 1923, Nernst had already vacated the directorship of the Institut to become president, if only for a short time, of the Physikalisch-Technische Reichsanstalt in nearby Charlottenburg, and Max Bodenstein succeeded him at the Institut. Bodenstein, it turned out, was perfectly happy to leave Simon to his own devices in continuing Nernst's low-temperature program, although a considerable amount of the administrative burden also fell on Simon. Somewhat fortunately, Nernst himself returned to the university, this time as professor of physics in the Physikalisches Institut adjoining the Physikalisch-Chemisches Institut; under this arrangement "Simon was able to keep in close touch with him, though not, as he said, in too close touch, for with advancing years Nernst grew more obstinate and dogmatic."[64] But the importance of the school for the superconductivity chronicle lies mainly in the role its principal members, Nicholas Kurti, Kurt Mendelssohn (Simon's first cousin), and Simon himself, would play in somewhat later years—and not in Berlin.

Among the many problems beginning to interest Simon and his fledgling team in 1924 was the electrical resistance of the alkali metals at liquid hydrogen and higher temperatures.[65] Their interest was not so much in resistance behavior as such, but in something even closer to Onnes's heart: confirming the law of corresponding states—this time for "atomic conductivities" (the specific conductivity per gram-atom of a substance, a concept introduced by Benedicks). For this purpose, Simon chose for corresponding temperatures those values of T for which $\beta\nu/T$ is constant. A survey had shown that the alkali metals exhibit very high atomic conductivities. With their low absolute $\beta\nu$-values as well, inferred from Lindemann's model (Eq. 2-7) based on their large atomic volumes, and few valence electrons, it occurred to Simon that these metals were particularly good superconducting candidates. As early as February of 1924, Simon dwelt on these matters in a letter to Onnes:

> If we seek a common property of substances that become superconducting at low temperatures, we find that they all have a very small ν-value compared to those of non-superconductors. Thus, for lead $\beta\nu = 88$, for mercury 96, thallium 100. The last two are average values since these substances are not regular crystals and, hence, the atomic heats are not described in terms of a single ν-value. For white tin we obtained recently $\beta\nu = 126$ between 9^0 and 14^0 absolute; for indium Lindemann's formula gives $\beta\nu = 115$. In contrast, $\beta\nu$ for copper is 315, for iron 453, cadmium 172, platinum 260, gold 170, and silver 215.
>
> Since it seems reasonable to suppose that substances with the lowest energy content become superconducting most readily [e.g., Onnes's relation between free path and oscillator energy] these facts are somewhat surprising. If we grant, however, the existence of a zero-point energy, then substances with the smallest ν-value are precisely those with the lowest energy content. We can also imagine that the relatively large atomic volumes exhibited by these elements, which are associated with small ν-values according to Lindemann's formula, is the primary cause for the occurrence of superconductivity.
>
> In any case, from this viewpoint we may search for new superconductors on a purely empirical basis. Candidates are bismuth, gallium, strontium, barium, radium, and particularly the alkalies sodium, calcium, rubidium, and cesium. As a result, I have been investigating the conductivity of the alkalies near 10^0 for half a year, though [the samples] were perchance badly contaminated. Presently we are occupied with producing the purest possible materials.
>
> I should like to ask, Herr Professor, whether the above substances have been investigated yet in your own institute. I should be extremely grateful for your kind information [in this regard].[66]

Evidently, then, Simon was unaware of Onnes's negative search among the alkalies down to 1.5 °K. Indeed, this apparent dilemma would remain a conspicuous rule governing the appearance of superconducting elements. Even today very few of the alkali metals show signs of superconductivity, and only under extreme conditions, in accordance with the anticorrelation rule alluded to by Mme Curie.

For the time being, the lack of liquid helium ruled out superconductivity as a topic for experimental pursuit in Berlin. To be sure, it would not be for long; a team—albeit not Simon's—was poised to close the gap. Contributions to the subject from either of the two low-temperature groups in Berlin, one at the university and one at Charlottenburg, were long overdue. But Leiden's monopoly was not yet broken. The course of the superconductivity program at Leiden and future lines of attack on the problem elsewhere were to be sealed shortly by an important development in Sizoo and Onnes's ongoing investigation.

CHAPTER 7

New Developments

7.1 Hysteresis

A hint of the new development appears in a note added in proof to Tuyn and Onnes's definitive paper on, among other things, tests to settle once and for all the validity of Silsbee's hypothesis.[1] The footnote is missing from the version reprinted in the United States based on a manuscript copy; this version, however, is introduced by a sad editorial commentary lamenting another milestone in the superconducting chronicle.

> While the manuscript for this paper was on the ocean in transit from Leiden to Philadelphia, news of the death of Dr. H. Kamerlingh Onnes appeared in the daily papers. His death takes a brilliantly successful Figure out of the world of experimental science, and deprives the realm of scholarship of an important contributor.[2]

This joint paper with Tuyn was not quite the last paper coauthored by Onnes on superconductivity (albeit, he was no longer senior author). As indicated earlier, his health had been declining for some years—and indeed, it was always delicate. In 1922 Leo Dana, who had recently arrived in Leiden from Harvard for a postdoctoral fellowship, had already experienced difficulty in meeting with Onnes because of the latter's prolonged absence from the laboratory. Even though the year had seen at least one memorable event in the Leiden physics department (Figures 7-2 and 7-3), the laboratory was "in disarray" just then[3] and Onnes was at home, ailing, and upset over the recent death of his close associate J. P. Kuenen—who had been his chosen successor as well. Onnes

Figure 7-1. Physics Department, Leiden, in 1922. Courtesy AIP Niels Bohr Library.

Figure 7-2. Displays prepared for the 50th anniversary of the founding of the Leiden Laboratory and in honor of Onnes's 40th anniversary as professor. Posters commemorate the discovery of superconductivity and Onnes's form of the equation of state of a gas. Courtesy AIP Niels Bohr Library.

Figure 7-3. Formal photo of faculty, graduate assistants and instructors (second, third rows), and students in Mechanics and Glass Blowing Schools (top three rows). Kamerlingh Onnes is in the center of the first row with his wife at his right. Flim is second from left. Plackard proclaims Onnes's famous motto. Courtesy AIP Niels Bohr Library.

retired as director and professor the following year amidst a continuing administrative vacuum. The primary reason for the lack of direction was an ongoing religious schism in the country between Catholics and Protestants that touched the university—a Protestant institution. After some wrangling the new directorship was filled by a joint appointment of Keesom (Catholic, and scientifically the presumed heir to the position) and Wander Johannes de Haas (Protestant, and son-in-law of Lorentz).

De Haas got his start as Onnes's assistant from 1905 to 1911, completing his doctoral thesis (on the compressibility of hydrogen) at Leiden in 1912. Before rejoining the Leiden laboratory in 1924, he served briefly (1913–1915) at the University of Berlin and the Physikalisch-Technische Reichsanstalt in Berlin, followed by professorships at Delft and Groningen. It was during his stay at the PTR that he, on Einstein's suggestion, performed a classic experiment on what is still known as the Einstein–de Haas effect (the torque induced in a suspended iron cylinder as a consequence of its being abruptly magnetized).

As a physicist, de Haas's approach was imaginative and intuitive, predicated on the use of simple but clever experimental apparatus. By contrast, Keesom had a much firmer grasp of mathematical physics, was highly organized, and excelled in laying out systematic experimental programs—in short, he was much more like Onnes. As we know, Keesom represented Onnes at Brussels in 1924. Some of the experimental material in the Tuyn–Onnes paper had actually been presented by Keesom at that conference, and the full investigation constituted Tuyn's doctoral thesis at Leiden in 1924.[4] Published in Dutch, the dissertation was not very accessible to the scientific community abroad, but it was subsequently republished in a slightly reworked English version as a regular Leiden *Communication.*

The new observation alluded to in the Tuyn–Onnes paper seems at first to have been treated as only an annoying technical impediment in the attempts to confirm Silsbee's hypothesis—a matter of high priority for Onnes. The various troubling threshold "phenomena" had stymied real progress for over a decade in understanding—certainly in utilizing—superconductivity. Silsbee's hypothesis offered the possibility of unraveling the mess to some extent, because "its correctness would make the current threshold value disappear."[5] Unfortunately, Silsbee himself had only scant data at his disposal when he sought to validate his hypothesis. For tin, but less clearly for lead, he had perceived a correlation between the measured threshold field and that computed from the threshold current. No magnetic measurements on mercury were available in 1916. In 1921, when Onnes and Tuyn took up the question in earnest, magnetic measurements were still lacking for both mercury and thallium. More convincing was the observation first made by Silsbee, that the surface value of the self-magnetic field calculated from the threshold current for wires of different diameters appeared to be a better constant than the current density. But by mid-1924 a considerable store of new data had been accumulated, mainly under Tuyn and "not exclusively in order to test the hypothesis of Silsbee, yet suitable to it."[6]

Communication 174a begins with a compilation of data on the temperature-dependence of the magnetic threshold ("critical field," by then a term also in common use) for lead, tin, and indium. These data, collected between 1922 and 1924, are presented in the form of tables of transverse or longitudinal resistance as a function of field intensity for various bath temperatures between 1.85 °K and 4.3 °K (the boiling point). The resistances were usually extruded wires wound bifilarly on glass tubes, insulated with paraffined paper or silk thread. The applied fields were supplied by one of several conventional, air-core solenoids, some small enough to fit within the helium cryostat. The field intensity was calculated from the ampere-turns and the coil geometry (the "Fabry factor"). For graphical representation, the authors restricted themselves to a "representative" plot of the transverse results for tin (Figure 6-3)—the sanctioned plot that Keesom had already shown at Brussels. To minimize spurious contributions to the observed breadth in the magnetic transition curve, corrections were applied for the strength of the measuring current, for the lack of homogeneity in the background field, for the influence of the terrestrial field, and for temperature oscillations in the helium bath. The aforementioned footnote introduces, quite casually, one additional effect suspected of contributing to the curve broadening. This effect, "a kind of hysteresis of the magnetic resistance of supra-conductors," was, according to the note, first noticed by Sizoo in 1925.[7] Finally, for each table of resistance versus field corresponding to a given temperature, the threshold field value, defined as $H_{(1/2)}$, was derived "graphi-

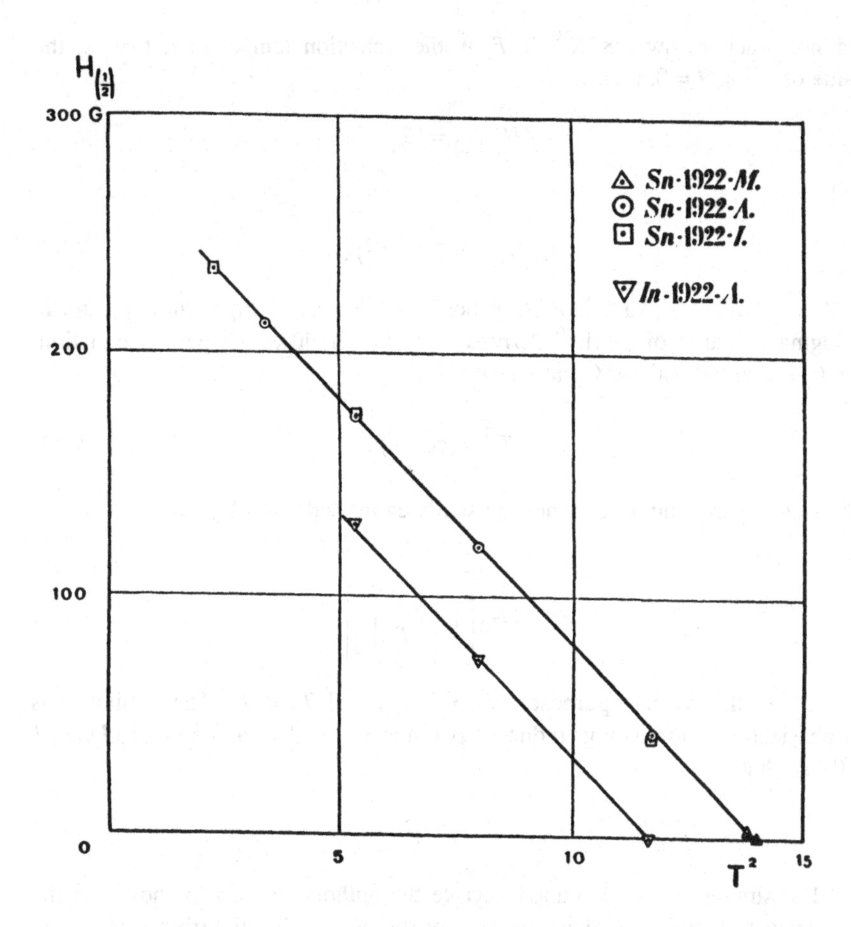

Figure 7-4. Plot of $H_{1/2}$ versus T^2 to test the parabolic relationship between threshold field and temperature. Tuyn and Kamerlingh Onnes, Franklin Institute, Journal, **201** *(1926), 398.*

cally." Values of $H_{(1/2)}$ defined in this manner for lead, tin, and indium in a transverse field were plotted as a function of temperature. The resulting plot is essentially that of Figure 6-4 (one also shown by Keesom at Brussels).

"The three lines show a great conformity," wrote Tuyn and Onnes. "Their concave sides are turned toward the T-axis and the curvature becomes stronger with decreasing T, at least to $T = 1.5$ °K."[8] The curves for tin and indium could be represented with considerable accuracy by the following parabolic relationship between field and temperature:

$$H_{(1/2)} = -hT^2 + H_{0(1/2)} \,. \tag{7-1}$$

That is, a plot of $H_{(1/2)}$ versus T^2 yields a straight line with intercept $H_{0(1/2)}$ at $T = 0$ °K (Figure 7-4). This was a qualitative advance over 1921's linear fit, seen in Eq. (6-4). For lead, such a fit was less assured, since the measurements did not reach below 3.8 °K.[9] If T_s is the transition temperature, that is, the value of T for $H = 0$, then

$$H_{0(1/2)} = hT_s^2,$$

and

$$H_{(1/2)} = h(T_s^2 - T^2)\,. \tag{7-2}$$

A somewhat similar relationship between field and temperature appears in Bridgman's paper of 1921,[10] derived from his modified Clapeyron equation, Eq. (6-6), with the aid of Curie's law:

$$T^5 - T_s^5 = (\text{const.})\,H^2\,. \tag{7-3}$$

The nearly-parabolic dependence is usually expressed nowadays as

$$H_c \approx H_0\left[1 - \left(\frac{T}{T_c}\right)^2\right], \tag{7-4}$$

where for all practical purposes $H_c \approx H_{(1/2)}$ and $T_c = T_s$. The similarity is emphasized by introducing 'reduced' parameters $t = T/T_c$ and $h' = H_c(T)/H_0(T = 0)$, so that

$$h' \approx 1 - t^2\,. \tag{7-5}$$

"The simple relation obtained," wrote the authors, "seems to show that the magnetic field will be of primary importance for the disturbing of supra-conductivity." Nevertheless, nothing definitive could be made of this: "The hypothesis of Silsbee can not be proved by magnetic measurement alone."[11] The missing element was more accurate measurements of critical currents and their temperature dependence. For the next round, both tin and lead samples were utilized. But, as was the case with the critical field, some practical ambiguity now arose in defining the critical current, because of the gradual onset of a potential difference across the sample as the current was increased (Figure 7-5). Strictly speaking, the true current threshold was required, not simply the current value at which the resistance is half restored. However, due to the gradual onset, and "if by the current threshold value the intensity of the current is meant, which restores at the extremities of the resistance the smallest, still measurable potential difference, this definition is not complete without making definite the amount of the potential difference and the dimensions of the resistance."[12] In order to ascertain this first restored potential difference, it was necessary to use galvanometers of maximum sensitivity. Also, much stronger currents (typically some amperes) were now resorted to compared with the few mA sufficing before. Because of the danger posed for the delicate apparatus and fragile sample, the measurements proceeded slowly and required great care.

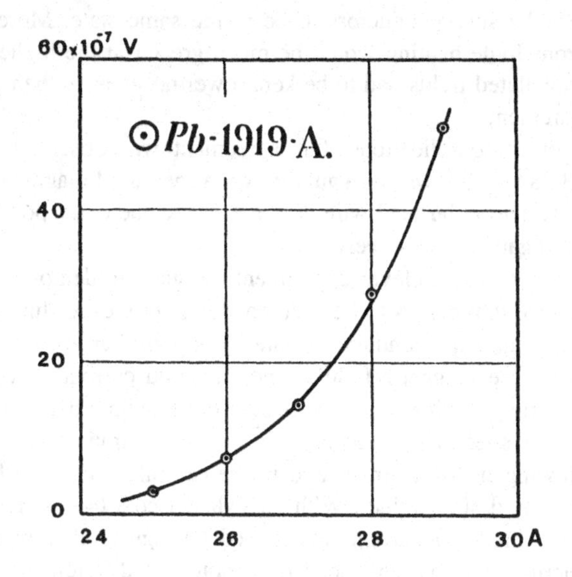

Figure 7-5. Gradual onset of voltage across sample as the current is raised. Tuyn and Kamerlingh Onnes, Franklin Institute, Journal, **201** *(1926), 401.*

Even so, excessive heating of the sample, usually evidenced simply by "a sudden ebulution of the helium bath," could be quite dramatic. In measurements on a particular bifilar lead sample, the current was slowly raised from 20 A past 24 A (the estimated threshold), and further raised in one-ampere steps.

> When the measurements with 29 A were almost ready, the [galvanometer] scale disappeared out of the field of vision in the telescope. The wire was heated by the restoring of the resistance and with fatal results. On dismounting the cryostat, the lead wire was found to have melted through. ... [13]

Three different tests were applied in attempts to corroborate Silsbee's hypothesis. There was nothing particularly novel about the first two; they merely supported what Silsbee had deduced from the prewar Leiden data. First, for lead the quantity $2i/r$ tended to be more constant, for a given temperature but for wires of different diameters, than the critical current density alone. (Or so Tuyn and Onnes claim; the correlation shown in *Communication 174a* is only marginally more convincing than Silsbee's earlier one.) The same had been observed earlier for thallium.[14]

A seemingly more direct test involved comparisons of these self-field values ($H = 2i/r$) calculated from the measured threshold current i with the measured threshold fields. This was not a simple matter, however, due to the order-of-magnitude difference in current necessary to reach the true magnetic threshold compared to the threshold current value. Because the first measurable restored potential differences were, by definition, equal in the field and current measurements, the corresponding restored resistances differed greatly, and there was no reason to expect the superconductors to be in the same state. Moreover, heat development from Joule heating would be much greater at the higher currents; therefore, the calculated fields had to be kept lower necessarily than those from magnetic measurements.

In spite of these complications, fair agreement was observed. For a tin sample at $T = 3.28$ °K, $2i/r$ was 44 gauss versus a measured magnetic threshold of 56 gauss. For a particular lead wire at $T = 4.21$ °K, the corresponding values were 356 and 430 gauss, respectively.

The third test involved a clever experiment designed to demonstrate explicitly that a resistance that ceases to be superconducting by exceeding the threshold current regains its superconducting state by applying an external magnetic field that weakens the magnetic field of the threshold current. A copper wire was stretched coaxially within a hollow glass cylinder with a band of tin wound on its surface. The specimen could be held below the transition temperature, and currents flowing in the cylinder and in the central wire could be adjusted independently to any desired values. With the cylinder cold ($\approx$ 3.7 °K), a current stronger than the threshold current was passed through it, thereby destroying the superconducting state as evidenced by a potential difference between the ends of the cylinder. A current of opposite polarity was then passed through the central copper wire, such that the magnetic fields of the two currents weakened or canceled each other. Assuming "the threshold current holds its power for disturbing the supra-conductivity at its magnetic field, then it must be possible to restore the supra-conductivity in [the cylinder]—in spite of the strong currents."[15] In a modified version of the experiment, devised by Flim, the central copper wire was replaced by a lead wire; the lead wire remained superconducting during the experiment.

Both versions of the apparatus were ready for a first round of experiments on 12 March 1924. With either arrangement, the disappearance of the resistance of the tin cylinder was observed with falling temperature, first with a constant current of 2.5 A passed through the outer cylinder only, and next with an additional current of the same intensity but with opposite polarity passed through the central wire. In both cases, the potential difference vanished as the bath temperature sank below the critical temperature for tin, about 3.72 °K.

The validity of these results was attested to by Silsbee himself within days after they appeared in print in the United States; Silsbee had lost no time in subjecting the results to close scrutiny, since "in view of the ingenuity of the method and the carefulness of the experimental work, their results seem[ed] to deserve a somewhat more quantitative analysis than that given by the experimenters themselves."[16] Assuming a perfect cylinder of homogeneous material and a definite critical field, H_c, Silsbee laboriously computed the potential difference for any given value of H_c and currents in the conductors. He concluded "that the potential difference should finally vanish at the same temperature in both experiments, and that for a tube of the wall thickness used, the potential difference under conditions intermediate between full resistance and supraconduction should not differ in the two cases by more than 4 percent of that corresponding to normal resistivity. The experiments," he added, "showed no difference so great as 4 per cent between the two cases."[17]

Further experiments were delayed for nearly a month, awaiting another rare helium day scheduled for 17 April. This time the temperature and the current in the tube were held constant while the current in the inner wire was varied; this experiment was repeated at different temperatures. A graph of potential difference versus current in the inner wire showed that the potential decreased to a minimum value, the lower the lower temperature, followed by an increase approaching the value corresponding to normal resistivity.

"On the faith of the results obtained up till now," concluded Tuyn and Onnes, "we think we may accept the hypothesis of Silsbee as being correct."[18] Silsbee was equally convinced:

> It may, therefore, be concluded that the results of these experiments can be completely accounted for by the assumption of a critical magnetic field, without making use of the concept of critical current.[19]

To be sure, this was an important conceptual conclusion. Potentially, an aspect of the experiments devised to test Silsbee's hypothesis was of even greater importance—something also vaguely appreciated, both by the group in Leiden and by Silsbee. Unfortunately neither party pursued the matter. Noting that, depending on the current ratio, there is somewhere in the cylinder a surface $H = 0$, its location depending on the unknown current density distribution in the superconductor, "such experiments can perhaps show something about this distribution," noted Tuyn and Onnes.[20] In the same vein, Silsbee's computations had shown "that there will exist in the cross-section of the tube a number of zones in some of which the material will have normal resistivity while in others it may be supraconducting. In many cases the superconducting layer shrinks to a current sheet of infinitesimal thickness but carrying a finite current."[21] In other words, the experiment had the important potential for revealing something about the magnetic field distribution in a superconductor—a problem that would soon take center stage in upcoming superconducting investigations.

The next logical step following Sizoo and Onnes's pursual of the effects of elastic deformation on the transition temperature was to check for similar effects on the threshold field, that is, on the magnetic transition curve. The new measurements, with contributions from de Haas, were summarized in two reports presented back-to-back at the Amsterdam Academy from November to December of 1925. (These were to be the last regular reports on superconductivity coauthored by Onnes.) The investigation had in fact been diverted from its original purpose almost from the start. During the very first measurements on tin, "besides the expected displacement a quite new phenomenon was discovered, which gave an unexpected turn to the inquiry" ... "something which caused [the authors] to accept the result with considerable reserve."[22] Indeed, they devoted two full helium days to the initial measurements comprising the first of the new *Communications*.

A hint of the new phenomenon had been obtained some months earlier with the observation of a slight difference in the ascending versus descending magnetic transition curve for tin. $H_{1/2}$ was higher in increasing than in decreasing fields, implying that "*the magnetic transition curve in reality is a hysteresis-figure* [*sic*]."[23] Before attaching much significance to this finding, one rather simple explanation would have to be ruled out: the possible presence of iron impurities in the tin. The remanent magnetism of the iron could easily account for the hysteresis. Once again the solution was mercury, since it could be obtained in very pure form by repeated distillation. Resistances were prepared in the usual way as mercury threads enclosed in glass capillaries, rather like the prewar samples but much shorter (now typically 10 cm long). The shortening was devised partly to accommodate as many resistances in the cryostat as the cramped space permitted, to maximize the homogeneity over the length of the resistance, and to decrease the chance of breakage from nonuniform cooling or heating:

> Even with very careful cooling the mercury threads of about half the number of the resistances generally broke. Besides, when after the measurements, the resistances were heated again from helium temperatures to room temperatures, some on [*sic*] the glass capillaries usually broke. Because of this we have not yet succeeded in measuring one resistance on two different helium days.[24]

The background field was normally supplied by the same small copper windings used earlier, some of which were small enough to fit inside the helium cryostat. Some transverse field measurements were made with the much stronger Weiss electromagnet.

The effect was still there, but now with an additional twist. "*The result of the investigation was not only that the hysteresis really exists, but besides that the graphs of the magnetic disturbance show very sharp discontinuities*."[25] The discontinuities were most conspicuous as "jumps" in curves obtained in decreasing transverse *or* longitudinal fields, whereas ascending curves tended to remain quite continuous, as in Figure 7-6. At first it was suspected that local inhomogeneities in the magnetic field were to blame, with part of the mercury thread becoming superconducting as the local field sank below the critical value. To ascertain this possibility, samples as short as 10 mm were employed; presumably they were too short for field inhomogeneities to have any influence. To thwart the possibility of other instrumental errors creeping in, they resorted to a Zernike galvanometer with a "sensibility" three times as high as those normally used and a time lag of only three seconds, to carbon rheostats that could be varied continuously, to various calibrated milli-amperemeters, and a variety of current leads. But neither the hysteresis nor the discontinuities were significantly affected.

Satisfied that the behavior was associated with the resistances themselves, thus "a property of the superconductivity," and not simply local variations in resistivity from iron impurities, variations in field uniformities, or otherwise associated with faulty apparatus, the *crystalline state* of the material seemed to be the next likely explanation; that is, pieces of single crystals formed in the

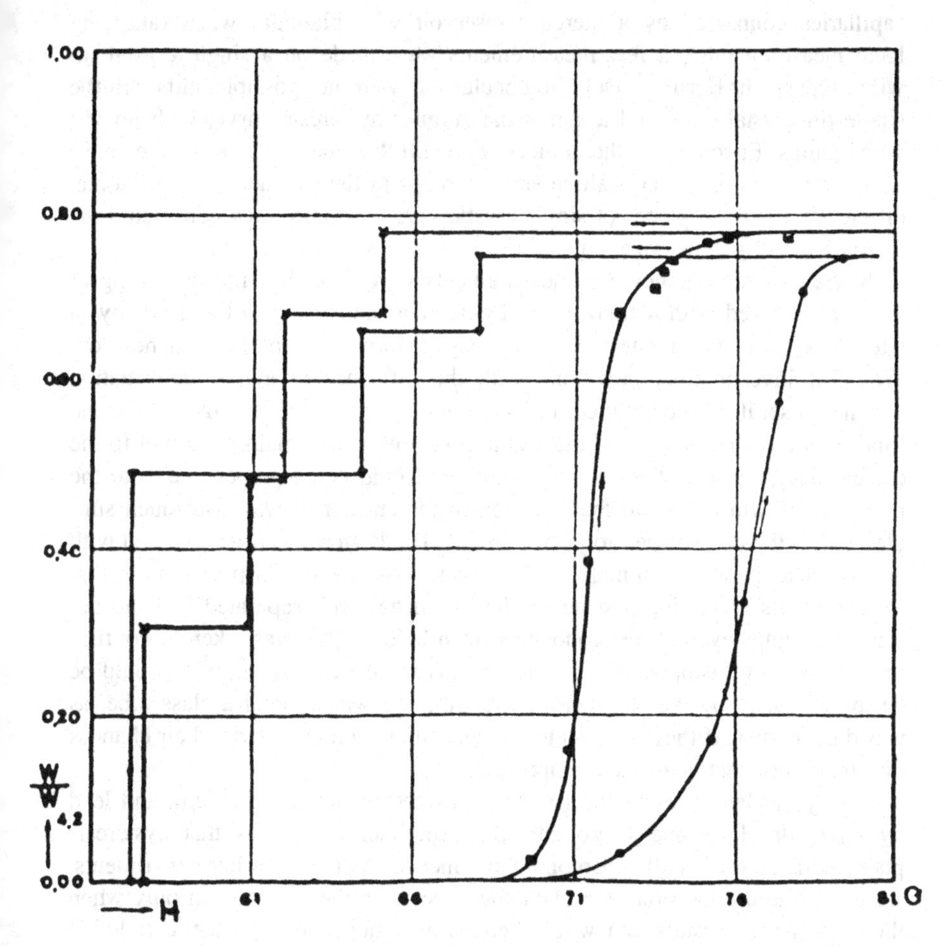

Figure 7-6. Hysteresis and discontinuities in magnetic transition curves for mercury. De Haas, Sizoo, and Kamerlingh Onnes, RN, **29** *(1926), 239.*

mercury thread by slow cooling, with each jump representing the onset of superconductivity in such a crystal. "The fact that different crystals lose their resistance at different values of the magnetic field might be due to the different orientation of the crystals in reference to the magnetic field" (i.e., to the directional field dependence of $H_{1/2}$). This suggestion was to be tested by (1) measuring the same resistance on different days (despite the aforesaid problem with thread breakage), since the state of crystallization brought about by the cooling was unlikely to be the same each time, (2) "the hysteresis figure of a real single crystal wire would probably show a single nature," and (3) "by measuring the resistance between different points of the thread it would be possible to decide if the change in resistance, which corresponds to a jump, is local, or is spread over the whole length of the thread."[26] Measurements intended to illuminate these questions occupied most of the second half of 1925. They were made mainly on mercury threads, some of which consisted of two capillaries connected by a mercury reservoir with platinum wires (allowing local measurements); a few measurements were made on a single crystal tin wire prepared in Berlin.[27] Definite conclusions were not possible, although the single tin crystal exhibited a somewhat simpler hysteresis curve, with no obvious jumps. Encouraged, the authors were still "inclined to look for the cause ... in variations of the crystalline state, perhaps in the existance [*sic*] of some, differently oriented, pieces of single crystal wire." In short, they considered the ongoing investigation "very urgent."[28]

In view of the urgency, the measurements were extended into the spring of 1926, and closed with a final report by de Haas and Sizoo to the academy in late May.[29] Onnes's name would be missing from the reports from now on. (The full investigation, beginning with the influence of elastic deformation which had set it off, constituted, in fact, Sizoo's dissertation in 1926.)[30] For the final round of measurements, the winding geometry was mainly parallel to the current distribution instead of the background field being perpendicular to the plane of the winding or to the direction of the current. It was also much simpler, and it thereby seemed to prove a point. The hysteresis curve obtained with an extruded tin wire noninductively wound along a mica strip was more distinct, with its ascending and descending branches well separated and the descending jumps even more pronounced than before. This was taken to confirm the single crystal supposition. Single crystals in an extruded tin wire would be destroyed easily by the strain from wrapping the wire around a glass tube, as was done earlier; if they were simply stretched along a mica strip, their chances of remaining intact were much improved.

Two years later these effects were confirmed in indium, thallium, and lead by Tuyn, de Haas, and J. Voogd: "the experiments ... show that hysteresis phenomena appear in all superconducting metals. As in the earlier experiments, we get the impression that the phenomena occur in the purest form only when there are large crystals and when there is a homogeneous magnetic field."[31] With these investigations into phenomena that defied a definitive explanation, the superconductivity program at Leiden had taken yet another sharp turn. The focus of interest had shifted over the years, generally narrowing its scope each time: from the properties of solenoids and simpler coils to bifilar hairpin samples, and now finally to single crystal specimens. The latter measurements had barely begun as the decade neared its end. They were stalled pending the development of in-house techniques for growing better single crystals, and they were also sidetracked temporarily by the sudden interest in superconducting compounds and alloys.

With regard to the production of single crystals, help was on the way with the impending arrival of a Soviet visitor (L. V. Schubnikow), as we shall relate in due course. Investigation of superconducting compounds, something neglected until now, was spurred by the collaboration of another outsider—the Belgian metallurgist E. van Aubel of Ghent. Meanwhile, the change in emphasis to experiments on single-crystal specimens would prove to be of great significance in the long run, mainly for the interest shown in them by new colleagues in Berlin and their, not Leiden's, interpretation of the results.

7.2 New Superconductors and New Laboratories

A renewed search for additional superconductors began in 1927 with a negative but important finding. As was pointed out at Brussels in 1924, tin offered a good test of the importance of the crystal lattice for the superconducting state[32] (despite the fact that its lattice was unaffected by the superconducting transition, as shown by Keesom shortly before the 1924 Solvay Conference). The structure of ordinary white tin, the known superconductor, is tetragonal whereas the structure of its modification, *grey* tin, is cubic (the structure of diamond). Grey tin is also a more difficult material to work with, available only in powder form. In the event, both simple resistance measurements and a more elaborate persistent current setup failed to reveal any trace of superconductivity in grey tin between the boiling point of liquid helium and 1.2 °K.[33] Not particularly surprised, de Haas, Sizoo, and Voogd allowed that

> ... while Kamerlingh Onnes came to the conclusion that a loose atom connection and a great distance between the atoms were to advantageous [*sic*] for the creation of the supraconductive state the above experiment shows that the phenomenon is more complicated.[34]

Gallium, with its low melting point, unique electronic structure (3 electrons in the outer shell, 18 in the next shell), and near-superconducting behavior at 1.6 °K, remained a prime candidate. In 1928 the measurements were extended down to 1.1 °K, still showing a small residual resistance.[35] Suspecting a trace of indium in the sample, the test was repeated with spectroscopically pure gallium in 1929. This time a new cryostat designed by Keesom was used, one able to reach temperatures below 1 °K.[36] The attainable minimum temperature was actually not lower than Onnes's record of about 0.8–0.9 °K in 1921, but the availability of new vacuum pumps of greater capacity allowed cooling a much larger volume of liquid helium (a capacity of ~300 cm^3 compared with the earlier ~1 cm^3) to this temperature. The new apparatus, in the hands of de Haas and Voogd, proved to be a success almost at once. On 31 May 1929, they confirmed superconductivity in gallium—the sixth superconducting element recorded at Leiden, with a critical temperature of 1.07 °K. A footnote to their report acknowledged, however, that gallium was not the only new superconductor, and that the Leiden Laboratory was no longer alone in the quest.[37]

It will be recalled that Canada had become involved in a wartime pilot project for extracting helium from natural gases under John Cunningham McLennan. In the course of operating a small plant near Calgary, Alberta during 1919–1920, considerable supplies of helium of high purity were collected. Consequently, McLennan proposed to the British and Canadian authorities that they exploit this war booty for scientific purposes. The proposal was approved and the project was funded by various agencies. Apparatus for liquefying air, hydrogen, and helium was procured (some of it gratis, courtesy of the Admiralty) and installed in the Physical Laboratory of the University of Toronto. McLennan's generous gift to Leiden in 1921 of a cylinder of helium gas was not given without ulterior motives. Onnes supplied the essential advice, even a complete set of drawings of the Leiden installation that were needed for constructing both the Canadian hydrogen and helium liquefiers. Helium was liquefied for the first time at Toronto in January of 1923, when the Cryogenic Division of the Physical Laboratory formally opened.[38]

Although the foundation for an important school in low-temperature physics had been laid in Toronto, of greater immediate significance for superconductivity was the commissioning two years later of a third helium liquefier, at the Physikalisch-Technische Reichsanstalt at Charlottenburg, a suburb of Berlin. The PTR already possessed a relatively seasoned cryogenic laboratory with a liquid hydrogen program in full swing, initiated under Emil Warburg's presidency in 1913. The German Physical Society had been consulted on where in Germany to establish a low-temperature institute.[39] It was agreed that Göttingen was the right place. However, with a small hydrogen liquefier a fait accompli at the PTR, Planck, as presiding chairman of the society, intervened and opted for the Reichsanstalt instead. The hydrogen liquefier, a copy of Nernst's, had been constructed by Walther Meissner, a former student of Planck.[40] (Meissner received his doctorate in theoretical physics under Planck in 1906.)

By 1923 Meissner's original hydrogen liquefier was no longer adequate for the many and growing demands made on it, and it was replaced with a larger plant. Preparations were also made for the next logical step, liquid helium. The influence of Nernst, during his brief and otherwise not very productive tenure as president of the PTR from 1922–1924,[41] helped in getting the helium program going. The availability of copious quantities of liquid hydrogen was one requirement for the liquefaction of helium by the standard method based on Linde's principle with liquid hydrogen precooling. Another was helium gas. German industry met the challenge, with gas provided partly from a neon–helium mixture made available from the Linde–Gesellschaft, partly through purification from raw helium that the Auer–Gesellschaft obtained from monazite sand. The liquefier, constructed under Meissner's direction,[42] was ready in 1925. The formal occasion of the first liquefaction at Charlottenburg, on 7 March, was duly marked in a letter from Meissner to Onnes written 2 days later:

> Having liquefied helium at the Reichsanstalt for the first time on March 7, I should like to express how I reflected with admiration on your own achievement in the liquefaction of helium; particularly because it was necessary for you initially to produce the helium (of which I obtained 700 liters by separation of a helium-neon mixture) by a quite different procedure, and, in contrast to [my own effort], you could not be certain of ultimate success in the liquefaction. I hope it pleases you that, in a field so dear to you, also in Germany the importance of your own work will hitherto be more fully appreciated.[43]

In 1927 a new building for the Cryogenic Laboratory was inaugurated. It housed a large liquid nitrogen plant, the new liquid hydrogen plant and, of course, the helium liquefier (capable of producing 10 liters per hour). Funding for the building and its machinery came mainly from the central government, but support for the plant as well as the research program also came from the Notgemeinschaft der Deutschen Wissenschaft (Emergency Association of Ger-

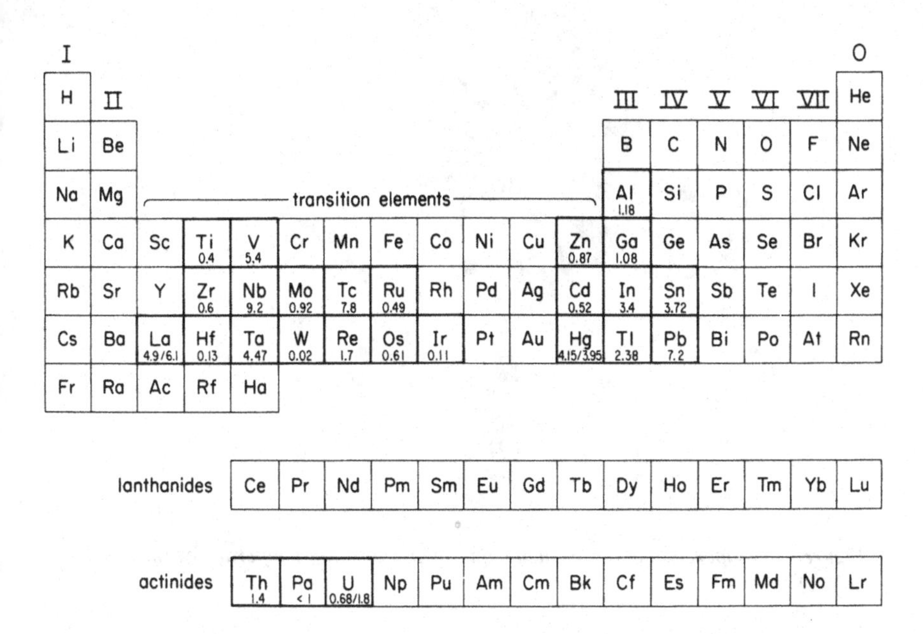

Figure 7-7. Walther Meissner. Courtesy AIP Niels Bohr Library.

man Science). The Notgemeinschaft was a postwar organization formed largely on Planck's initiative for the purpose of channeling Reich funding to scientific institutions and individuals in basic research.[44] The superconductivity program under Meissner was launched to complement the ongoing low-temperature work nearby at the University of Berlin under Simon. Meissner's group would concentrate on solid-state problems, according to the official accounts, with Simon's team primarily focusing on continuing its thermodynamic investigations. In point of fact, Simon generously refrained from initiating an active program of his own in superconductivity to avoid the appearance of stepping on his colleague's turf at the PTR.

Metallic resistance was by no means a new topic at the PTR,[45] and now superconductivity took high priority. No sooner had liquid helium been demonstrated than an ambitious program of systematic measurements on high-purity single and polycrystalline samples down to 1.6 °K was under way.[46] For a starter, many of the "beautiful results of Kamerlingh Onnes and co-workers" were verified. However, in view of the "apparent lack of a principle for the sequence of the transition points,"[47] Meissner took the liberty of concentrating on a different group of elements in the periodic table from those traditionally pursued at Leiden (Figure 7-8). His search centered on transition metals, elements with atomic numbers relatively close to those of the known superconduc-

Figure 7-8. Superconducting elements and their transition temperatures (modern values).

tors but different in an essential respect, being "hard" with *high* melting points. The first superconductor uncovered, announced at the Hamburger Physikertagung in 1928, was tantalum, which had the surprisingly high transition temperature of 4.4 °K.[48] Thorium followed in 1929 with a critical temperature of approximately 1.4 °K.[49] As if to underscore the lack of any simple relationship between T_c and βv-value, niobium came next, in 1930.[50] It was found to have a critical temperature near 8 °K (the modern value is about 9.2 °K)—a record T_c for known superconducting elements that has not been broken even today. Niobium had actually been measured by McLennan's group a year earlier. They had obtained strong indication of superconductivity, correctly blaming an observed rapid decrease in resistance near the boiling point of helium on a trace of (superconducting) tantalum in the sample.[51]

Although the periodic table failed to provide much guidance, a far more bountiful source of superconductors was uncovered at Leiden in 1928. That year de Haas and Voogd, collaborating with the Belgian metallurgist E. van Aubel, turned their attention to binary metallic mixtures, paying special attention to compounds in which one of the components is a superconductor. Compounds, compared to alloys, were known to exhibit extremely high values of specific electrical conductivity at ordinary temperatures, and a rather regular crystal lattice; however, their electrical resistance at low temperatures had scarcely been touched. A representative collection of samples, for example Ag_3Sn, Cu_3Sn, Bi_5Tl_3, SbSn, were prepared by van Aubel in his laboratory at

Ghent in Belgium and measured at Leiden, one by one. Between 0 °C and the temperature of liquid hydrogen no surprises were encountered, as might be expected from Matthiesen's rule: The resistances showed little dependence on temperature in keeping with their high specific resistance. In liquid helium definite surprises were in store, however. Not only did SbSn and Sb_2Sn prove to be superconducting with T_c slightly higher than for pure tin, but a rod of Bi_5Tl_3 exhibited superconductivity well above the boiling point of helium—somewhere between 4.2 °K and 9 °K—representing an enormous elevation in the transition point compared to T_c for pure thallium (2.4 °K).[52]

This tendency of two-component systems was not altogether novel, having been observed by Onnes back in 1913 with amalgamated tin foil (a solid solution of mercury in tin), but it was clearly not a general rule. Thus, Cu_3Sn behaved like a perfectly normal conductor with a near-constant residual resistance all the way down to 1.3 °K. Nevertheless, the impression was gained that a combination of a superconductor with a nonsuperconductor was likely to become more easily superconducting than the superconducting component, providing that the atomic weights of both metals are close. The same was found for certain other eutectic *alloys* with tin and with lead and thallium; as a rule, the fall in the thermal transition curves for such alloys became less steep than in the case of pure or "true" superconductors.[53] Why not, then, go one step further? Taking a clue, the eutectic mixture of gold and bismuth, of which *neither* components are superconductors, was carefully studied, and sure enough, the solid solution of about 4% bismuth in gold became superconducting at 1.9 °K.[54]

As it happened, Meissner had been similarily inspired later in the same year, and he was able to show that copper sulfide also becomes superconducting, in this case at 1.1 °K.[55] Sulfur is manifestly an insulator, and copper, albeit a fine ordinary conductor, had shown no evidence of superconductivity to the lowest temperatures attainable at Leiden or Berlin, whereas the superconductivity of bismuth remained an open question at the time.[56]

Although these developments were of great intrinsic interest, revelations of potentially far greater practical importance awaited de Haas and Voogd's first attempt to determine the magnetic threshold value for Bi_5Tl_3 in the early fall of 1929.[57] The threshold proved too high for the 1250-gauss solenoid usually employed for this purpose, and required a Weiss–Oerlikon electromagnet instead. In fact, at the boiling point of helium $H_{1/2}$ exceeded 4 kG! Moreover, the slope of the rise of $H_{1/2}$ with decreasing temperature, 1830 gauss per degree Kelvin, was also very much steeper than, for example, lead (~200 gauss/ °K), as is apparent in Figure 7-9. (This figure gives, by extrapolation, approximately 6.5 °K for the transition point of Bi_5Tl_3.)

> We found, that at 3.4 °K a field of 5.3 kilogauss did not yet disturb the superconductivity of Bi_5Tl_3. We have not yet extended our investigations to still lower temperatures. An extrapolation however of the data would predict that at a temperature of 1.3 °K a field of about 9 kilogauss would not yet disturb the superconductivity. *This would render possible the production of magnetic fields of this*

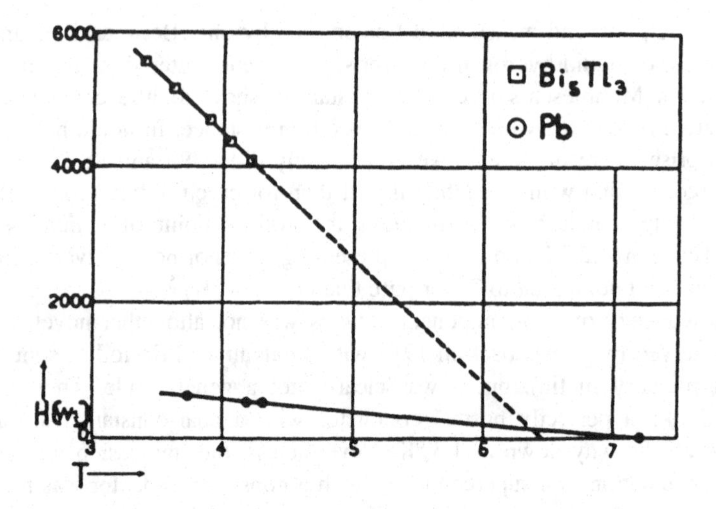

Figure 7-9. Comparison of threshold fields for lead and the compound Bi_5Tl_3. De Haas and Voogd, RN, 32 (1929), 876.

> *order of magnitude with a solenoid from Bi_5Tl_3 wire without production of heat.* It is evident however, that besides this eventual application, the phenomenon itself is of the highest importance.[58]

Suddenly Onnes's old dream of exploiting superconductors in high-field magnets seemed within reach. Prospects appeared even brighter when de Haas and Voogd turned to the eutectic alloy Pb–Bi in the spring of 1930.[59] They knew beforehand that this alloy was superconducting at 4.2 °K,[60] but that was all (as measurements near the transition point of pure lead were notoriously difficult). Again, "a preliminary investigation showed that the magnetic disturbance would require considerable fields."[61] Undaunted, they borrowed a special cryostat, placing it between the pole shoes of a very powerful electromagnet. This setup, normally used by Jean Becquerel and de Haas for cryogenic magneto-optic research, allowed measurements as a function of temperature in magnetic fields up to 30 kG. The results, as seen in Figure 7-10, speak for themselves.

> The required magnetic fields are evidently very high. If a solenoid were made of the saturated solid solution of bismuth in lead, we should be able to generate magnetic fields of 14000 gauss at the boiling point of helium without development of heat and at 2 °K even fields of 19000 gauss.[62]

Naturally, Meissner followed these startling developments at Leiden with keen interest, as shown by his New Year's letter to de Haas written at the end of 1930, which dwelt almost exclusively on the lead-bismuth results.

> Recently in a physics colloquium at the University I had occasion to once again refer to your highly interesting work in the field of superconductivity, in particular

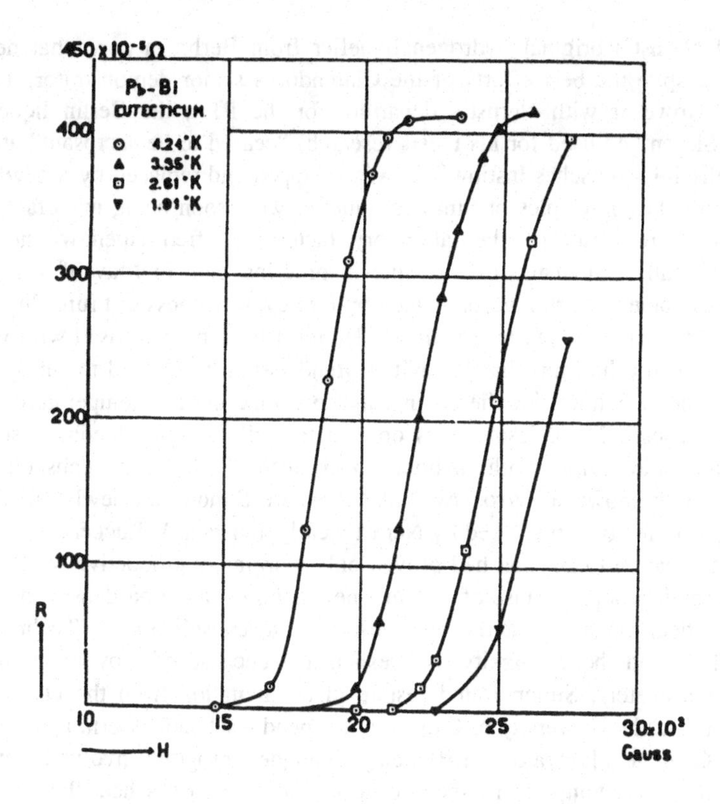

Figure 7-10. Magnetic transition curves for the eutectic alloy Pb–Bi at four different bath temperatures. De Haas and Voogd, RN, **33** *(1930), 267.*

to the work with Schubnikow [visitor from Kharkov, of whom we will hear more later] and with Voogd. This remarkable curve for bismuth is really extraordinarily interesting. I also hear that you have obtained a similar curve for the susceptibility of bismuth. Perhaps you have by now made practical use of the higher threshold value of the lead-bismuth alloy, and prepared a magnetic coil in liquid helium? It may interest you that recently I myself undertook measurements on single crystals of lead and gold down to 1.1° abs. in fields to 5,000 gauss, without uncovering any indication of the saturation demanded by the theory of Frank. Moreover, your curve for bismuth reveals no indication of the saturation at 30,000 gauss, even at 11° abs.[63]

Not surprisingly, a full 3 years elapsed before the first recorded attempts were made to incorporate the new alloy materials in actual magnets. Their intricate metallurgy and their notoriously poor handling characteristics proved to be formidable barriers. The first and apparently only attempt was made in Oxford's Clarendon Laboratory—then a fledgling low-temperature institution but one that would soon grow in stature.

The low-temperature school at Oxford owed it origins to Lindemann and his efforts to rejuvenate the aging Clarendon and get a program going on low-temperature specific heats.[64] We will recall that Lindemann had imported a copy of Nernst's original hydrogen liquefier from Berlin, a copy that never worked despite the best efforts of the Clarendon's senior demonstrator, T. C. Keeley. However with Nernst's departure for the PTR, the Berlin liquefier (which Simon had used for his thesis research) "ceased to be sacrosant" at the Physikalisch-Chemisches Institut;[65] it was scrapped and replaced by a new one of Simon's design. Copies of Simon's liquefier were soon being manufactured by a small firm run by the laboratory factotum Alfred Hoenow and his brother,[66] and Lindemann lost no time in ordering one. In May of 1931 he personally came to Berlin, accompanied by Keeley, to witness test runs directed by Kurt Mendelssohn and take delivery. He must have been pleased with what he saw, because he immediately invited Mendelssohn to Oxford to set up the machine, get a helium liquefier going, and organize the low-temperature program in general. Mendelssohn was obliged to stall for time, having just accepted a position (along with Nicholas Kurti) at the Technische Hochschule in Breslau (now known as Wroclaw, Poland) where Simon had newly filled an important professorship vacated by Nernst's early assistant A. Eucken.

The relocation to Breslau had proven only marginally productive for Simon and his small group, because of Simon's new administrative burdens stemming from the decision by the authorities, just then, to consolidate the Technische Hochschule and the University of Breslau, and because of growing political unrest. Fortunately, Simon found respite in an invitation from the chemistry department at the University of California to spend a sabbatical term at William Francis Giauque's laboratory in Berkeley. Giauque's program involved, among other things, low-temperature specific heats tied to Nernst's heat theorem—a subject dear to Simon's heart. It also involved a long-term effort since 1926 (pursued independently by Peter Debye at Leipzig) to produce lower temperatures ($\ll 1$ °K) than those possible by simply pumping on liquid helium, by utilizing a novel method based on the adiabatic demagnetization of paramagnetic salts.[67] For carrying out this scheme in practice, liquid helium was still needed (for precooling) but it was not available at Berkeley (nor at Leipzig, for that matter). On this score Simon, arriving at the end of 1931, saved the day. As noted earlier, Simon was already adept in concocting cryogenic apparatus, having constructed a variation on Nernst's old hydrogen liquefier and, some time later in Berlin, a modified version of the helium cascade liquefier based on pumping off under adiabatic conditions helium gas absorbed on charcoal instead of relying on high pressure as in the Linde process. Now he devised an ultrasimple scheme for liquefying helium by Cailletet expansion in a small strong-walled copper container.[68] Helium gas is compressed; the resulting heat is absorbed by liquid (or solid) hydrogen; the copper vessel is isolated from the hydrogen bath; the helium gas, cooled by expansion, liquefies as a result.

Because of its great simplicity Mendelssohn, while he was at Breslau, decided to adopt this method for the Clarendon, and carried a small helium liquefier-cryostat to Oxford in January of 1933 in a suitcase. No sooner had he returned to Germany than Hitler came to power. With conditions at Breslau rapidly deteriorating, and for added reasons of health, Mendelssohn was back in Oxford in April for a permanent stay. He was to be one of the many Jewish refugee scientists sponsored by Lindemann, who entertained no doubts about the Nazi menace. At the same time, Lindemann saw the opportunity for assembling at the languishing Clarendon Laboratory a team of crack continental scientists on par with those of Cambridge's Cavendish Laboratory. There were no regular positions open at Oxford, but thanks largely to Lindemann's influence, research grants were made available by Imperial Chemical Industries to the scientists from abroad, and Mendelssohn was among the first.

Staying in Germany was anathema to Simon even though, as a front line fighter in the war and holder of the Iron Cross, First Class, he was exempt from the decree dismissing Jews from academic positions. After some negotiations, he followed Mendelssohn later in the year, bringing with him Kurti, two helium liquefiers, and some electrical equipment. Before long Heinz London, his doctoral thesis at Breslau completed, joined the relocated low-temperature group at Oxford, of which Simon was the acknowledged senior member. Gradually the deplorable conditions at the Clarendon, which was housed in a Victorian Gothic edifice dating from 1872 and had a woeful dearth of funds, were overcome, and a scientific program got under way. In 1936 Simon was also appointed University Reader in Thermodynamics at Balliol College. Many years later, in 1956, he succeeded Lord Cherwell, nee Lindemann, as Dr. Lee's Professor of Experimental Philosophy—a month before his untimely death.

Mendelssohn's first experimental task at Oxford required the use of a high-field magnet, and he opted for putting the new alloy superconductors to work. The magnet was needed for use in magnetic cooling—Simon and Kurti's research priority at the Clarendon, once its feasibility had been demonstrated at Berkeley and Leiden. An optimist, Mendelssohn viewed heat conduction through the current leads as the chief remaining technical challenge of high-field superconducting magnets. Full of clever ideas as well, he proposed eliminating the problem by transferring the energy by induction instead of by current transfer. For this he devised a type of dc current transformer known nowadays as a "flux pump." It employed a normally conducting primary winding and a superconducting secondary circuit consisting of a secondary with a few turns of large radius and a solenoid with many narrow turns for producing the high field.[69] The optimism turned out to be highly premature. A prototype constructed in September of 1933 from lead-bismuth[70] failed utterly to generate the high field levels expected on the basis of the high critical fields reported at Leiden and Silsbee's rule.[71] (Kurti and Simon at first managed to make do with an "extremely ancient" iron-copper magnet that was a mainstay for many decades at the Clarendon. They also used the Paris Academy's large Bellevue magnet that was capable of producing up to 75 kG—to be sure, with a pole tip-spacing of a few mm!)[72]

Foiled as well was Keesom's nearly simultaneous determination in November to construct a coil wound from Pb–Bi wire. Keesom's report on the frustrated project is vague as to whether an actual coil was made. Apparently the attempt was abandoned when, in a short sample test at 4.2 °K, a current of 23 A corresponding to a wire surface field of 308 gauss proved to be sufficient to restore the resistance.[73] This field is of the order of the threshold field values for *pure metals*—a far cry from the threshold field (14 kG at 4.2 °K) measured for the alloy by de Haas and Voogd. "The conclusion [wrote Keesom] is that for an alloy the disturbance of supraconductivity by an electric current does not depend in the same way on the value of the magnetic field due to the current as Silsbee's hypothesis prescribes for pure metals."[74]

The abject failure of these projects remained inexplicable in 1933, although some insight into its causes was not long in coming. In light of what we know today about the Pb–Bi alloys, there is reason to believe that either coil should have done relatively better. The Oxford magnet may have been compromised by having been machined from a cast cylinder,[75] not wound from elaborately drawn wire like that prepared by Keesom.[76] And much might have been learned had Keesom taken the trouble of making measurements on the current-carrying capacity of his wires in external fields in the kilogauss range,[77] not merely in fields of a few hundred gauss as he reported. Opines the father of modern high-field superconducting magnets:

> It is quite conceivable that if such measurements had been made, and the information successfully disseminated at the time of the early Pb–Bi studies, our state of technological development today might be quite different, possibly to an extent that is hard to envision.[78]

7.3 Magnetic Transition on Closer Inspection

Discussion of the short-lived flurry of activity over the practical application of alloys has brought us somewhat ahead of the main thread of developments. At Leiden the bulk of the magnetic investigations continued to be those carried out by de Haas and Voogd on the magnetic disturbance of both alloys and pure superconductors, with concentration on the influence of the state of crystallization on the magnetic transition curve. They concluded their experiments with a series of final reports to the academy in January of 1931. Foremost among these was a report on the magnetic transition curves for *single crystals* of pure tin.[79] This *Communication* remains a somewhat classic paper on superconductivity, mainly for the inspirational role it was to play in the train of events leading to Meissner's breakthrough in 1933. The single crystal samples were prepared by techniques discussed in Section 10.1—basically by slowly cooling liquid metals in thin-walled glass tubes. They were typically 0.2 mm in diameter (small enough that accurate resistance measurements were possible with weak currents) and 15 mm long (short enough that the field of a solenoid was reasonably homogeneous over the whole length). The magnetic disturbance was monitored with the external field parallel or perpendicular to the crystal (wire) axis—the "longitudinal and transverse disturbances," respectively.

As expected from the most recent trend in the measurements, the hysteresis effects and discontinuities were much more dramatic than for polycrystalline materials.[80] Typical results for the longitudinal and transverse transition curves at some particular temperature are well summarized in Figures 7-11 and 7-12, respectively. In the longitudinal case they found that

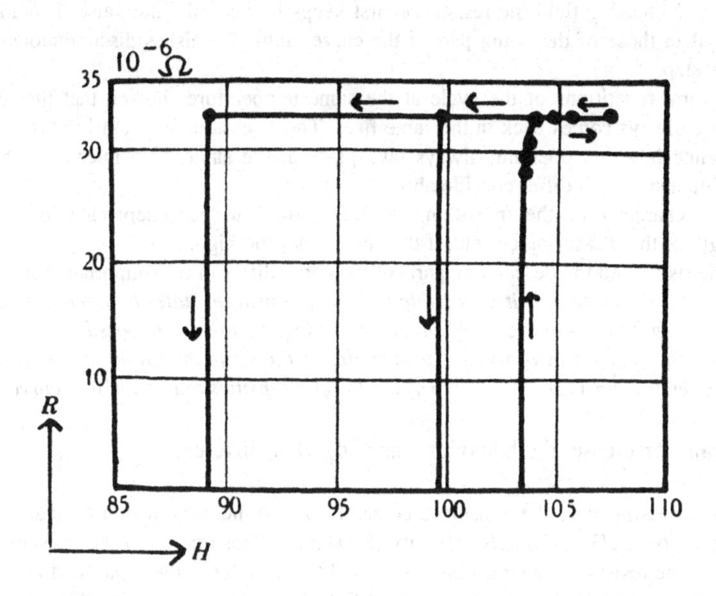

Figure 7-11. Magnetic transition curve for a single tin crystal in a parallel magnetic field for increasing and decreasing fields. De Haas and Voogd, RN, **34** *(1931), 65.*

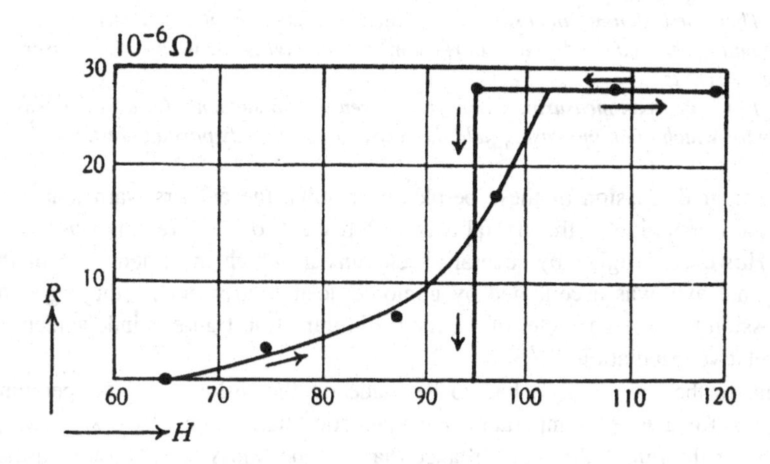

Figure 7-12. Magnetic transition curve for a single tin crystal in a traverse magnetic field. De Haas and Voogd, RN, **34** *(1931), 66.*

...when the field is increased the resistance always comes back for a definite value of the field intensity within a small magnetic range. The rising line is therefore a steep curve, which begins straight and is somewhat curved at the upper end. When the resistance is restored to its normal value a further increase of the field does not change it anymore (but for the increase of the resistance in much higher fields).

> In a decreasing field the resistance first keeps its normal value, also in fields lower than those of the rising part of the curve, until it vanishes discontinuously in one step.
>
> Several repetitions of the cycle at the same temperature showed that the resistance always comes back in the same field. The discontinuous vanishing of the resistance however does not always take place at the same field intensity. The latter intensities can differ considerably.
>
> The character of the transition curve is proved to be independent of the strength of the measuring current, if the latter is not too high.
>
> The rising and the descending parts of the curve differ in two important points:
>
> 1st. *On the rising part it is possible to realize transition states between normal resistivity and supraconductivity; on the descending part this is impossible.*
>
> 2nd. *The rising part of the line always lies at the same intensities of the field, the descending part not, though always at lower intensities than the rising curve.*

In the transverse case, the behavior was altogether different.

> In an increasing field the resistance comes back continuously within a magnetic range of about 2/3 of the field intensity for which the restoration of the resistance begins. The resistance has regained its normal value at about the same field intensity as has been found for the longitudinal field. In first approximation the character of the rising line is independent of the measuring current, becoming less curved only with higher measuring current.
>
> This is not the case for the descending branch. In a descending field the resistance first keeps its normal value.
>
> *Then for a definite intensity of the field the resistance falls suddenly and the amount of this sudden decrease in resistance is the higher the weaker the measuring current is.*
>
> *For very weak measuring currents can even vanish suddenly for a field intensity for which in an increasing field the resistance is already partially restored.*[81]

In their discussion of these perplexing results, the authors warn that, as a rule, the orientation of the crystal axis with respect to the wire axis was uncertain. However, judging by several check runs in which the orientation of the tetragonal axis was ascertained by goniometric measurements, "[they] got the impression that the character of the magnetic transition figure is independent of the [relative orientation]."[82]

One further comparison had to be made on the influence of temperature. Hitherto, for a given temperature the magnetic "half value" $H_{1/2}$ was always higher for the longitudinal disturbance than for the transverse. Those measurements were always made on polycrystalline wires. But in order to determine the temperature dependence for single crystals, the definition of magnetic threshold needed clarification, in light of the pronounced hysteresis effects. De Haas and Voogd now proposed as an alternative characteristic field intensity the magnetic half value for the rising curve of the longitudinal disturbance, because in this case the restoration of resistance is so sharply defined as to blur the distinction between half height and threshold. Defined thusly, the only way to compare quantitatively the longitudinal and transverse behavior—for example their respective temperature dependences, as seen in Figure 7-13—was in terms of

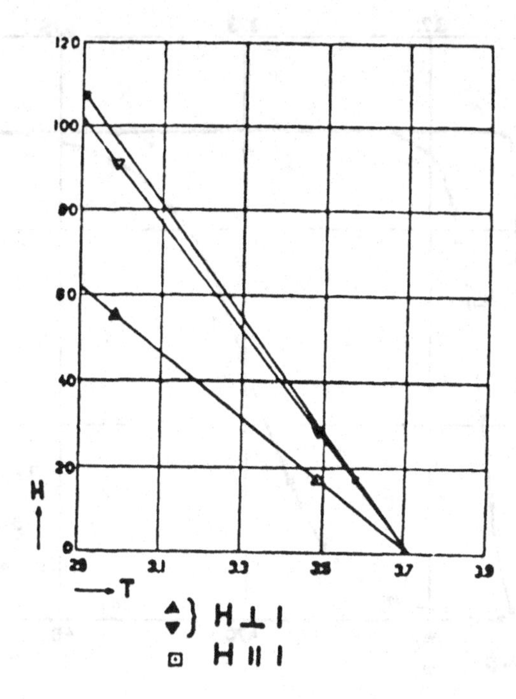

*Figure 7-13. Temperature dependence of longitudinal and transverse magnetic disturbance. De Haas and Voogd, RN, **34** (1931), 69.*

the field intensity for the first onset of resistance. The new data show "that in a longitudinal field the resistance comes back in about the same field in which in a transverse magnetic field the resistance has just regained its normal value," and that "for the transverse disturbance the magnetic half value of the rising curve will be smaller than for the longitudinal disturbance, which has always been observed in the case of polycrystalline wires with small hysteresis."[83]

In spite of the comparison ambiguity, the authors rightfully maintain that

> the principal point found is the characteristic difference between the longitudinal and the transverse disturbance. In further experiments we shall try to learn something more of the mechanism of the magnetic disturbance. ... Hence we must find out *how far in the different stages of the magnetic transition curves the magnetic field has intruded into the wire* [italics added].[84]

This they did, albeit three years later and long after these Leiden results aroused the attention of Max von Laue. Indeed, it was von Laue who would draw attention to their true significance.

One clear-cut result accrued quickly from de Haas and Voogd's studies, however. While single crystal wires yielded dramatic results but did not shed much light on the nature of the magnetic disturbance per se, they prompted reexamination of the perennial question: Is the *thermal* transition curve truly discontinuous under ideal conditions? For a long time a dependence on sample

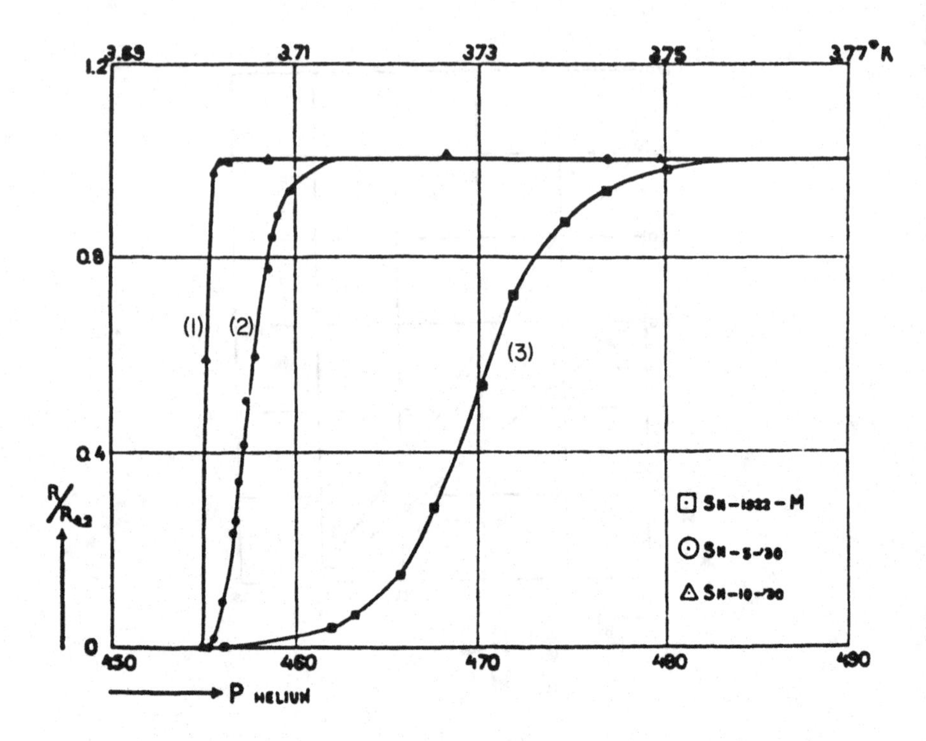

Figure 7-14. Influence of purity on sharpness of transition: (1) pure tin single crystal; (2) pure tin polycrystal; (3) less pure tin polycrystal. The abscissa indicates helium vapor pressure, proportional to temperatures. De Haas and Voogd, RN, **34** *(1931), 196.*

purity of the observed breadth of the step to normal resistance had been suspected, even though metals in a state of extreme purity were not necessary for exhibiting the superconducting state in the first place. (The curve was certainly sensitive to the strength of the measuring current as well.) The most recent evidence on this included new measurements by Meissner and B. Voigt at the PTR.[85] The exceptionally broad thermal transition curves usually observed by them were believed to be caused by physical and chemical impurities in their samples, including Hg, In, and Tl samples. Explanations for this behavior of polycrystalline materials came readily to mind. Internal stresses formed at the boundary faces of the crystallites from asymmetries when the wire is cooled was one possibility; these had, after all, been convincingly shown by Sizoo and Onnes to shift the transition curve toward higher temperatures. The possibility also remained that the crystallites would become superconducting one by one, with the transition temperature for each grain depending on the orientation of the current vector with respect to the crystal axis.

Whatever the mechanism responsible for the broadening, the availability of strain-free single pure crystal wires of white tin possessing a highly anisotropic tetragonal crystal lattice offered the best chance to resolve the purity question experimentally once and for all. De Haas and Voogd devoted considerable ad-

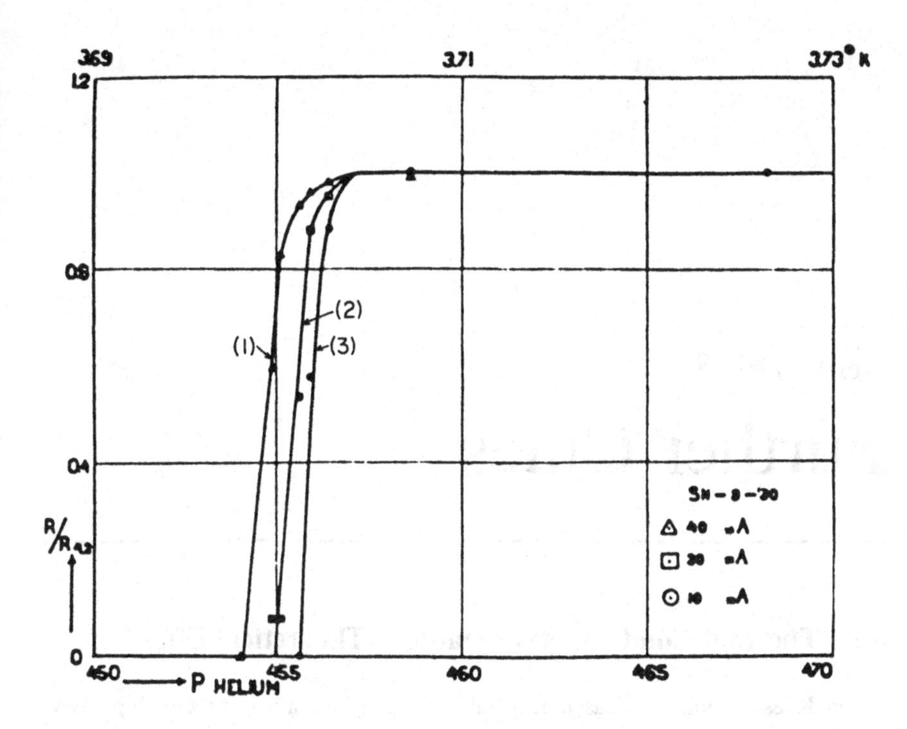

Figure 7-15. Influence of strength of measuring current on sharpness of transition for pure tin single crystal: (1) 40 mA; (2) 20 mA; (3) 10 mA. De Haas and Voogd, RN, **34** *(1931), 197.*

ditional effort to refining methods of preparing wire samples in glass capillaries treated in a vacuum furnace, controlling irregularities by x-ray analysis, determining the axial orientation by various means, and mounting them carefully with leads for the measuring current and potential measurements attached by a hot nichrome wire technique. Using very weak (mA) commutated measuring currents and with the terrestrial magnetic field compensated by a Helmholtz coil, they obtained the clean results depicted in Figures 7-14 and 7-15.[86] Figure 7-14 shows quite clearly that the steepness of the transition curve increases with the purity of the crystalline state, even if the shift of the curve as a whole is too great to be attributed to stresses approaching the breaking value.[87]

It is also rather evident from Figure 7-15 that the transition curve depends on the strength of the measuring current. In the limit of zero current, and using single crystals of the highest purity, they estimated the width of the transition to approach 0.001 °K. That the culprit is the magnetic field of the measuring current was demonstrated by showing that increasing the current lowers the transition temperature by an amount such that the increase of the critical field equals the increase in maximum self-field of the measuring current.

CHAPTER 8

Further Clues

8.1 Thermal Conductivity; Quantum-Theoretical Efforts

When Keesom and de Haas jointly inherited the laboratory directorship, they divided the research program somewhat loosely into two broad areas. Keesom, in addition to being responsible for the cryogenic plant, would concentrate on liquid helium, molecular physics, and thermal properties of solids. De Haas took overall responsibility for research on electrical, magnetic, and optical properties of matter at low temperatures.[1] If anything, this division—which was occasionally plagued by minor friction between the directors on competence in borderline subjects[2]—had positive benefits as far as the superconductivity program went. While de Haas and his colleagues were focusing on the perplexities of the superconducting transition in a magnetic field, the first convincing indication that the superconducting state is indeed characterized by properties different from those in the normal state came from *thermal* measurements, and mainly from Keesom's section of the laboratory.

We should recall that among Onnes and Holst's last joint experiments before Holst departed for Zürich and Philips were two studies that involved highly preliminary measurements on the specific heat and thermal conductivity of mercury in liquid helium. "With respect to the specific heat," they had reported, "nothing peculiar happens at the point of discontinuity."[3] More surprisingly—downright astonishingly—the same appeared to hold for the thermal conductivity. "The thing that immediately strikes us, is that there is no distinct discontinuity as was found at 4.19 °K in the electrical conductivity, although the thermal conductivity becomes much larger, when the temperature decreases."[4] This negative finding was in fact highly revealing: The implied failure of the Wiedemann–Franz relation cast strong doubt on the adequacy of explaining superconductivity in terms of classical mechanisms and free electrons.

Renewed measurements on the thermal properties of superconductors were started at Leiden in the early 1930s. Despite the division of labor, measurements on thermal conductivity were carried out under de Haas's supervision, with H. Bremmer as his principal assistant. Bremmer's experimental career had a somewhat inauspicious start; Casimir relates several incidents in which he precipitated serious laboratory accidents with delicate glass vacuum systems. ("He later distinguished himself in mathematical physics.")[5] Measurements on lead and tin were reported to the academy in late March, 1931.[6] Fortunately they involved more robust apparatus (a brass vessel replacing the usual glass dewars), with the exception of a gas thermometer. A rod of lead or tin was mounted in an evacuated metal vessel. It was attached at its cold end to a heat sink, usually at liquid helium temperatures, and was heated at the other end by a concentric electric heater. The thermal conductivity was determined by measuring the temperature at two places along the rod; correction for the loss and gain of energy by convection was determined by a separate experiment. (Radiation losses were negligible at helium temperatures.) A gas thermometer was employed because, unlike a resistance element, it was not influenced by a magnetic field.

The results for lead showed that the thermal resistance passed through a broad minimum roughly in the vicinity of the transition point, and rose sharply at still lower temperatures. However, Grüneisen and Goens had found as a rule similar maxima in thermal conductivity for impure metals at higher temperatures;[7] therefore it could not be ruled out that the low-temperature minimum was due to "the last traces of impurities." According to a contemporary theory of thermal conductivity developed by Rudolf Peierls, dating from 1929,[8] which clearly explained the well-known asymptotic behavior of typical thermal resistance curves at ordinary and high temperatures, the thermal resistance should be proportional to $1/T^2$ at the lowest temperatures. In that case, one would expect a point of inflection, but no minimum. Close examination of the thermal resistance curve near the transition point revealed no obvious point of inflection; "if it shows any, it surely is only weak."[9]

The measurements on tin brought out new surprises, a hint of which was given in the earlier lead measurements. A single run had been made with the lead rod exposed to a longitudinal field well above the threshold field for the particular sample temperature. In that case the thermal resistance diminished significantly, but increased once more when the field was switched off. This contradicted the behavior of normal conductors, whose thermal resistance *increases* in a magnetic field. Consequently, the new tin samples[10] were systematically measured, first with and then without a magnetic field. The results, shown in Figure 8-1, show that the thermal resistance curve forks at the transition temperature. The curve obtained below T_c with a tin sample in zero background fields, or in fields below the threshold value, is essentially a smooth continuation of the curve above the transition point. (As with lead, an increase in thermal resistance is seen with decreasing temperature in the helium temperature range, following a broad minimum between the helium and hydrogen temperatures.) However, the curve for fields greater than the

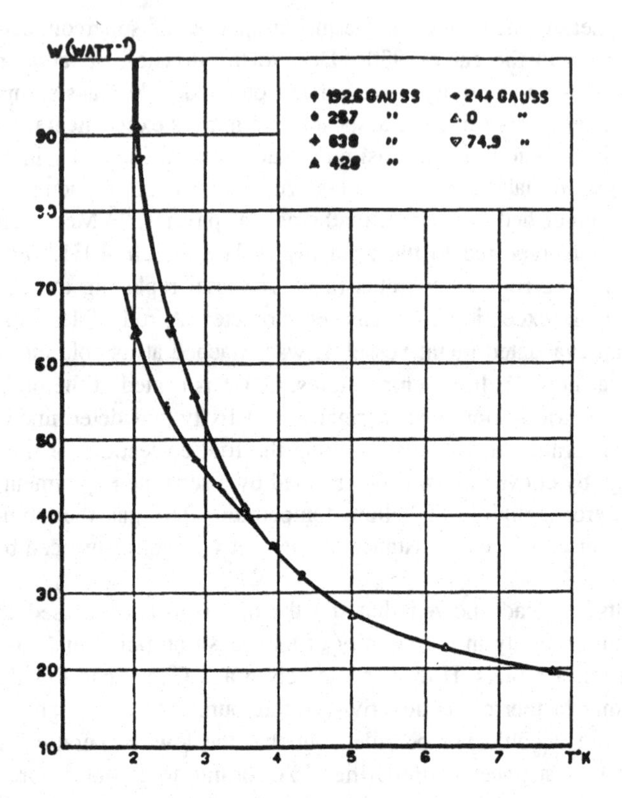

Figure 8-1. Thermal resistance of pure tin versus temperature. Upper curve corresponds to zero background field or $H < H_c$; lower curve obtained with $H > H_c$. T_c for tin is approximately 3.72 °K. De Haas and Bremmer, RN, **34** *(1931), 334.*

threshold value (H_c depending, of course, on T) shows a noticeable discontinuity at T_c.

Curiously, one might have expected the continuity for *normal* tin, not the reverse. The lack of a discontinuity below the threshold field troubled many investigators in 1932; one was Meissner. Thus, in a letter to Ralph Kronig in Groningen that November, he wrote:

> ... permit me to make the following remarks. According to experiments at Leiden (compare my report on superconductivity, Ergebnisse der exacten Naturwissenschaften, Vol. II [1932]), no change in thermal conductivity occurs at the superconducting transition point. It seems to me extremely difficult to reconcile this fact with your hypothesis for the origin of superconductivity; unless it can be established with certainty that thermal conductivity of good conductors at low temperatures for the most part acts through conduction electrons, and only for a small part through elastic waves of the lattice. If the rigid electron lattice moves through the atomic lattice without exchanging energy with it, there is, as far as I can see, no possibility of a transfer of heat to the atomic lattice. ... In addition, in the majority of cases the superconducting current in all probability is a surface

current flowing on the surface of the conductor, so that only the outermost electrons can be displaced by the electron lattice.[11]

Meissner is referring to one of several renewed attempts at that time to treat the problem of superconductivity theoretically. By then quantum theory had progressed to a stage where finding a solution seemed a more realistic goal. Felix Bloch, who was Heisenberg's first graduate student at Leipzig, had accepted the challenge in the fall of 1928, soon after the publication of his landmark thesis on normal conductivity and upon his arrival in Zürich to begin an assistantship under Wolfgang Pauli. His attempt, never published, was based on the assumption that there is a close connection between ferromagnetism and superconductivity; that is, persistent currents below the critical temperature are analogous to permanent magnetization below the Curie point. His essential idea was that the thermodynamically most probable superconducting ground state is one that supports a finite spontaneous current below the transition point.[12] However, despite repeated calculational attempts and encouragement from Pauli, he invariably found that the state of minimum energy was unable to carry a persistent current—a finding known at the time as "Bloch's theorem on superconductivity." (Another, facetious, theorem proposed by Bloch echoes his mounting frustration in this attempt: "The only theorem about superconductivity which can be proved is that any theory of superconductivity is refutable.")[13]

Nor did Lev Landau, Bloch and Peierls's fellow assistant at Pauli's Institute in 1929, who made a similar attempt, have much more success, though he did not disdain publication.[14] He, too, sought to exploit the analogy between superconductivity and ferromagnetism as his guide; his superconducting model contains local "saturation currents" flowing in different directions in different parts of the superconductor that produce net currents only when organized by an applied field. While Landau hinted that his theory was, in fact, unsatisfactory (for one thing, it predicted a critical field dependence that was at variance with de Haas and Voogd's newly published results on thallium), it contained rudiments of the eventual Ginzburg–Landau theory of superconductivity and anticipated the Meissner–Ochsenfeld experiments.[15]

The physical significance of the impossibility of Bloch's and Landau's models (i.e., of Bloch's first theorem), was provided by Leon Brillouin some years later in a 1935 discussion on superconductivity at the Royal Society in London, to which we shall return in Section 11.2. Suppose, with Brillouin, that a current flows through a metal. If we apply a potential difference between the ends of the conductor of proper sign, the energy of the system can be decreased; therefore the energy cannot have been a minimum. The only way around the problem, within the frame of classical theories, is for the current to be metastable, something made doubtful by Onnes's persistent currents and certainly the Meissner effect (which was unknown to Bloch and Landau in 1929).

In any case, an alternative approach to the problem was taken up several years later in independent efforts by Bohr and by Kronig. Both of their models were essentially a (quantum-theoretical) resurrection of Lindemann's old lattice model. Kronig's version was the more specific of the two. It supposes the existence, as a result of the interaction between electrons, within a few degrees of absolute zero, of a rigid electron lattice intermeshed within the ionic lattice of the metal. The transition point is the "melting point" of the electron lattice, determined by the thermal motion of the electrons—presumably sharply defined like the melting point of an atomic crystal.

In his first attempt, the results of which were submitted for publication in August of 1932,[16] Kronig assumed that the electron lattice could move as a whole through the metal; being rigid, it could not give up energy to the ionic lattice. An enthusiastic supporter of Kronig's approach to this problem was McLennan, who had recently relocated to England for an indefinite stay. "Now that I have given up my laboratory my thoughts turn to theory," he wrote Kronig in mid-1933.[17] On receiving a copy of Kronig's paper in late August of 1932 McLennan had concurred that "the idea of the electron lattice looks to be just the thing that is wanted. It fits in with many of the observed effects and it will be interesting if the Debye–Scherrer diagrams bring out the required lines."[18] However, heated discussions with Bohr and Bloch in Copenhagen that October soon persuaded Kronig that the potential barrier between lattice sites would prevent bodily displacement of the electron lattice. Instead he submitted a new paper near the end of the year, based on a modified model in which the electron lattice as a whole remains fixed in position, but single chains—a one-dimensional electron lattice—can move freely along their length.[19] In this case, the potential barrier problem is greatly reduced. Writing to Bohr in late November of 1932, Kronig acknowledged Meissner's letter of 11 November, adding that Meissner's conception of a surface current "would be in accord with the chain picture. I have asked Mr. Meissner to tell me in detail how he comes to this conception, but have up to now received no answer."[20]

In the event, Meissner's conception—not Kronig's—would provide the missing clue. Neither Kronig nor Bohr made substantive progress with their respective models. Bohr had taken up the problem of superconductivity in the spring of 1932, inspired with the notion that "superconductivity concerns a coordinated motion of the whole electron lattice." He assumed that a satisfactory many-electron wave function carrying the supercurrent could be constructed by slightly modifying the normal ground state wave function, and expressed confidence generally that the solution to superconductivity lies in the new quantum mechanics.[21] A lively but short discussion then ensued from June to July between Bohr, Leon Rosenfeld (one of Bohr's assistants in Copenhagen), and Bloch, who remained skeptical, objecting that Bohr's conception was too vague. As a result, Bohr soon developed second thoughts about publishing his own manuscript on the subject, partly on experimental grounds; he found it difficult to reconcile his own theory with Canadian superconducting experiments with electric fields of radio-frequencies despite McLennan's own enthusiasm for the Bohr/Kronig approach. Moreover, there were too many uncertainties attached to the moving lattice concept.[22]

Bohr settled for viewing superconductivity as "a purely quantum problem, which quite escapes visualization by means of basically mechanical pictures" and "[found] it difficult ... to agree with details of [Kronig's] second treatment."[23] Bloch, in his continuing role of critic, was more specific in his objections. "The more I have thought about it [he wrote Bohr in December of 1932], the more I consider the Kronig work to be in error. The last hope which Kronig places in ... the one-dimensional lattice is unjustified, because the condition, which according to Kronig would cause the overtaking of the potential hill [e.g., zero-point motion of the electrons], simply means that the zero-point oscillations are so big that this lattice cannot exist."[24]

All of this has distracted us from de Haas and Bremmer's measurements. Weak evidence for a discontinuity in thermal conductivity for zero field was subsequently reported in their renewed and still more accurate measurements, made this time on indium. Reducing the results in the form of a log-log plot of thermal resistance versus temperature, they "got the impression that at the transition-point a discontinuity occurs, as well for the supraconductive state as for the non-supraconductive state."[25]

Their evidence has not withstood the test of time, as modern measurements have been unable to verify any distinct discontinuity at the normal transition temperature.[26] However, the broad features of the Leiden curves are qualitatively explained by the modern theoretical picture of thermal conductivity in normal metals at low temperatures. To first approximation the onset of superconductivity in relatively pure metals is merely a perturbation of the normal conduction mechanism, which is primarily due to free electrons.[27] The electronic thermal resistance is rather well represented by:

$$W_e = \alpha T^2 + \beta/T \,. \tag{8-1}$$

The first term corresponds to scattering by the thermal lattice vibrations (phonons), and the second term to scattering by impurities and imperfections. The coefficient α involves various parameters, in particular the Debye temperature; β is related to the (temperature-independent) residual electrical resistivity. Near absolute zero, the thermal conductivity rises linearly according to β/T. It reaches a maximum and then drops again as the quadratic term takes over. The location of the maximum thus depends on the relative values of residual electrical resistance and Debye temperature. For a given metal, the higher the purity the lower the residual resistance and, consequently, the lower the temperature of the maximum. Thus, for tin and indium αT^2 is small compared to β/T; for lead the two terms are comparable.

8.2 Reversibility at Last?

While de Haas's group was preoccupied with measurements of thermal conductivity, Keesom was engaged in a parallel experimental program on low-temperature specific heats of solids—a venerable program dating back to 1914. Preliminary measurements on lead in liquid helium had been made by Keesom and van den Ende in 1926, but their accuracy was rather marginal, prompting renewed measurements in 1930.[28] These were carried down to 2 °K with various improvements in the Nernst–Eucken calorimeter and in the measuring procedures. The results agreed reasonably well with Debye curves over the entire temperature range, with no evident untoward departure near the transition point. Unfortunately, however, uncertainties in the constantin thermometer calibration in the vicinity of this very temperature rendered this particular conclusion suspect.

More positive results were obtained with tin and zinc early in 1932, with the help of an improved (phosphor bronze) thermometer and further improvements in the calorimeter:

> Attention is drawn to the fact that at 3.7 °K the atomic heat [of tin] shows a particularity, in this sense that for a lower temperature the atomic heat is larger than for a higher one. This particularity cannot be due to an inaccuracy of the thermometer, as in this region not a trace of an irregular change of the resistance could be detected. ... Moreover we ought to have found then an analogous particularity for zinc, which was not the case. It seems as if there is a sudden jump in the value of θ [the Debye temperature] at about 3.7 °K. This temperature coincides with or is very near to the transition point at which tin becomes supraconductive. This coincidence suggests that there is a connection between the particular behavior of the atomic heat and the phenomenon of supraconductivity.[29]

The measurements, still rather preliminary, were repeated without delay, assisted now by Anna Petronella Keesom, Keesom's daughter. The final results,[30] reported in June, leave no doubt (Figure 8-2). The atomic heat falls from a value of about 0.0078 cal/ °K to 0.0054 cal/ °K over a temperature interval estimated to be less than 0.012 °K at a temperature coinciding with the transition point to within 0.01 degree. That the discontinuity is indeed connected with superconductivity is corroborated by the odd data point marked [❑] in Figure 8-2, obtained by accident. The intention was to postpone investigations on the influence of a magnetic field, a more complex undertaking. Inadvertently, however, during one experimental run a magnetic field was applied in a neighboring cryostat. It produced a perturbing field at the calorimeter of a few gauss, sufficient (with the unshielded ambient field) to lower the transition point below the bath temperature. This particular measured value agrees—more or less—with the normal curve.

An effect of the magnitude observed with tin would not have been detected easily with lead. The slight absolute jump in specific heat would have been swamped by the enormous lattice (normal) specific heat at the transition point—which is 50 times that of tin. Simon and Mendelssohn, in fact, had also searched in vain for a discontinuity in lead late in 1931.[31]

One other conclusion was reached in the latest round of measurements: "Transition to the supraconductive state is not connected with a transformation heat." The same was found in subsequent measurements on thallium, again without a magnetic field, by Keesom and Kok.[32] Thallium, like tin and unlike lead, was another promising sample material, possessing a moderately high normal specific heat in liquid helium. It showed no detectable latent heat but a jump in specific heat at the transition point. In fact, the observed jumps, for

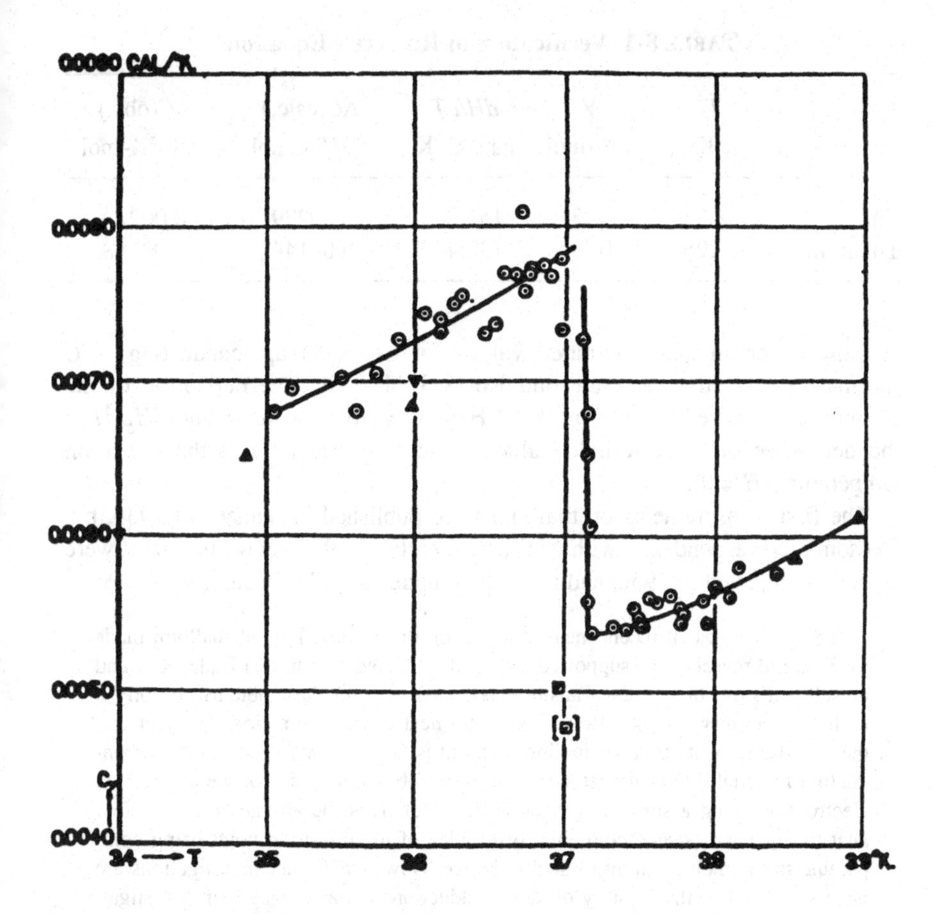

Figure 8-2. Specific heat discontinuity of tin at the transition point. The measurements are for zero background field, except for point marked [□] which was accidentally obtained *with a magnetic field higher than* H_c. *Keesom and Kok, RN,* 35 *(1932), 744.*

thallium as well as for tin, were in excellent agreement with a formula for the difference in specific heat at T_c in the absence of a magnetic field proposed at that time by Arend Joan Rutgers, Ehrenfest's first doctoral student and professor at the Technical University in Delft.[33] Rutgers was motivated by Keesom's somewhat earlier analogous application of Gibbs functions to the transition between helium II and helium I where a similar jump in specific heat had been discovered the same year by Keesom and K. Clusius.[34]

The formula, first presented by Rutgers at one of Ehrenfest's regular colloquiums, is the following:

$$\left(\frac{dH_c}{dT}\right)^2_{T=T_c} = \frac{4\pi\Delta C}{T_c V}, \tag{8–2}$$

TABLE 8-1 Verification of Rutgers's Equation

	T °K	V cm^3/mol	dH/dT gauss/ °K	ΔC(calc.) cal/ °K-mol	ΔC(obs.) cal/ °K-mol
Tin	3.71	8.37	151.2	0.00229	0.0024
Thallium	2.26	16.9	137.4	0.00144	0.00148

and the observed and calculated values for $\Delta C = C$ (superconducting) − C (normal) are listed in Keesom and Kok's Table 8-1. (The derivation of the formula is discussed in Section 10.1.) Here V = atomic volume and dH_c/dT is the derivative of the threshold value of the magnetic field at the transition temperature ($H = 0$).

The first measurements on thallium were published in January of 1934, but Keesom's correspondence with McLennan early in 1933 shows that they were actually completed in 1932, and more were under way. In January, he wrote:

> … I may remark that recent measurements on the specific heat of thallium made by Kok and myself have supported the result obtained by van den Ende, Kok and myself with tin, that at the transition point the specific heat obtains abruptly a higher value, there being however no latent heat of transformation. The fact that the specific heat of supraconducting material is larger than that of non-supraconducting material seems at first sight not to corroborate the idea of the conduction electrons forming a somewhat rigid lattice. Nor does the absence of any latent heat to my mind [seem] favorable to the idea of the transition point being comparable to the melting point of such a lattice. However, further investigations are necessary to clear the mystery of superconduction. We are engaged in investigating the specific heat of supraconductors under the influence of magnetic fields, and I think interesting data will result from them.[35]

The new round of experiments referred to here, on thallium in a constant magnetic field,[36] had in fact been largely completed when the first results on thallium went to press. Their publication was delayed pending calibration of the new phosphor bronze thermometer in a magnetic field. The results would prove to be highly relevant to the general problem of understanding the behavior of superconductors in a magnetic field—a problem poised to take center stage in all of this.

Of more immediate relevance to Keesom's thermodynamic program was the fact that the latest helium results provided the long-sought after experimental answer to the old question posed in turn, by Keesom, Bridgman, and Langevin: Can the transition from the superconducting to nonsuperconducting state be treated thermodynamically as a reversible process? Figure 8-3 shows specific heat results for thallium obtained with a constant external magnetic field of 33.6 gauss, plotted as atomic heats versus temperature. The broken curve represents the atomic heat of thallium with zero external field, derived from the earlier measurements. The sharp peak in the curve with a field occurs at a

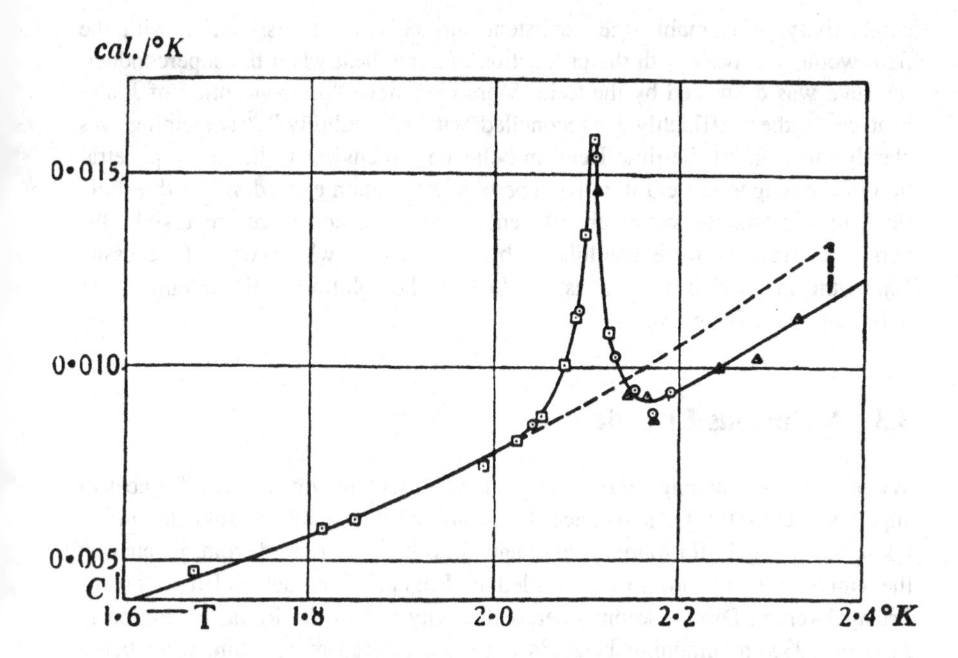

Figure 8-3. Specific heat of thallium in a constant field, providing evidence for a latent heat. The broken curve is for zero background field. Keesom and Kok, PH, **1** *(1934), 506.*

temperature of about 2.11 °K or close to the transition temperature for the applied field, as deduced by de Haas and Voogd.[37] "The suggestion presents itself," wrote Keesom, "that the sharp increase of the apparent atomic heat at about 2.1 °K in reality is due to a latent heat in connection with a gradual vanishing of supraconductivity by increase of temperature in an external magnetic field."[38]

The validity of Rutgers's formula, derived with the aid of an infinitesimal reversible cycle, would appear to have demonstrated the applicability of reversibility for the superconducting transition in the *absence* of an applied field, despite the lack of a latent heat under those conditions. Now the statement could also be generalized to transitions *in* a magnetic field, because what amounts to a comparison of the latent heat deduced from the measured specific heat with that calculated from Keesom's old formula, Eq. (6-6), showed agreement to better than 2%. The gathering experimental evidence for what had until then—in the absence of a latent heat or change in crystal lattice—been accepted essentially ad hoc, was seriously weakening Casimir's first "superstition" of superconductivity: that the phenomenon is a quirk of the free path of the conduction electrons, not a special state of the electron gas. On the other hand, Casimir's second superstition (the assumed validity of Maxwell's electrodynamics for superconductors) still held sway when these measurements were made, and presented a new dilemma for the transition in a magnetic field. Because superconductivity was believed to be merely a state of perfect ($R = 0$)

conductivity, presumably the persistent surface currents associated with the field would die away with the production of Joule heat when the superconducting state was destroyed by the field. Moreover, since "the conception of Joule-heat can rather difficultly be reconciled with reversibility,"[39] something was clearly missing. At the time Keesom believed the answer to lie in the penetration of the magnetic field into the superconductor upon exceeding the threshold field. In this process, which absorbs energy and was considered reversible, the persistent currents were annihilated by induction,[40] which was of necessity *before* the material became resistive. In fact, the solution to the dilemma was only a few months away.

8.3 A Curious Episode

While Keesom was engaged in his final measurements on the specific heat of superconducting tin, his assistance was requested in a small but amusing undertaking now largely forgotten. McLennan, who had just retired from his chair at the University of Toronto and settled in England, was scheduled to give a Friday Evening Discourse on superconductivity before the Royal Institution in June of 1932, maintaining Dewar's tradition. It occurred to him, what better way to demonstrate the phenomenon than with a persistent current? The ensuing correspondence in this connection, most of it between McLennan and Keesom,[41] documents a curious episode in the history of superconductivity that deserves a brief recapitulation. It also underscores the considerable logistics involved in the little project.

The celebrated Friday Evening Discourses of the venerable Royal Institution, founded by Count Rumford (Benjamin Thompson), originated in about 1825 when evening meetings of the dues-paying members of the institution began on a regular basis.[42] On these occasions, a lecture was given, accompanied by experiments or demonstrations if the topic was of a scientific nature. Michael Faraday, who came on board the institution at that time, was from the beginning largely responsible for arranging the lectures. Combining "fluency and animation as a speaker" with "a remarkable faculty in devising experiments for the lecture table to illustrate his theme," he frequently gave the lecture himself before a packed audience.[43] The lectures have continued ever since, and have become one of the great traditions of the Royal Institution. James Dewar's expositorial skills, as noted earlier, were perfect for this splendid forum. Nor was McLennan a slouch when it came to popularizing science, as he would amply demonstrate with Keesom's cooperation.

McLennan to Keesom, 26 March 1932: McLennan explains his desire to exhibit a persistent current in a lead ring. Unfortunately, liquid helium won't be available in England for another year [when Mendelssohn's liquefier is expected to be operational]. Is there a possibility of help from Leiden?

Keesom to McLennan, 8 April: Keesom replies enthusiastically, "prepared to do what I can for it." According to his information, transporting a dewar by auto from Leiden to Amsterdam, from Amsterdam to Croydon aerodrome by aeroplane, and finally by auto to London should take about 4 1/2 hours. He reckons at least 5 1/2 hours between topping off the dewar at Leiden and the end of the experiment. Allowing an evaporation rate twice as large as when the apparatus is "in a quiet condition," the dewar must be capable of retaining an adequate liquid helium level for 10 hours. Keesom is prepared to offer one of his own dewars, but inquires as to the particulars of the intended experiment. In the interest of safety, a liquid nitrogen-shielded dewar would be preferable to hydrogen, in which case the liquid helium glass will have to be totally silvered except for a small window.

> As there is always some possibility—however small it is—that a glass cracks, and as any panic under passengers absolutely must be avoided, I think it necessary that the apparatus shall be transported by a separate aeroplane with no strange passengers. Moreover, in order that I can bear the responsibility against our aeroplane company, if you will use one of their aeroplanes, I think it must be one of my men [Who else but Flim?] who carries the apparatus filled with liquid helium and nitrogen (or hydrogen) to London.[44]

McLennan to Keesom, 10 April: McLennan is delighted by Keesom's generous offer of assistance. Lord McMaster of Sempill,[45] an old wartime friend, has kindly offered to fly the precious load to Croydon in his personal aeroplane, along with Keesom's courier. The experiment is intended to: (1) demonstrate the presence of a persistent current in a lead ring by its effect on an external magnetometer, and (2) to provide an opportunity of seeing liquid helium. Since McLennan has to attend the forthcoming planning for the Physikertag later scheduled at Bad Nauheim, he proposes to stop over in Leiden and go over the details.

After some further exchanges, they agree to meet in Leiden on 26 April.

McLennan to Keesom, 26 April: McLennan writes from hotel in the Hague, thanking Keesom for the opportunity of meeting with him that morning and seeing all of his interesting researches. He is making the necessary arrangements with the office of the Minister of Finance, customs, and the Master of Sempill. The latter arrives at Schiphol on 3 June; it remains to be determined if his aeroplane "is open or closed and whether it is a 'two-seater'."[46]

McLennan to Keesom, 20 May:
Everything is in order for the experiment.

McLennan to Keesom, telegram, 29 May:
Everything now in order for June third four o'clock at Amsterdam's Schiphol.

McLennan to Keesom, telegram, 29 May:
Will send one hundred cubic feet gaseous helium back with Flim unless you cable me immediately it is required before.

Keesom to McLennan, 31 May: As a contingency measure, Keesom has arranged with KLM to have an aeroplane on standby at Schiphol in the event that the Master of Sempill does not arrive by four thirty. Mr. Flim would prefer to return by steamer, allowing him time to "see something of London at [*sic*] Saturday."

Keesom to McLennan, telegram, 31 May:
All prepared here.

McLennan to Keesom, telegram, date unclear:
All ready in London for Friday.

KLM to Keesom, 2 June:
Herewith we confirm our telephone discussion of this afternoon in which we informed you that Colonel the Master of Sempill, AFC, AFRAES is a very well known pilot. This gentleman is Chairman of the Royal Aeronautical Society, Director of the London Light Aeroplane Club and the National Flying Service Ltd., and holds numerous other important functions in the field of aviation.

The aeroplane G-ABJU is a closed-fuselage Puss-Moth and belongs to the Master of Sempill himself.

In view of the flying capabilities of this pilot, we have no doubt that the trip from London to Amsterdam and back will proceed to your satisfaction. In the eventuality of G-ABJU failing to arrive at Amsterdam, we will keep, in accordance with your request, an extra aeroplane available at Schiphol.

We trust that the fast and pleasant method of travel by air will induce your assistant, Mr. Flim, to return by KLM to Amsterdam.

The persistent current, a little over 200 amperes, was started by Keesom about three o'clock on the afternoon of 3 June, induced by an external magnetic field. Earlier in the day, the liquid helium had been produced and collected in a glass dewar flask supported in a second glass dewar containing liquid air. Both dewars were supported in a third metallic dewar, also filled with liquid air. The ring of lead, mounted with its plane vertical at the bottom of the inner helium dewar, measured about 2.5 cm in diameter and was about 3 mm in cross section. Everything went without a hitch. The portable dewar was brought by motor car to Schiphol aerodrome, and at the appointed hour Flim, hefting the cryostat dewar, and the Master of Sempill were ready to go. Keesom and A. Plesman, Director of KLM, saw them off (Figure 8-4). The flight took about three hours.

Flim to Keesom, telegram, 3 June:
Transport successful.

McLennan to Keesom, 3 June: He recapitulates the events of the day. Flim had arrived at the Royal Institution at about 8:30 p.m., a half hour before the lecture, with the helium level in the dewar flask well above the lead ring—no small tribute to the instrument makers at Leiden. When the lecture began at 9:00 p.m.,[47] the current had persisted for over 6 hours and was still running.

Figure 8-4. Portable dewar prior to boarding at Schipol aerodrome, Amsterdam. Left to right, W. Keesom, G. J. Flim (with cryostat), Master of Sempill (pilot), and A. Plesman of KLM. Courtesy Museum Boerhaave, Leiden.

"The experiment was great and a brilliant success." Indeed, it must have been an enjoyable affair, since McLennan was a public lecturer second to none. "It was a sight for the gods to see him beam with enthusiasm and joy at his helium super-cooled ring ..."[48]

The Master of Sempill to Keesom, 3 June:

Just a short letter to thank you very much indeed for coming to Amsterdam on Friday, and to say how delighted I was to have had the honor of meeting you after having known your name for so many years.

Mr. Flim and I had quite a good flight although when we got over England the air conditions were somewhat disturbed and a certain amount of the liquid was lost. Nevertheless, there was sufficient for the demonstrations to be made before the Royal Institution, and everyone heartily applauded the mention of your name.[49]

McLennan later shed some more light on the near-disastrous accident aloft that was merely alluded to by the pilot. At a meeting of the Student's Mathematical and Physical Society in Toronto the following November he explained that

> ... we provided against every cause of failure except one. The accident happened when the aeroplane encountered an upstream current on the return trip. Due to the diminishing air pressure the helium and its jacket of liquid air both started to bubble and the cork was blown out. Fortunately the pilot recovered the stopper and plugged up the flask before the helium could evaporate.[50]

Keesom to McLennan, 8 June: He was glad to hear of the success, and also begged McLennan to accept his best thanks for the manner he entertained Mr. Flim, "who very much enjoyed his journey."[51]

Keesom to Colonel the Master of Sempill, 8 June: Keesom thanks him for his kind letter and for his important role in the episode.

McLennan to Keesom, 4 July: Informs Keesom of having arranged with British Oxygen to ship his cylinder of helium gas to Leiden. Again, Flim will accompany the shipment, taking particular caution to protect the cylinder's valve during the flight to minimize the loss of helium gas.

Keesom to McLennan, 9 January 1933: The next time we hear of the matter is in Keesom's letter to McLennan the following January reporting on his results with Kok on the specific heat of thallium.

> I duly receeived through the hands of Dr. Crommelin the separate copy of your supplement to Nature on supra-conduction for which I sincerely thank you. It reminds me in the first place that, due to all sorts of business after my journey to Buenos Aires for the Refrigeration Congress, I did not yet thank you for your generously sending us an amount of helium vastly suppasing [*sic*] the quantity we spent for the first experiment of June 3rd. So I beg leave to tender you my most coordial [*sic*] thanks for it.
>
> I may add that I was very much upset when I learned that during my absence they had bothered you for a charge in connection with transport from here to Schiphol. I hope you will allow me to return same, and I beg you to forget after that bother they called you.
>
> May I ask you what address we can return the cylinder in which the helium you sent us was transported? Or do you prefer that we remit you the cost of it?[52]

Keesom refers to a *first* experiment on June 3. Indeed, shortly after his latest letter to McLennan the latter staged two encores of the performance. The first was a demonstration of liquid hydrogen before the Royal Institution on 24 January; another involved a persistent current in liquid helium the following week. Both demonstrations are the topic of discussion in further voluminous correspondence between Keesom and McLennan from mid-January to mid-February.[53] The liquid hydrogen went by steamer via the Hook of Holland in the care of Flim and McLennan's lecture assistant, Mr. Minns. The helium dewar was flown to Croydon with the Royal Air Mail, and once more entrusted to Flim.

It should be added that Simon, too, demonstrated a persistent current, of his own making, later that very spring to a packed audience at the Technischen Hochschule in Breslau.[54] A feature of that lecture was a small version of his new and highly portable helium liquefier. The temperature of the helium bath, monitored by a direct reading gas thermometer, could be read from the auditorium. The 400 A current was induced in a lead ring somewhat larger than McLennan's with the aid of a solenoid, and the constancy of the current was monitored by both a magnetic needle and a small search coil. Simon eventually repeated the demonstration, including the liquefier, at his own Friday Evening Discourse before the Royal Institution in February of 1935.[55]

CHAPTER 9

Experimentum Crucis

9.1 Von Laue Steps In

The evidence that the superconducting transition is a thermodynamic change of state came as the culmination of an eleventh hour systematic search for thermal properties that had been virtually taken for granted for well over two decades, save for Dana's inconclusive tests for a latent heat accompanying the transition in the early 1920s and Keesom's negative search for a change in crystal lattice at about the same time. This dogmatic faith in the applicability of thermodynamics had its counterpart in the conventional laws of electrodynamics. Unlike the members of Keesom's calorimetry team, however, and particularly Keesom himself, who on the whole remained open-minded on the question of thermodynamics, none in de Haas's group, or anyone else for that matter, bothered to put the question of whether a superconductor does indeed obey Maxwell's equations, albeit with infinite conductivity, to a serious experimental test. That is, a test to verify the applicability of Lippmann's rule, which holds that the magnetic flux linked by any closed loop inside a body of infinite conductivity cannot change, but is "frozen in," unaffected by external changes. To be sure, the one and only, albeit spurious, test performed, with Onnes and Tuyn's lead sphere (Section 6.3) appeared to confirm this long held viewpoint; Lorentz's analysis of the misleading results (misleading because of the sphere being hollow) in terms of Maxwellian electrodynamics undoubtedly carried considerable weight in perpetuating the classical view. In addition, however, it was tacitly assumed that the only essential property of a superconductor is infinite conductivity; thus as late as 1931 de Haas and Voogd proclaimed: "We therefore regard the vanishing of the resistance within a few hundredths degree as the most characteristic phenomenon of the supraconduction in pure metals."[1] In other words, the *magnetic* properties were seen as secondary or at least predictable from Maxwellian theory. Therein lies a serious contradiction, since the classical magnetization behavior of a perfect ($R = 0$) conductor (according to Maxwell) is *irreversible*, as we will recapitulate shortly, whereas a thermodynamic process is of necessity reversible. One of the most curious aspects of the development of superconductivity during the period 1920–1930 is that this striking inconsistency was so slow in inspiring a closer look at the magnetic behavior. This was the case despite the overwhelming evidence for the special role of a magnetic field in the phenomenology of superconductivity: as inducer of persistent currents; as the cause of a jump in thermal resistance; as a prerequisite for a latent heat; as a threshold impediment to the very superconducting state itself.

Gavroglu and Goudaroulis have examined this matter from a more formal, methodological perspective. They discuss the apparent contradictions and the persistence of old "concepts out of context" in terms of "paradoxical situations" that arise when "an observed unexpected phenomenon is expressed by the descriptive language of the dominant explanatory schema [prevailing theories] and which leads to the formulation of a physical problem amenable to solution."[2]

> What eventually established a paradoxical situation was the basic assumption behind the initial efforts to find an explanation for the new phenomenon. A superconductor was considered to be a *perfect* conductor and, even though Kamerlingh Onnes from the beginning suspected that quantum effects caused such a peculiar behavior, the phenomenon was not regarded as a "pure quantum phenomenon," but rather as the manifestation of quantum effects in an otherwise classically perfect conductor.[3]

However, the requisite unexpected phenomenon was slow in coming, since "the scientific community was working on the wrong problem," that is, on infinite conductivity. Only when attention shifted from the electrical properties to magnetic ones, still with the notion of infinite conductivity as the guiding principle, was the stage set for the creation of such a "paradoxical situation," as indeed we shall see in due course.

Because it is central to subsequent developments, we are not amiss at this point to review the firmly entrenched picture of the irreversible behavior associated with a perfect conductor when it is cycled in a magnetic field. Suppose first that the conductor is made superconducting by cooling it below T_c in the absence of a magnetic field—that is, it undergoes a transition from state a to state b as indicated in Figure 9-1. If subsequently an external magnetic field, less than the critical field, is applied ($b \rightarrow c$), circulating persistent currents are induced on the surface of the specimen, by virtue of the rule of Lippmann, in such a manner as to create a magnetic flux density in the interior that cancels the flux density due to the applied field. The result is zero net flux density in the interior, or perfect diamagnetism. Though the flux density created *inside* the body by the surface currents is equal and opposite to the flux from the applied field, this is not the case *outside* the body; there the net distribution of flux resulting from the superposition of flux from field and currents is as depicted

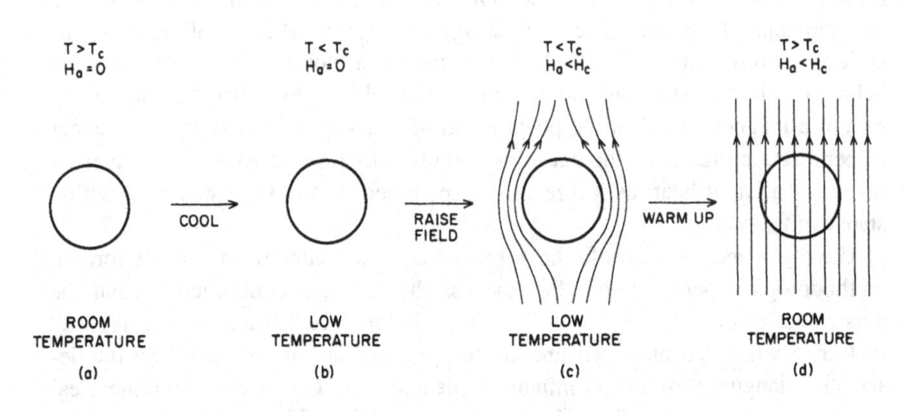

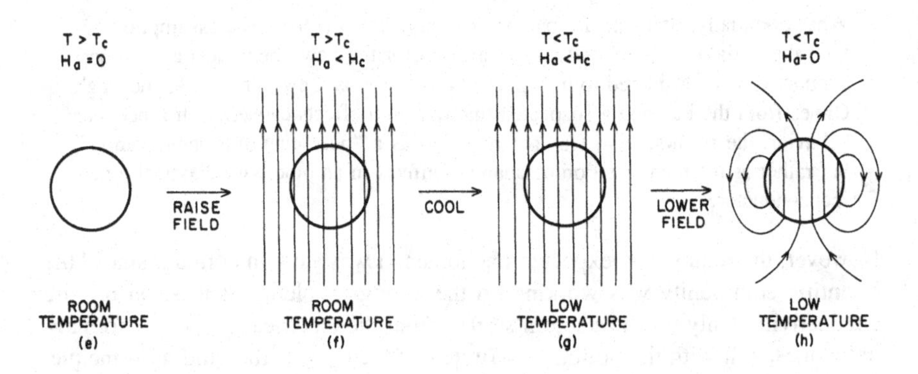

Figure 9-1. Presumed behavior of a "perfect" conductor in a magnetic field. Top sequence: exposing a sample, already superconducting, to a magnetic field. Bottom sequence: cooling a sample, initially in the normal state, in a constant magnetic field. Dahl, HSPBS, **16** *(1986), 32.*

(*c*). If the body is warmed with the field held constant, the screening currents decay and the external flux penetrates (*d*). Or, if the applied field is reduced to zero with the temperature maintained below T_c (not shown in Figure 9-1), the body is left in its original unmagnetized state.

A quite different pattern results if the body is cycled in the reverse sequence. If the conductor at room temperature is first exposed to a magnetic field less than H_c, the flux will penetrate (assuming the specimen is not ferromagnetic), since the resistance is finite (sequence *e* to *f* in Figure 9-1). Cool it now below the transition point while maintaining the external field ($f \rightarrow g$); there should be no magnetic effect since $H = 0$ and the internal flux distribution remains unchanged. Thus, the state of magnetization (*g*) is quite different from the state (*c*), although the sample is exposed to the same applied field and maintained at the same temperature in both cases. If now the applied field is reduced to zero, the flux density inside the perfectly conducting metal cannot change; persistent currents are again induced, maintaining the flux inside, with the result that the body retains a permanent magnetization—exhibiting a "frozen-in" permanent dipole field. Again, we have a different magnetization state (*h*) from state (*b*), even though external conditions are identical. The state of magnetization of a "perfect" conductor is not uniquely determined by the external conditions, but depends on the cyclic history leading to those conditions.

In September of 1932 Max von Laue gave a lecture on superconductivity at the eighth Deutschen *Physiker Tages* at Bad Nauheim, on the occasion of receiving the Planck Medal of the German Physical Society. Formerly Planck's leading student and his closest disciple,[4] Laue, in addition to his academic duties at the University of Berlin, was theoretical adviser to the PTR at Charlottenburg and followed the superconductivity research programs with keen interest. "It is like the time of Faraday," he was fond of remarking to Meissner, "quite new investigations and quite new considerations are necessary to explain the new phenomena."[5] Although he was close to Meissner, Laue was to remain a theoretical loner in the flagging efforts to understand superconductivity between the wars. He consistently pursued the subject within the framework of classical electrodynamics and thermodynamics, in contrast to the quantum-theoretical approach of Heisenberg, the London brothers, and others who came later. Perhaps Laue's most important contribution to superconductivity was his role as chairman of the physics committee of the Notgemeinschaft, in which role he directed available financial resources to worthy projects, ensuring continuance of research even in the bleakest economic times of the Weimar Republic.

Laue's Nauheimer lecture was entitled "An interpretation of some experiments on superconductivity,"[6] and dealt with the measurements recorded by de Haas and Voogd the year before on the magnetic disturbance in monocrystalline tin wires,[7] which had recently come to his attention. He was struck by the fact that the magnetic field strength at which the first trace of resistance is observed in a rising transverse field is approximately *half* the threshold field H_c in a longitudinal field, and that the resistance is only fully restored at a field approximately equal to H_c. This, he pointed out, is as to be expected on the basis of the classical picture of a perfect conductor (Figure 9-1*c*). In a transverse field the diamagnetic behavior causes the field lines to be distorted and crowded together at the transverse diameter of the cylinder. The magnetic disturbance will set in as soon as, at some point on the surface of the wire, the tangential component of the field equals the threshold field. Quantitatively, consider a superconducting cylinder of radius R_0 in a uniform magnetic field, with its axis perpendicular to the direction of the field (Figure 9-2). Simple magnetostatics show that the distribution of field lines is analogous to that of the stream lines in an incompressible, frictionless fluid flow directed perpendic-

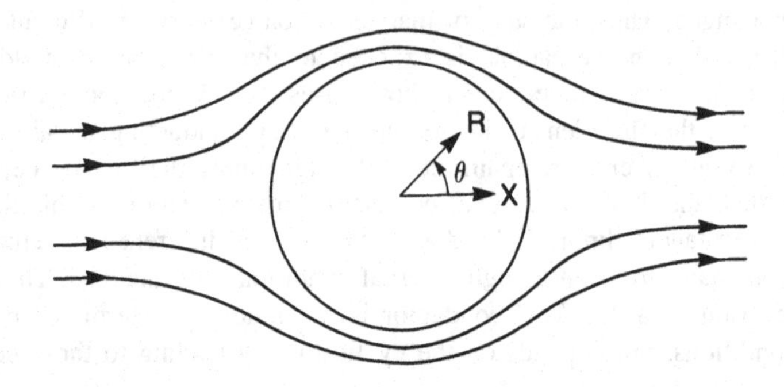

Figure 9-2. Superconducting cylinder in a uniform transverse magnetic field.

ular to a cylinder.[8] The radial and tangential field components outside the cylinder are:

$$H_R = H_0 \left(1 - \frac{R_0^2}{R^2} \right) \cos \theta$$

$$H_T = H_0 \left(1 + \frac{R_0^2}{R^2} \right) \sin \theta$$

where H_0 is the strength of the uniform field well removed from the cylinder. At the surface the radial component vanishes and the tangential component is:

$$H_T = -2H_0 \sin \theta \,. \qquad (9\text{-}1)$$

Thus, the maximum value for H_T occurs at $\theta = \pm\pi/2$ and equals $2H_0$. In the longitudinal case, the presence of the cylinder does not distort the applied field, and the field reaches its critical value H_c simultaneously throughout the cylinder, giving a sharp transition from the superconducting to normal state. In the transverse case, the field near the cylinder reaches the critical value when the external field is only half the longitudinal critical value, so the first trace of resistance should occur at a field $H = H_c/2$ (neglecting self-field effects of the measuring current). Thereafter, the resistance should rise approximately linearly as the field gradually penetrates into the cylinder, until it has returned completely at about the same field strength as in the longitudinal case.

For a sphere, Laue added, similar analysis predicts the initial restoration of resistance at a field value 2/3 H_c. For an ellipsoid the situation is more complicated; now the "demagnetizing coefficient"[9] depends on the orientation of the principal axis with respect to the field.

In examining de Haas and Voogd's data more closely, particularly their Figure 3 (Figure 7-13), which plots as a function of temperature the field strength for which the resistance begins to return in the longitudinal and transverse cases, respectively, Laue noted that between 2.9 and 3.7 °K both $T - H$

curves are approximately linear. They show a ratio of $H_c(\perp) / H_c(\parallel)$ independent of temperature and approximately 0.60, not 0.50 as expected on the basis of Laue's own interpretation. The discrepancy did not disturb Laue, however. For one thing, the shape of the magnetic transition curve for a rising transverse field did not rule out the possibility that measurements with higher sensitivity might extend the onset of resistance to lower fields. There was the question of how nearly circular the wires were, and the accuracy in determining the field direction seemed open to question as well. The importance of the orientation of the crystallographic axis was unknown. Moreover, Laue noted, Meissner himself had pointed out in private discussions that even in the longitudinal case some demagnetization can be expected, because of the finite length of the wire.

With regard to the *hysteresis* effects, Laue suggested that these, too, could probably be accounted for in terms of the highly irreversible magnetization curve associated with the proverbial frozen-in moments produced in lowering the field from above the critical value; such behavior would negate any simple relationship between resistance and field at any given temperature. He pointed to Onnes and Tuyn's old experiment on a superconducting lead sphere, which, in his opinion, demonstrated Lippmann's permanent magnet and "proved plainly the constancy of the moment." We encountered this spurious experiment earlier, and we will have more to say about it shortly. It suffices to note that Laue's analysis of de Haas's experiments can hardly be considered a rigorous depiction of the magnetic behavior in view of subsequent findings. However, it was to be invaluable in stimulating heightened interest in this aspect of superconductivity generally and was directly responsible for determining the course of the next round of experiments on the subject at Leiden.

9.2 Opportunity Missed

At Leiden, de Haas's group lost no time in following up on the arguments presented by von Laue. The matter was tackled anew in a fresh round of investigations under way in the spring of 1933 by de Haas and Voogd, now joined by Joshina Jonker. The goal was partly to extend their earlier measurements on the shape of the thermal transition curve as a function of crystal purity and strength,[10] partly to reexamine their even earlier measurements of the magnetic disturbance of monocrystalline tin wires[11] in light of Laue's deceptively simple interpretation of those results. Final publication of the new findings was delayed for a full year—a delay that proved costly. However, preliminary results were reported by de Haas in February of 1933 in his Leipziger Vorträge read at the *Physiker Tagung* at Leipzig that month (organized by Debye).[12] This lecture provided the first, succinct summary of the ongoing measurements at Leiden. Historically, it is important for a particular reason, as we shall see.

Special importance was attached to remeasuring accurately the magnetic transition curve for the transverse magnetic disturbance. Whether Laue's interpretation was quantitatively correct or not, this curve, with its broad ascending

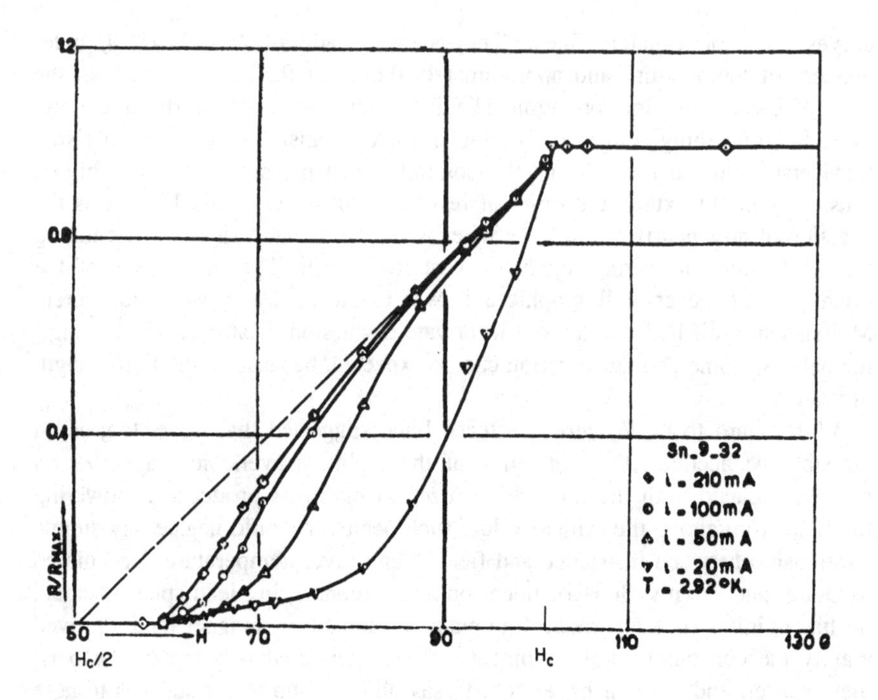

Figure 9-3. Effect of measuring current on steepness of magnetic transition curve in a transverse field. The dashed curve, superimposed on the original curve by de Haas et al., illustrates von Laue's $H_c/2$ relationship. De Haas, Voogd, and Jonker, PH, **1** *(1934), 284.*

slope from initial to full resistance, clearly held a clue to the mechanism of field penetration into the wire. However, first the effect of the measuring current—downplayed by Laue—had to be determined. As suspected, the shape of the curve was found to be highly sensitive to the strength of the measuring current (Figure 9-3), very much akin to the effect on the thermal transition curve (cf. Figure 7-15). Only for large measuring currents (several hundred mA) did the step in the curve approach the vaguely expected linear rise to normal resistance. However, the field strength corresponding to the onset of the first measurable resistance appeared to be independent of current strength and about 58% of the longitudinal critical field H_c, very much as Laue had already deduced from the earlier data. Insisting their wire "surely was circular,"[13] de Haas considered this discrepancy from Laue's ideal 50% too large to be explained by deviations from a circular cross section.

To clarify the matter, de Haas's team took Laue's hint, turning next to the influence of the geometry of the wire cross section. They measured the shape of the transverse ascending magnetic transition curve for single crystal wires of near-elliptical cross section, with the field variously directed parallel or perpendicular to the major axis. Not too surprisingly, restoration of resistance began at different field levels, depending on the orientation—that is, on different magnetizing coefficients—in qualitative if not in quantitative agreement with Laue.

The deviation from the expected behavior was about as that observed in the case of a circular cylinder. As a case in point, for a particular wire sample the onset of resistance began at 67.2 gauss with the field parallel to the major axis and at 54.5 gauss with the field perpendicular to the axis; the theory predicted 65 gauss and 40.59 gauss, respectively. Despite the discrepancy, this was surely further confirmation that the return of resistance was indeed connected with the penetration of the magnetic field.

The experiments up to this point, particularly those involving the "0.58 mystery," have been the subject of considerable attention over the years since they were performed in Leiden, by David Shoenberg among others. Thus, the dashed curve superimposed on the experimental curves in Figure 9-3 (not found in de Haas's original figure) suggests that Laue's $H_c/2$ relationship might be applicable to very large measuring currents. The solution to the 0.58 mystery, Shoenberg points out, was eventually shown to be a size effect associated with the "intermediate state" mechanism introduced much later.[14] The fact that the Leiden experiments always gave a value close to 0.58, not 0.50, is explained by the fact that the wires used were nearly always of about the same diameter (~0.2 mm). It was shown by E. R. Andrew that the ratio H_d/H_c, where H_a is the applied field at which the resistance first appears, depends on the wire diameter in a roughly linear manner, approaching 0.50 for very large specimens, as suspected from the dashed curve in Figure 9-3.[15]

Thus far, however, the measurements taken by de Haas, Voogd, and Jonker had produced nothing particularly new. But something new was not long in coming. The last set of measurements involved tracing the *thermal* transition curve by establishing first a constant transverse magnetic field (below H_c) while the wire was still warm, and then cooling it below T_c; by this procedure the transition curve was mappped for decreasing and increasing temperatures. From the Leipziger lecture we learn that the measurements, with what de Haas still termed in the current vernacular "eingefrorenem Felde," were barely under way at that time, and de Haas warned that the results should be viewed with "reservation." Now, according to the classical theory (sequence $f \rightarrow g$ in Figure 9-1), the behavior under this particular powering and cooling sequence should be qualitatively different from that under the earlier sequence. Though preliminary, the resultant curve, Figure 9-4, leaves little doubt that this in fact is *not* the case. The curve shows a strong resemblance to the magnetic transition curve at constant temperature, again exhibiting hysteresis between the descending and ascending branch and is reproducible only for increasing temperatures. "*The character of the curve*," reported de Haas, "*develops as in the case of changing the field at constant temperature*."[16] [Emphasis added.]

Here, then, was the all-important clue to the magnetic behavior. It may come as somewhat of a shock that strong evidence for the analogous behavior, whether it was cooling in a constant field or cooling and *then* introducing the field, was close at hand but went unheeded as early as February of 1933. This revelation should have alerted de Haas and colleagues that something was seriously wrong with the application of classical electromagnetic theory to superconductors. In hindsight, it should have prompted further, nowadays obvious,

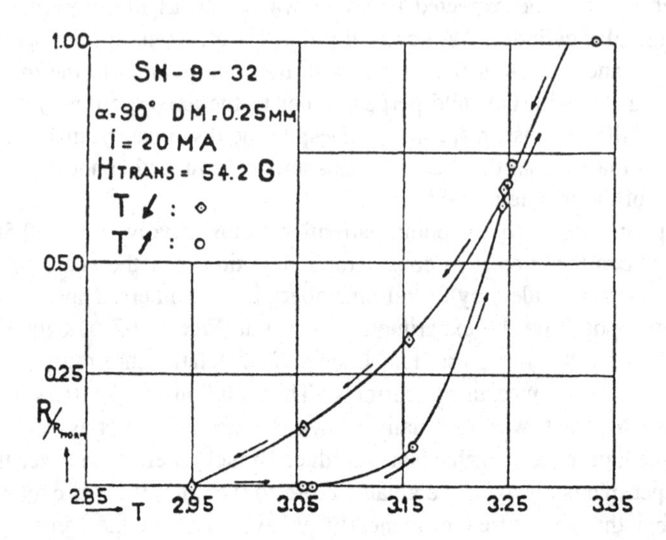

Figure 9-4. Preliminary thermal transition curve for tin sample in constant external field: clue to magnetic behavior in early 1933. De Haas, Leipziger Vorträge 1933, 69.

measurements that could have clarified the superconducting behavior immediately. Their failure to recognize the compelling clue must be partly blamed on lingering unease and preoccupation with the distracting "0.58 mystery," and partly on their concentration on resolving the issue of whether the hysteresis effects were fundamental or due to spurious, secondary causes tied to the persistent currents. At the time of the *Physiker Tagung* de Haas was still not convinced that they were fundamental, because the results were strongly affected by the method of attaching the potential leads: soldering them resulted in a broad hysteresis curve, whereas carefully spot-welding them to the sample produced much narrower curves.[17] On the other hand, the occurrence of "thermal hysteresis" under conditions when induced currents were not expected (cooling in a field) "point[ed] to a real hysteresis."[18]

At this juncture, de Haas refrained from offering a definite opinion on this point, warning merely that "experience teaches one to be most cautious in drawing conclusions from experiments on supraconductivity."[19] He added that Meissner was onto the same hysteresis problem, having measured the thermal transition curve (in zero background field) without the usual precaution of commutating the measuring current: that is, without reversing the current direction to eliminate thermoelectric effects. Under these circumstances, the hysteresis was even more pronounced. Meissner and Simon, de Haas noted, were inclined to attribute this result to a gradual redistribution of the current over the wire cross section during the transition.

The measurements at Leiden continued into mid-1933, confirming the analogy without doubt. Thus, not only did the more accurate thermal transition

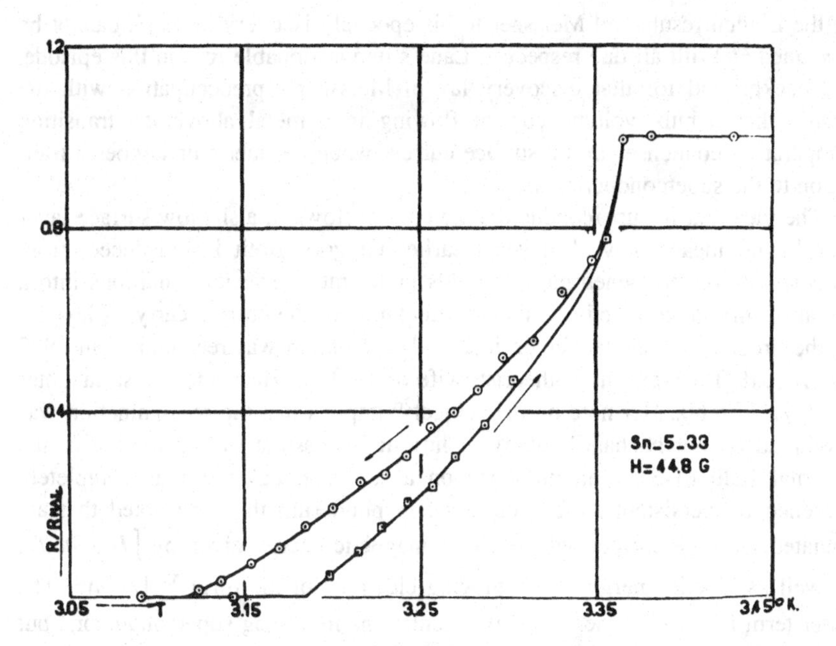

Figure 9-5. More accurate thermal transition curve for tin, circa mid-1933. De Haas, Voogd, and Jonker, PH, **1** *(1934), 288.*

curve obtained later in the spring (Figure 9-5) exhibit marked hysteresis but, like the magnetic transition curves, its shape was similarly sensitive to the measuring current and the wire cross section. By the fall, the conclusion that something was definitely wrong could no longer be avoided. In fact, the resolution to the problem was "in the air." New experiments by younger staff members, Casimir recalls, were ready at Leiden to settle the matter.[20] Alas, the experiments were too late; the solution came from Berlin. Wrote de Haas tersely, as the final version of the paper by Voogd, Jonker, and himself went to press:

> This analogy between thermal and magnetic transition curves is not intelligible from the standpoint of von Laue, unless one postulates that by cooling in a field the field is expelled from the supraconductor. Meissner and Ochsenfeld, in fact, have now observed a similar effect.[21]

"Two of us," de Haas added simply, "have confirmed the results of these measurements [by Meissner and Ochsenfeld] by still another technique."[22]

9.3 A State $B = 0$

Meissner, of course, was well aware of von Laue's arguments, as he was in close and regular touch with him. Although Meissner considered them both "important and interesting,"[23] Laue's claim in later years that his interpretation of the Leiden results led Meissner to his epochal discovery in 1933, cannot be sustained.[24] With all due respect to Laue's unquestionable role in this episode, the background for that discovery lay in Meissner's preoccupation with the notion that a bulk volume current flowing in a metal above the transition temperature condenses into a surface current when the metal undergoes a transition to the superconducting state.

The idea that in superconductors the current flows in a shallow surface layer was by no means new. Ten years earlier Gregory Breit had deduced, from calculations on the penetration of fields under rather special conditions into a metal of infinite conductivity, that the thickness of the current-carrying layer is of the order of molecular dimensions.[25] The problem was reexamined in 1925 by G. L. de Haas-Lorentz—theorist, wife of W. J. de Haas and eldest daughter of H. A. Lorentz. Her note on the subject[26] implies that the reexamination was prompted by her husband's query of how it is possible for a variation of the external field to exert an influence on a superconductor if it is completely screened by persistent surface currents. In pondering this, she noted that associated with the current will be electromagnetic energy given by $\int H^2/8\pi \, dV$, as well as kinetic energy of the moving electrons of the form $\sum 1/2mv^2$. The latter term is normally neglected in calculations involving superconductors, but de Haas-Lorentz found that the two quantities are actually of the same order of magnitude when the current layer has a thickness given by

$$\lambda = (mc^2/4\pi ne^2)^{1/2} . \tag{9-2}$$

Here n is, as usual, the number of electrons per unit volume, and m and e the electron mass and charge, respectively. The penetration depth is of the order of 500 Ångström. If the external magnetic field changes, surface currents of depth λ are set up and shield the interior so that at every point here the magnetic field remains constant.

Since de Haas-Lorentz's note was published in Dutch in an inaccessible Dutch journal, *Physica*, it received little notice. However in the late summer of 1933, Richard Becker of the PTR, with his assistant G. Heller and student F. Sauter, took up the problem anew, and pointed out that because of the inertia of the electrons, an electric field cannot lead to an infinitely strong current.[27] Simplified versions of their derivations have been reproduced frequently.[28] The argument proceeds from what was previously taken for granted generally, that a body of infinite conductivity requires curl $\mathbf{E} = -\dot{\mathbf{H}}/c = 0$ (assuming for the moment $\mathbf{B} = \mathbf{H}$), because an electric field would accelerate the electrons steadily, giving rise to an infinite current. The requirement $\dot{\mathbf{H}} = 0$ does not apply near the surface, observed the authors, because here we have associated with the electrons an electric field and hence an acceleration given by:

$$m\dot{\mathbf{v}} = e\mathbf{E} . \tag{9-3}$$

Since the current density is $\mathbf{J} = ne\mathbf{v}$,

$$\mathbf{E} = \frac{4\pi\lambda^2}{c^2}\dot{\mathbf{J}} \tag{9–4}$$

where λ is given as before by Eq. (9-2). Taking curls on both sides of Eq. (9-4) and using curl $\mathbf{E} = -\dot{\mathbf{H}}/c$, we have

$$\frac{4\pi\lambda^2}{c}\,\text{curl}\,\dot{\mathbf{J}} + \dot{\mathbf{H}} = 0\,. \tag{9–5}$$

Substituting Maxwell's second equation curl $\mathbf{H} = 4\pi\mathbf{J}/c$ leads finally to

$$\nabla^2\dot{\mathbf{H}} - \dot{\mathbf{H}}/\lambda^2 = 0\,. \tag{9-6}$$

Thus, instead of simply $\dot{\mathbf{H}} = 0$ from ignoring the electric field, the solution to Eq. (9-6) shows that $\mathbf{H}$ decreases exponentially inside the surface of the perfect conductor. If the number of superconducting electrons is of the order of the number of atoms, the distance in which $\mathbf{H}$ decays is about 10^{-6} cm.

Casimir recalls Pauli's suggestion, which was put forth at a colloquium at Zürich in the winter of 1932–33,[29] to apply Eq. (9-3) to a superconductor. Nobody present took up the challenge but according to Casimir, Pauli also put the problem to one of his students, who remains unnamed. "However, this was a rather poor student and nothing came of that."[30]

Meissner himself, too, first addressed the question of the current distribution in 1925.[31] In 1928 he even performed an inconclusive experiment that was designed to throw light on whether superconductivity is basically "a volume or a surface effect"; Meissner also thought it possible, but not very likely, that supercurrents involve electrons in the metal-insulator interface at the conductor surface. He reported the results at the Deutsche Naturforscherversammlung in Hamburg that year.[32] The experiment involved resistance measurements on tin samples, some of which were clad with new silver, some not. (It was basically a repetition of Onnes's old, equally inconclusive experiment in 1913 on mercury in a steel capillary, mentioned in Chapter 4. That experiment, on the possible influence of contact with a normal conductor, had been defeated by the uncertain contact between mercury and steel.) The new results, showing the tin to be superconducting below T_c with or without the silver cladding, appeared to favor a volume effect.

> This [he hastened to add] does not exclude the possibility that the supraconducting current is unevenly distributed over the cross section of the conductor, or even flowing in the outermost layer of the conductor. What the experiment does show is that a conductor-insulator surface layer is not necessary for the maintenance of a supraconducting current.[33]

The question was still not resolved 4 years later, as we know from Meissner's correspondence with Kronig in the late fall of 1932. Although in November Meissner had argued that "in the majority of cases the superconducting current in all probability is a surface current,"[34] he remained unconvinced of its validity in general. Thus, in December he wrote Kronig,[35] in reply to Kronig's query of how he came to this conclusion about surface currents:

> Concerning my remark that in all probability the supracurrent is a surface current, this is not quite correct or is in any case only somewhat superficially the case. On page 248 of my report [on superconductivity in *Ergebnisse der exakten Naturwissenschaften*, volume 11] is indeed emphasized that the supraconducting experiments do not furnish positive evidence for how the current is distributed in a supraconductor. It may well be that if one proceeds from the *non*-supraconducting state, then in the supraconducting state the uniform distribution is maintained over the cross sections. This is by no means certain, however. I am occupied with preparations for elucidating these questions experimentally. If, however, one produces a current in the reverse order, say by induction, in an already supraconducting metal, then a surface current undoubtedly is involved, since the supraconducting current induced in the surface prevents a change in the magnetic field in the supraconductor. ... Since in most cases the current is generated or interrupted while in the supraconducting state, it seems most probable to me that in the majority of cases the current in a supraconductor is a surface current. ... As noted, we [intend to] attack this question by direct experiment.[36]

This seems to be Meissner's first public reference to preparations that were then under way in Berlin for explicit experiments on the penetration of magnetic fields in a superconductor. The experiments must have been in an advanced stage of planning as early as January of 1933, judging from the tone of Meissner's next letter to Kronig (and as the actual laboratory records now confirm). Here he expresses his appreciation for Kronig's kind transmission of a copy of his latest work on the one-dimensional lattice model, adding that "I find it very interesting in view of the experiments on the current distribution which I am engaged in."[37]

Meissner's next public reference to the subject appears in his lecture during the *Physikalische Vortragswoche* at the ETH, Zürich, in late June of 1933.[38] On this occasion, in contradistinction to his initial intention to confine his talk exclusively to superconducting experiments at Charlottenburg,[39] he dwelt particularly on the thermal hysteresis effects uncovered at Leiden and confirmed by himself—effects that he for one did *not* doubt were real. "This [hysteresis]," he insisted, "is not intelligible if the current [distribution] in the supraconducting state is the same as in the non-supraconducting state."[40] We can also glean from the ETH lecture some specific details of the forthcoming experiments that were designed to settle the question once and for all. The experimental principle, he explained, would rely on a pair of parallel superconducting cylinders connected in series with an external current source. The magnetic field between the cylinders, to be measured with a small coil connected to a ballistic galvanometer, would depend on the current distribution in the conductor. Von Laue and F. Möglich, he added, had recently calculated the expected field under some "specific assumptions." However, "these experiments should allow a decision on whether the uniform distribution is preserved at the onset of the supraconductive state."[41]

In fact, the experiments, conducted with Meissner's assistant Robert Ochsenfeld, were already under way. Meissner's laboratory notebooks in this period show that the first three months of 1933 were occupied with various low-temperature calibrations.[42] From the second week of April through the third week of June there is an unaccounted gap in record keeping, except for a sketch in Meissner's hand illustrating the initial experimental principle with supporting design calculations.[43] This material is not dated, but from contemporary correspondence with von Laue we can surmise that it was jotted down in mid-April; we will return to it in Section 10.2. The first data entries for the new experiments proper are dated 22 June, or just prior to Meissner's ETH lecture. The subsequent record shows measurements and data reduction continuing at regular intervals, with the last entries dated 13 October 1933. Yet, the only public discussion of the experiments was a brief progress report read by Meissner at the *Physikertagung* at Würzburg in the third week of September.

In retrospect, the preliminary and highly puzzling results uncovered in the very first round of measurements must have occupied nearly all of Meissner's attention. Unfortunately his presentation at Würzburg is absent from the published proceedings of the *Physikertagung*.[44] Virtually the only reference to it occurs in the subsequent exchange of letters between Meissner and Peter Debye, the conference organizer and editor. On 2 November Debye reminded Meissner that "a small number of manuscripts from the Würzburger Vorträge have not yet reached us, among them yours. Allow me to inquire whether we can expect to receive something for publication within the framework of your lecture and the discussion at the meeting?"[45] To which Meisner promptly replied:

> Following my return from Würzburg I have, in accordance with the wish of [Reichsanstalt] President Stark, continued the investigation without delay, with the intention to bring out additional material in a forthcoming publication. Concerning the results reported in Würzburg and those obtained subsequently, confirmed results are contained in a short, preliminary communication in the November 3 issue of *Naturwissenschaften.* Only when all the experiments have been brought to a stage of completion will a detailed report be submitted for publication. In light of this, a printed version of my Würzburger lecture by itself appears to me inappropriate, or at least unnecessary.[46]

The letter to *Die Naturwissenschaften*[47] ranks high in historical importance, and is about as sparse in details as Onnes's original paper announcing superconductivity in 1911. It was submitted over Meissner and Ochsenfeld's signatures on 16 October—2 months after the paper by Becker, Heller, and Sauter went to the publisher. In view of its urgency and excessive length (equivalent to two printed columns, or twice the allowance for contributions to *Die Naturwissenschaften*'s *kurze Originalmitteilungen*) the letter was accompanied by a brief explanatory note from Meissner to A. Berliner, the journal editor.

> Due to a trip of President Stark lasting several days, I am only today able to transmit to you a contribution for the section "short original communications" of the *Naturwissenschaften.* I am hopeful that nevertheless it can be published in

November's issue number 3. Should this not prove possible, I would be thankful if you could notify me by telephone.

Since this communication is already somewhat longer than intended, I have omitted all figures. I trust ... that the present length is acceptable.[48]

The experiments, begins the half-page letter, were specifically designed to test the prevailing assumption represented by sequence $f \rightarrow g$ in Figure 9-1—that is, that in cooling a superconductor in a constant external field the conductor remains permeable to the magnetic lines of force. "On the contrary," they reported with no further comment, "our experiments on tin and lead have given the following results":

1. In cooling below the transition point the distribution of lines of force in the region exterior to the supraconductor is altered and becomes essentially as expected for a body of zero permeability, or of diamagnetic susceptibility [per unit volume] $-1/4\pi$.

2. In the interior of a long lead tube, in spite of the change of magnetic field in the exterior as described under 1, in descending below the transition point the previously established magnetic field is essentially undiminished.[49]

Thus, two basic experiments were involved here. The rest of the letter provides only the barest of details. The first experiment included measurements apparently discussed at Würzburg. It utilized two adjacent and parallel cylindrical single crystals of tin. Each cylinder, we are informed, was 3 mm in diameter, 140 mm long, and the two cylinders were spaced 1.5 mm apart (4.5 mm between centers). The magnetic field between the cylinders, as noted earlier, could be measured with a small search coil capable of being flipped 180° about an axis parallel to that of the cylinders and connected to a ballistic galvanometer. With no transport current but with an external uniform field of approximately 5 gauss applied transverse to the cylinders, the ballistic "throw" was observed, first with the cylinders at room temperature, and then after they were cooled below the transition temperature. The ratio of the ballistic deflection below and above the transition point, proportional to the ratio of the flux of the magnetic field under the two conditions, was determined to be 1.70. For a pair of polycrystalline lead cylinders (of unspecified but presumably nominally identical geometry), the corresponding ratio was 1.77. The latter value, the authors noted, was in fact in excellent agreement with the ratio deduced from von Laue and Möglich's aforesaid calculations assuming that the permeability of the superconductor is zero below the transition point. The small deviation in the case of monocrystalline tin was easily accounted for by the uncertain spatial distribution of the winding of the measuring coil and by the not quite cylindrical cross section of the crystals.

Next came an important variation on the basic experiment. The intention had been to leave out the homogeneous external field, connecting instead the two tin cylinders in series electrically. Thus modified, the experiment was repeated: With a small constant current from an external supply flowing (presumably a few amperes), the ballistic deflection was again measured, first at room temperature and then after the cylinders were cooled down. Invariably, a larger ballistic deflection was observed below the transition point, interpreted to signify the expulsion of the magnetic self-field from the conductors at the onset of the superconducting state.

We now know that the latter measurements, which were apparently in an early stage when the letter was submitted to *Naturwissenschaften*, produced a positive result quite fortuitously. It was standard practice to compensate for the Earth's field with a larger Helmholtz coil surrounding the helium dewar. However, this precaution was inadvertently omitted in the very first set of measurements with a weak transport current. What was observed, then, was the fortuitous expulsion of the Earth's field at the onset of the superconducting state.

Less clear-cut was the outcome of the second basic experiment, which utilized a single *hollow* tube of polycrystalline lead. Its outer diameter was 3 mm, the inner diameter was 2 mm, and it was 130 mm long. The measuring coil was small enough that it could be operated inside the cylinder hollow or outside, next to the cylinder. As before, the tube was cooled in a constant field. On reaching T_c the field in the *interior* was observed to increase by approximately 5%, while the external field was again altered by roughly the amount to be expected for zero permeability. On switching off the background field with the sample still superconducting, the field in the interior remained unchanged, but the external field did not completely decay, indicating instead a small residual field of typically 5 to 10% of the original homogeneous field close to the outer surface of the tube.

Pondered Meissner and Ochsenfeld:

> The representation of the results in terms of changes in a microscopically defined permeability may possibly give rise to a difficulty for the behavior in the interior of the lead tube, since no longer can there be said to exist a unique relationship between induction and field strength. Alternatively we can presumably attempt to represent the results in terms of microscopic or macroscopic currents in the supraconductor, under the assumption of permeability equal to unity in current-free regions. These currents appear to change or occur spontaneously at the transition to supraconductivity following the new effect.[50]

"Perhaps," ended the letter, these effects invite "an analogy to ferromagnetism." Indeed, the analogy between the transition temperature and the Curie temperature was something Onnes had speculated on more than once.

Evidently, the diamagnetic behavior uncovered by Meissner and Ochsenfeld was fundamentally different from the diamagnetism of Lippmann for perfect ($R = 0$) conductivity. Indeed, that conception had suggested the von Laue effect, predicated on the cooling *preceding* application of the field—the invariable laboratory practice until 1933. Rather, the new effect is the spontaneous expulsion of an *existing* field, which does not follow from perfect conductivity. It is illustrated schematically in the bottom sequence of Figure 9-6. In cooling a superconductor below the transition point in the presence of an applied field lower than H_c, step $f \rightarrow g$, the previously established interior flux density is suddenly expelled at the transition point. Now state (g) is identical to state (c).

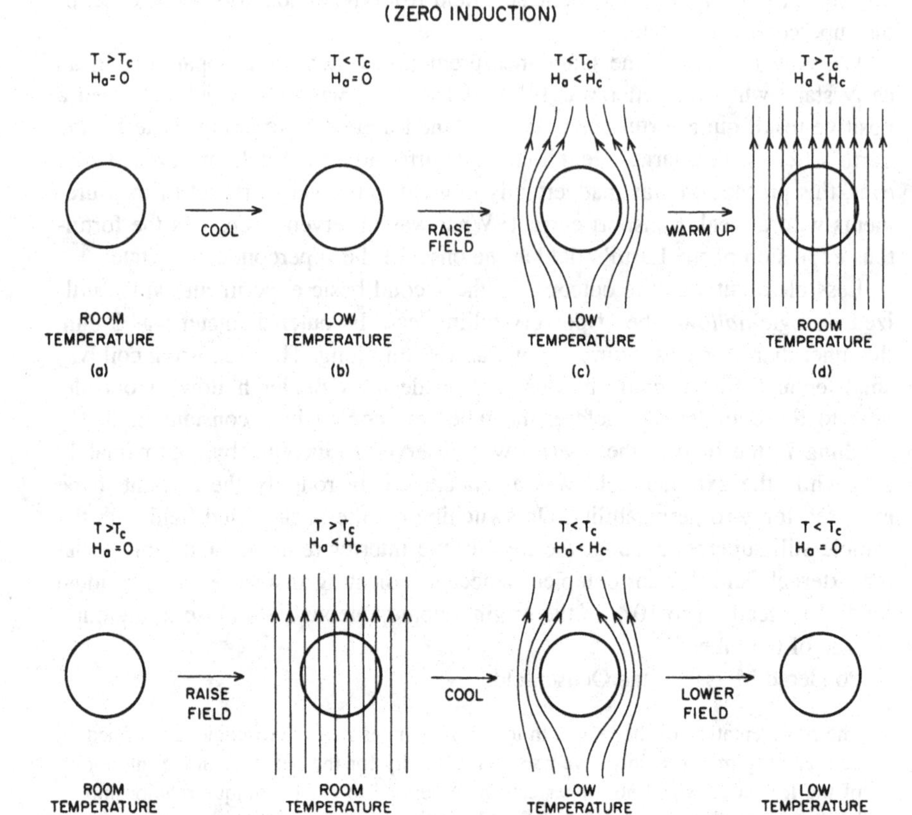

Figure 9-6. Actual magnetic behavior of a superconductor in a magnetic field. Note that top sequence is identical to the behavior expected classically for a perfect (R = 0) conductor. The bottom sequence, cooling in a constant field, represents the Meissner effect. Dahl, HSPBS, **16** *(1986), 37.*

Then, by switching off the external field, no trapped flux and no induced dipole remain. State (*h*) is identical to state (*b*). The state of magnetization depends only on the external conditions of the field and temperature, *not* on the sequence leading to those conditions. To be sure, the new property of superconductors in the presence of a static field, or condition $B = 0$, can be described alternatively by postulating zero permeability with a negative bulk magnetization given by:

$$\mathbf{M} = -\mathbf{H}/4\pi \,, \tag{9-7}$$

or described by persistent currents flowing in or near the surface. However, it is no longer a consequence of infinite conductivity.

In the present round of experiments, Meissner had gone to the trouble of explicitly demonstrating the difference between the classical and the new behavior, of necessity utilizing for this the recalcitrant lead tube (since the tin cylinders would have produced an ambiguous result). The check was vital to ensure the validity of his experimental procedures, in view of the dramatic new results. It showed that cooling the tube and *then* applying the homogeneous field produced zero field strength in the hollow of the tube.[51] This was as expected, because the induced diamagnetic screening currents circulating on the outer surface should generate a flux density that cancels the flux density due to the applied field in the hollow as well as in the interior of the metal. The distribution of field lines in the exterior was also as expected from the conditions of zero permeability prevailing within the body of the lead tube.

The observation of an essentially persistent field within the hollow of the tube after cooling in a field, even after switching off the external field, was less easily reconciled with pure diamagnetism in the hectic weeks of 1933. Particularly troubling was the observation of a small net residual external field corresponding to a finite paramagnetic moment when the field was switched *off*—a matter that was to become the focus of attention at Berlin and the laboratories abroad. What *was* immediately obvious was the vindication of the original suspicion that prompted the experiments of 1933 in the first place. If, indeed, $\mathbf{B} = \mathbf{H}$ and moreover $\mathbf{B} = 0$ inside the body of the superconductor, it follows from curl $\mathbf{B} = 4\pi\mathbf{J}/c$ that curl $\mathbf{B}$ is also zero, and hence the bulk current density $\boldsymbol{J}$ is zero as well.[52] The new effect also ruled out Tuyn and Onnes's interpretation of their old experiment with a lead sphere—that is, the notion of persistent currents rigidly bound to the superconducting metal. The vector relationship Eq. (9-7) could not be maintained when the sphere in the external field $\mathbf{H}$ was turned.

CHAPTER 10

Precarious Affirmation

10.1 Initial Reaction to Meissner's Discovery

The Meissner–Ochsenfeld letter created something of an immediate stir—for its extreme brevity, seeming contradictions, and obvious importance. Among the first to respond was McLennan, now retired but still a keen observer of developments in low-temperature physics. On 12 November he wrote Meissner from his adopted domicile in Surrey, England that

> ... I am very interested is your article [*sic*] in *Die Nat.* of Nov. 3 and would be so glad if you could let me have some further details of the experimental arrangement. You add a few of the results you have obtained. The latter appear to me to be very important, but I confess that owing to the brevity of your communication I cannot make out exactly what you did. Would you please let me know what you mean by "und wird nahezu so, wie es bei der Permeabilität 0, also der diamagnetischen Suszeptibilität-1/4π, des Supraleiters zu erwarten wäre." I shall be grateful for your kind help in making it clearer for me.[1]

Meissner obliged him with a short explanation by return post, assuring him that "it gives me great pleasure that you take such interest in our experiments on the new effects occurring during the onset of supraconductivity."

> A reply to your question is most simply provided by the enclosed sketches. Sketch 1 [no longer extant] covers the case of the parallel cylindrical supraconductors, sketch 2 that of the lead tube. [Figure] a depicts the distribution of lines of force above the transition point, b the corresponding distribution below it. At the onset of supraconductivity the lines of force appear to be almost completely crowded out of the supraconductor, so that the permeability is approximately zero. Less easily interpreted is the experiment with the lead tube, owing to the internal

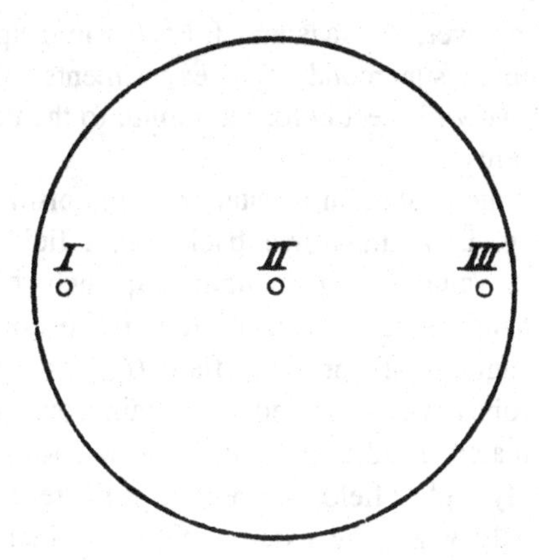

Figure 10-1. Location of bismuth probes in cylindrical tin crystal. De Haas and Casimir-Jonker, PH, **1** *(1934), 291.*

> remaining field. We nevertheless intend to pursue the phenomena which arise in this case with the aid of thicker tin tubes, and as soon as the experiment is brought to some degree of completion I will naturally prepare a detailed publication on all the measurements.[2]

The following week found McLennan, indefatigable lecturer, en route to Utrecht to give yet another talk. He paused in Leiden and paid a visit to de Haas. "De Haas," he reported to Kronig that evening in a letter posted from the Hague, "thinks the magnetic effect observed by Meissner and described recently in *Die Naturwissenschaften* very interesting."[3] Actually, although Meissner beat de Haas and co-workers into print on the new effect, a fresh set of measurements as revealing as those by de Haas and J. M. Jonker[4] was already completed when the Meissner–Ochsenfeld letter appeared. De Haas's experiment was simple in concept, unambiguous, and timely, involving direct measurements of the field penetration in a superconducting rod with the aid of bismuth probes. Unfortunately, de Haas disdained advance publication of the results by the expedience of a journal letter. The paper appeared eventually in *Physica's* February 1934 issue, along with the companion paper by de Haas, Voogd, and Jonker.[5]

The conductor utilized for this purpose was a single cylindrical rod of monocrystalline tin. It was much thicker than the one used earlier in the Leiden investigations, measuring 7 mm in diameter. It was pierced by three fine holes lined with glass capillaries. One was located on the axis of the cylinder (II in Figure 10-1). The two others were parallel to it, on the diameter transverse to the applied field, spaced about 1 mm inside the cylinder surface. Mounted in the capillaries were delicate bismuth wires, approximately 1 cm long and supplied with current and potential leads of copper. The strong magnetoresistance effect in bismuth, especially at low temperatures, was well known; the use of bismuth probes for determining small field changes was equally well known, if hardly standard laboratory practice. (Kohlrausch treats them in the 1901 edition of his famous text).[6] However, de Haas and J. M. Casimir appear to have been the first to utilize them in superconducting experiments. They had a visitor from the USSR, whom we will meet before too long, to thank for implementing these elegant measurements.

They first monitored the change in resistance of the bismuth wires, with the tin rod superconducting, in a transverse background field as the field was slowly increased past a value strong enough to quench the superconducting state. Note that the change in resistance of the tin specimen itself was not measured, since the (longitudinal) threshold field H_c was well known. The resistance in all three probes was observed to remain constant up to a certain field strength (approximately $H_c/2$); above it the resistance of probes I and III rose rapidly. At a slightly higher field, the central probe reacted as well, and its resistance also rose. Only when the external field reached the value H_c (36 gauss in this particular run) did the curves of probe resistance versus field coalesce with a curve exhibiting the normal H^2 dependence expected for bismuth exposed to the uniform background field. De Haas and J. M. Casimir took the results to mean that at a definite external field the superconducting state was first destroyed in the peripheral region (I, III) of the interior (when the initially induced persistent screening currents disappeared, allowing entry of the field); subsequently the field gradually penetrated to the center of the rod.

In an increasing field directed parallel to the plane of the bismuth wires, on the other hand, the initial penetration took place almost simultaneously in all three capillaries, as expected.

Then comes the pièce de résistance of the experiment. For reasons of his own, de Haas was very tight-lipped in print about the final measurement. Even more reticent than Meissner, he offered no actual data whatsoever.

> Finally [ends the paper] we have performed a measurement in which the tin rod was cooled in a constant field. It appears that the field in the neighborhood of the outer Bi-wires vanishes, whereas near the central wire an increase of the field was found to occur at the transition point. Qualitatively, these results are in agreement with the measurements of Meissner and Ochsenfeld. ... These observations may indicate that the expulsion of the magnetic field from a supraconductor is a complicated phenomenon; however, in general the measurements are still somewhat uncertain. We intend to continue our investigations with greater accuracy.[7]

The extreme caution against premature conclusions exercised by de Haas did not extend to two of his younger fellow scientists who were privy to the fresh results before publication. One was Cornelius Jacobus Gorter. Gorter had obtained his doctorate under de Haas in 1931; subsequently he was appointed to a staff position in the Natuurkundig Laboratorium at the venerable Teyler's Foundation in Haarlem (H. A. Lorentz's institution from 1911 until his death in 1928). While still at Leiden his interest was aroused in the thermodynamics of the superconducting transition, and he decided to go over the application of Ehrenfest's phase transition in the derivation of Rutgers's equation (Eq. 8-2) in greater detail. Consider the cycle *ABCDA* in the *H–T* plane (Figure 10-2): cooling in zero field, followed by an isothermal transition to the normal state in a

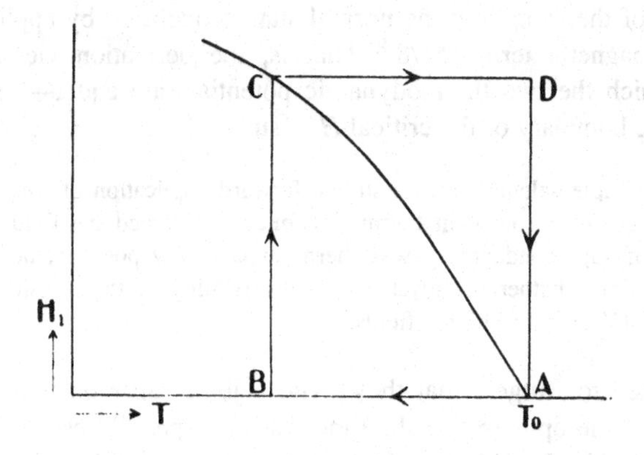

Figure 10-2. Gorter's thermodynamic cycle.

field exceeding the critical value, and finally warming the body above $T = T_c$. Applying the first law of thermodynamics for a body in a magnetic field, $dQ = dE - Hd\sigma = cdT$, and noting that, according to Keesom and Kok, the heat of transformation vanishes in zero field, Gorter wrote

$$-\int_{T_2}^{T_0} \Delta C\, dT + Q_2 = -\int_0^{\sigma_2} H_1\, d\sigma + H_2\sigma_2 .$$

Here $\Delta C = C_s - C_n$ is the difference between the specific heats in the superconducting and normal states, the index 2 indicates the transition point at C, and Q_2 and σ_2 are the heats of transformation and magnetization, respectively at C. For a long cylinder parallel to the field, $\sigma = -H_1 V/4\pi$ per gram-mole of material (neglecting the susceptibility in the normal state). Therefore,

$$Q_2 = \int_{T_2}^{T_0} \Delta C\, dT - \frac{H_2^2 V}{8\pi} ,$$

where $H_2 \equiv H_c$ is the (longitudinal) critical field at point C. Now Gorter made the ad hoc assumption that the second law of thermodynamics is valid for the transition at C, in effect postulating zero induction in the superconducting state. If so, $\int dQ/T = 0$ and consequently

$$\int_{T_2}^{T_0} \Delta C\, dT - T_2 \int_{T_2}^{T_0} \frac{\Delta C}{T}\, dT = \frac{H_2^2 V}{8\pi} . \qquad (10\text{–}1)$$

Double differentiation of Eq. (10-1) yields Rutgers's equation for zero field at $T = T_c$. This important expression, Gorter added, can also be derived in terms of thermodynamic potentials $U - TS + pV - \int \sigma\, dH$, since it results from equating the difference between the Gibbs free energy of a superconductor in zero field and that of the sample in the normal state brought on by application of a field to the magnetic term $H_c^2/8$.[8] That is, the derivation yields the field strength at which the two thermodynamic potentials are equal—the condition that defines the boundary of the critical H–T surface.

> Making this simple calculation as a straightforward application of Keesom's excellent undergraduate course in thermodynamics, I obtained the field at which disturbance of superconductivity was thermodynamically possible, not entering into a discussion whether the mysterious irreversibility of the transition would considerably shift it to still higher fields.[9]

Gorter suggested to Rutgers that they write a joint paper on the subject, but Rutgers "was of the opinion that the time was not ripe."[10] Less patient, Gorter went ahead and published his calculations in the archives of the Teyler's Foundation as soon as he arrived in Haarlem.[11]

Actually, the "mysterious irreversibility" still nagged the thermodynamic derivation. Nor could the convincing experimental verification of Rutgers's equation (making use of Keesom's specific heat data), which seemed to justify the assumption of reversibility in the derivation, get around the embarrassing fact (Section 8.2) that the decay of the persistent currents ought to be an irreversible phenomenon associated with Joule heating. Not only that, but all of the experimental data pointed to the fact that, except for very long, thin cylindrical or flat bodies ('needles' or 'razor blades') parallel to the field, the resistance is gradually restored when the field (or temperature) is raised. Some parts of the body may then be in the normal state, while other parts remain superconducting.

The letter by Meissner and Ochsenfeld appeared just as Gorter was pondering these questions. Gorter pounced on the news. Here was obviously the missing element—the elusive justification for reversibility needed to reconcile the inconsistency between the thermodynamic and magnetic depictions of the transition. Gorter lost no time. In a letter to *Nature*, dated 22 November 1933, he boldly suggested that the condition $B = 0$ is a *general* characteristic for superconductors.

> This last assumption throws new light upon my results. ... It appears that the condition $B = 0$, made in the thermodynamical treatment, does not cause loss of generality, since supraconductive states with $B \neq 0$ do not exist. If hysteresis may be neglected for a moment, the transition to the supraconductive state and back again is literally reversible (in spite of the 'persisting currents') so that it is obvious that $dQ/dT = dS$.[12]

Two aspects of the problem remained troubling, however. One was the nagging hysteresis behavior, both magnetic and thermal. Gorter inclined to attribute it, with Meissner, to the sudden redistribution of the magnetic field at the transition. The transition is presumably accompanied by eddy currents in other nearby conductors and in nonsuperconducting parts of the body itself. These currents represent energy that is lost when the currents decay, that is, hysteresis loss. "It may be expected that the transition *to* the supraconductive state especially will be retarded by the necessity of starting the eddy currents."[13] The other perplexing observation was that the condition $B = 0$ appeared not to apply inside the hollow of Meissner's lead tube. "It seems highly probable," Gorter added, "that parts of the tube were not supraconductive; this allowed the line of induction to pass." Even so, this point was sufficiently troubling to warrant a letter sent posthaste to Meissner on 27 November:

> Permit me to send you a carbon of a note which I submitted to *Nature* a few days ago, mainly concerning your latest measurements. If we assume $B = 0$ for the supraconducting state then, in my opinion, a satisfactory overall picture emerges. However, your finding that the field within a lead cylinder remains unchanged when the field is switched off is problematic. To my knowledge, neither is the finding in accord with the Leiden observations. ... [14]

As a postscript, he added in a somewhat smug note, the condition $B = 0$ came, in fact, as no surprise, in light of his own and Casimir's thermodynamic analysis. Meissner replied without delay.

> Unfortunately my further experimental investigation of the alteration of the magnetic field at the onset of supraconductivity has met with delay: My colleague in this, Dr. Ochsenfeld, found it necessary to suddenly abandon our collaboration when he was called away to a teaching position. Moreover, the preparation of new single crystals of larger dimensions and new experimental apparatus has also suffered delays. In lieu of new precision measurements ... I am, unfortunately, unable to comment on your explanation and views. In particular, the field in the interior of a supraconducting tube must await more accurate measurements. I can only add that the lead tube was almost completely immersed in liquid helium, so for the time being I consider it quite unlikely that parts of it were non-supraconducting. However, the difference in behavior of single crystal and polycrystal materials still remains to be investigated in greater detail. For now I must question the general assumption $B = 0$; perhaps conditions similar to those of ferromagnetism are involved here, where again no clear connection exists between field strength and induction. In this respect, your explanation of the hysteresis phenomena strikes me as somewhat implausible.[15]

Similar views are voiced in Meissner's subsequent, protracted correspondence with Werner Braunbek of the Technische Hochschule in Stuttgart. The two had engaged in a long discussion at the Würzburg *Physikertagung*. Braunbek had, independently of Becker, Heller, and Sauter and without their advantage of being close to the superconducting scene in Berlin, examined the electromagnetic behavior of superconductors in the classical limit of constant permeability. Taking into account the inertia of the electrons, as did they, he developed a model of electromagnetic waves propagating within a certain penetration depth in the conductor. (He also suggested an experiment, based on these ideas, for determining the density of superconducting electrons.) The Meissner–Ochsenfeld letter forced him to reexamine the model just as his journal paper on the subject[16] was in the proof stage. It also prompted the ongoing exchange with Meissner, and a rather fruitful one at that judging by the attention Meissner devoted to it at this important juncture in his affairs. The exchange focused on the question of zero versus constant permeability vis-à-vis the operational sequence of cooling and introducing the field. Meissner had no problem with the notion of a penetration depth. However, he also drew Braunbek's attention to Gorter's thermodynamic treatment, newly transmitted from Haarlem, while cautioning against uncritical acceptance of Gorter's sweeping conclusion with respect to the new effect.[17]

Shortly before news of the Meissner–Ochsenfeld results broke, H. G. B. Casimir was called back from Zürich to Leiden by Ehrenfest, his former teacher. Casimir had obtained his doctorate under Ehrenfest in 1931 and had just spent a year abroad as Pauli's assistant. It seems likely that the recall was forced by Ehrenfest's anticipation of his own tragic demise as a result of self-doubt and various personal problems. He committed suicide late in September, the very week of Casimir's return.[18] Be that as it may, H. G. B. Casimir's interest had been awakened in low-temperature physics in this period. To begin with he assisted his wife with her measurements in the cryogenic laboratory, "doing simple auxiliary measurements and providing food and coffee."[19] He also joined Gorter in a further consolidation of the phenomenological treatment of superconductivity based on thermodynamics and with the proviso of zero magnetic induction inside a superconducting body. Their joint paper appeared in *Physica* early in 1934,[20] accompanying quite fittingly the two seminal papers by de Haas and co-workers.

This paper contains an excellent précis of the central issues in superconductivity perceived by the principal players, as the eventful year of 1933 drew to a close. It begins with a brief survey of the four recent landmark experiments: by Keesom, van den Ende, and Kok; by de Haas, Voogd, and Jonker; by Meissner and Ochsenfeld; and finally by de Haas and J. M. Casimir. The picture emerging from all of this could be most simply understood in the context of hypothetically cycling a body of simple shape, either initially cooled in zero field and then exposed to a field or cooled in an existing external field. In either case, assuming the body has a shape other than a *ring*, the accumulated observational evidence invariably point to the condition $B = 0$—irrespective of whether the condition is attributed to persistent currents or a negative susceptibility. (If, however, a superconducting *ring* is involved, they cautioned, then in the first transition sequence the additional condition must be stipulated that the total flux threading the ring is zero; in the second sequence the total flux threading the ring is constant but not necessarily zero.) The authors first repeat Gorter's thermodynamic derivation, including that of Rutgers's equation, for the simplest case of a long needle in a parallel field. This treatment is extended to an arbitrary body or systems of bodies in an external field, emphasizing as always the importance of the magnetic energy from the vantage of thermodynamic potentials. In general a continuous transition must be expected. The disturbance of superconductivity in a cylinder by a current is treated next, confirming and "defining more precisely" Silsbee's hypothesis. (The distinction between the longitudinal and transverse disturbances was unknown when Silsbee formulated his rule.) Finally, the authors return to the two principal unsettled questions:

> An apparent difficulty is furnished by Meissner and Ochsenfeld's result, which is confirmed by de Haas and Mrs. Casimir's observation, that sometimes far below the transition temperature and even in the absence of an external field, regions in the body may exist, where certainly $B = 0$. It seems to us, that this may be ascribed to supraconductive rings inside the body, as certainly the conditions in these experiments favoured the formation of such rings. Inside the rings the field due to the current along the rings may prevent the transition to the supraconductive state. Probably also Kamerlingh Onnes and Tuyn's observations on persisting currents in a sphere may be ascribed to such rings.
>
> A second problem, which is not yet completely solved, is the mechanism of the gradual disturbance of supraconductivity (e.g., in a transversal magnetic field). One could imagine, that, as soon as the tangential magnetic field will surpass the critical value, supraconductivity will retire from the surface to deeper regions of the body. This picture, however, is not satisfactory, for then, in general, the field at the surface of the body would decrease, so that, at least if the transitions are really reversible, at the surface new supraconductive regions would be formed. So it seems better to imagine, that part of the supraconductor will be perforated or reduced to pieces, rather than to suppose that there will exist a sharp retiring boundary between the two phases. The mean magnetic field in larger regions will then have some value between zero and the maximum value, in agreement with de Haas and Mrs. Casimir's results. In practice the condition $B = 0$ may thus perhaps lose its rigour in the neighbourhood of the transition line.[21]

The various experimental groups abroad were not far behind the Dutch in reacting to Meissner and Ochsenfeld; 1934 brought forth a veritable rash of communications on the subject. The first of many letters to *Nature* was submitted from Oxford by Kurt Mendelssohn in February,[22] coinciding with the publication of de Haas's extensive measurements. Immediately after the appearance of Meissner's letter, Mendelssohn, with the Canadian Rhodes Scholar J. D. Babbitt, had set out to repeat his observations using solid and hollow tin spheres. The Simon liquefier brought to Oxford the previous January was ready for the task. It had been made operational in short order with the help of Babbitt—and even of Lindemann himself. Mendelssohn relates a memorable occasion when, late at night, Lindemann doffed his dinner jacket and threw himself into heating soldering irons. "Only later did I realize that I had witnessed a unique sight."[23]

In these experiments the tin sphere under test, 1.5 cm in radius, was cooled by conduction through a copper rod from a liquid helium reservoir. The background field was produced by a solenoid enclosing the dewar vessel; any residual induced dipole moment could be measured by a search coil connected to a ballistic galvanometer.

The experiments neither disproved the classical theory nor confirmed the Meissner effect with any certitude. On cooling the solid sphere in a field of about 70 gauss and then switching off the field, the sphere retained a magnetic moment of approximately one-sixth of the classically expected frozen-in value. (The expected moment could be most conveniently calculated by an old procedure derived by Lorentz and long regarded as prima facie support for the classical explanation of the Onnes–Tuyn experiment.)[24] The moment for a hollow sphere was considerably stronger—two to three times greater than for the solid sphere.

> As a result of these experiments [Mendelssohn and Babbitt concluded], it seems certain that the effective permeability of substances when they become supraconducting decreases, as observed by Meissner and Ochsenfeld. On the other hand, it appears clear that under our experimental conditions the permeability does not vanish entirely, as might be expected in view of the almost infinite conductivity, or if it does vanish, it only does so in certain regions and not throughout the whole volume of the supraconductor.[25]

In early April the Toronto group, too, reported similar cautious conclusions in a letter to *Nature* under the principal authorship of E. F. Burton, who had succeeded McLennan as head of the newly renamed McLennan Laboratory in 1932. Their experimental arrangement, sketched hastily for the preliminary announcement that was rushed to *Nature*,[26] is depicted in Figure 10-3. The superconductor was a hollow tin cylinder, and small search coils, five in all, were wound around the tin in various ways or placed near it, as shown in the sketch. The coils were connected to a flux meter, and the deflections read by means of a lamp and a scale.

From Meissner's results on cooling in a field, Burton's team expected to observe various search coil responses, depending on coil placement: a decrease, increase, or no change in flux. In actuality, they observed a range of responses, considerably less than expected in principle but in the right direction. Moreover, because "coil no. 3 projected about 5 mm from the surface where there was undoubtedly a magnetic field of high gradient, and ... coil no. 5 of necessity enclosed a considerable space where the field was not theoretically zero, but only relatively weakened," only "fair" agreement with the results of Meissner was to be expected.[27] The reason is that search coils used by either party have the drawback compared to, say, bismuth probes, of giving only an average field over the area of the coil winding.

April of 1934 was also the month in which Keesom and Kok's first paper on the latent heat of thallium in a magnetic field finally appeared in *Physica*.[28] By the time this paper went to press the experimental results of both Meissner and Ochsenfeld and of de Haas and J. M. Casimir were available, and Keesom agreed heartily "that one must be cautious as to the appreciation of the value of the magnetic field in the neighbourhood of a supraconductor."[29] The basic measurements on thallium had actually been performed as early as November and December of 1932. The inordinate delay in publication of the results had, as noted earlier, been caused by the need to calibrate in a magnetic field the phosphor bronze wire that served as the thermometer. The calibration itself was apparently not a simple matter, but was duly performed. Although the calibration enabled a reliable estimate of the temperatures under the same conditions that prevailed during the original calorimetric experiments, it unexpectedly threw light in another direction. Performed with the calorimeter block submerged in liquid helium, the calibration involved measuring the resistance of the phosphor bronze wire, mounted in the interior of the calorimeter core, as a

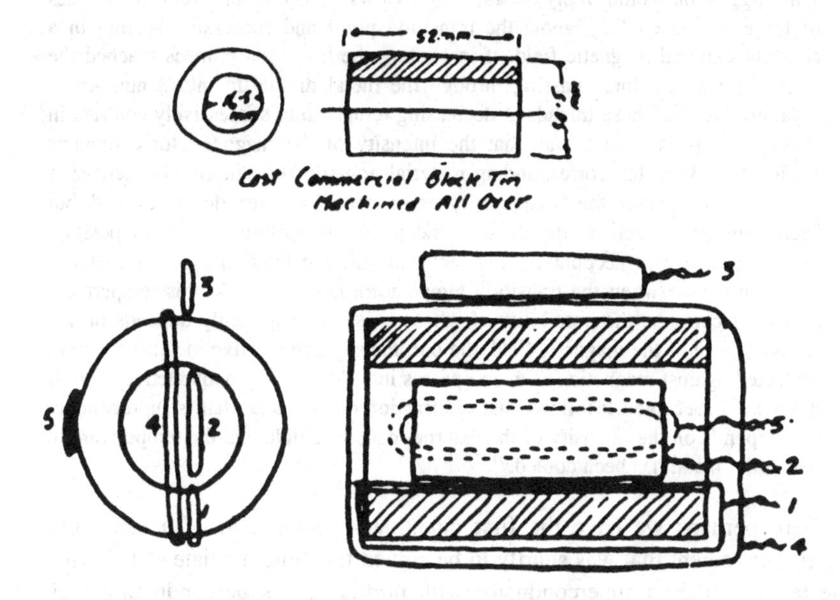

Figure 10-3. Experimental arrangements at Toronto for verifying Meissner's results, sketched hastily for a letter to Nature. Burton, Wilhelm, and Tarr, NA, **133** *(1934), 684.*

function of bath temperature (derived from the vapor pressure of the liquid) in rising temperatures from below to above T_c (for thallium), with and without a constant field. Each series of measurements yielded a continuous curve with no hint of a discontinuity at T_c. Keesom and Kok interpreted this to mean that the field in the interior did not change appreciably when the block passed into the resistive state.

> This result corresponds with that obtained by de Haas and Mrs. Casimir in one of their experiments. If we combine it with the results obtained by Meissner and Ochsenfeld, according to which the lines of magnetic force are pushed out from a metal when it becomes supraconductive, we are lead [*sic*] to the view that in our block when its temperature is decreased till below the transition temperature to the supraconductive state the lines of magnetic force are continuously pressed together, so as to prevent a certain part of the metal surrounding the geometrical axis of the block to remain non-supraconductive.[30]

Furthermore, the latent heat deduced from the area under the hump in the curve of atomic heats versus temperature (Figure 8-2) appeared to be somewhat smaller than the value calculated from measured specific heats in a constant field assuming a particular magnetization cycle (the one first treated by Gorter in his paper in the *Teyler Archives*). Mindful also of the phosphor bronze results, it seemed quite likely to Keesom that only part of the thallium block had become superconducting. How large a fraction was not obvious, but Keesom guessed from the specific heat discrepancy that it was in the range 90 to 95%.

> This suggestion would imply the following views as to the behaviour of the lines of force during cooling below the transition point and successive heating in a constant external magnetic field. If in cooling the transition point is reached the lines of magnetic force passing through the metal divide up into a number of separate threads. These threads at decreasing temperature successively contract in cross sections in such a way that the intensity of the magnetic force remains sufficient to keep the corresponding material non-supraconductive. In increasing again the temperature the threads or tubes of force, however, do not expand, but keep their cross sections they had at the lowest temperature. This supposition seems necessary to account for the fact that still the fraction z of the metal is supraconductive when the transition temperature is reached. At this temperature persistent currents, the possibility of whose existence apparently depends on the "gross" value of the magnetic field, die out, the supraconductive metal is no more protected against magnetic force, and passes into the non-supraconductive state. If this view is accepted it follows that the fraction of the metal that is supraconductive depends on the intensity of the external magnetic field and the temperature to which the metal has been cooled.[31]

Here perhaps we have the first quantitative discussion of a catch-all but fruitful expedient that was shortly to be coined the "intermediate state." That is, the term denotes a superconductor with normal and superconducting regions coexisting in parallel while the specimen undergoes a gradual transition (over a finite field range if at constant temperature or over a finite temperature range if at constant field). It was, by now, evident to all that the problem is exasperated in specimens of irregular shape (any shape other than long cylinders or ellipsoids, such as Keesom's or Meissner's hollow tube).

One more laboratory remained to be heard from: the fledgling Ukrainian Physico-Technical Institute in Kharkov, which was then the capital of the Ukraine. This was literally a brand new institution, born in the Soviet revitalization after World War I. Prior to the war physics in Russia had languished: "the 'scientific work' of the research students of the Leningrad University consisted in reproducing the experiments published in the last issue of a foreign review."[32] After the revolution, things picked up. "Physics was one of the first [areas] to feel the vivifying influence of the Proletarian Revolution."[33] A host of new physical and technical institutes sprouted up in Leningrad and other metropolitan areas; the first opened in 1918. Beginning in 1929, new physical centers arose in the principal republics and provinces of the USSR as well. Their leading cadres were supplied by the Leningrad Physico-Technical Institute. Each became an All-Union scientific center in its own field. One of these was the Institute in Kharkov, specializing in low temperatures and high electrical tensions.

The new cryogenic laboratory in Kharkov had been founded by the director of the Ukrainian Institute, I. V. Obreimow. Formerly of the Polytechnical Institute in Leningrad and an authority on crystal physics, Obreimow soon passed the directorship of the cryogenic laboratory on to his former student, Lev Vasil'evich Schubnikow—an ultimately tragic personage in the unfolding events of superconductivity. Schubnikow, with his teacher, had received early recognition for a method of growing single crystals from a molten metal still in use today. It has become known as the Obreimow–Schubnikow method, though it was developed independently by Bridgman. In it, molten metal is poured into a test tube terminating in a fine capillary and placed in a furnace somewhat above the melting point of the metal. A single crystal of metal is grown from a seed crystal in the capillary by slowly cooling it from one end, and the glass is dissolved in hydrofluoric acid.[34]

Schubnikow's subsequent collaboration with Obreimow had centered on the plastic deformation of crystals. Upon his graduation from the Polytechnic, the Narkompros (National Educational Commissariat) sent him to Leiden where he served a 4-year stipend under de Haas in the company of his wife and physicist Olga Trapeznikova. There, Schubnikow's experience was put to good use with de Haas's suggestion that he tackle the problem of producing perfect single crystals of bismuth for the ongoing low-temperature resistivity measurements in external magnetic fields, and in particular for pursuing the magnetoresistance of bismuth itself. Bismuth was one of the first metals to be obtained in large crystals, and was considered one of the substances to crystallize most easily. However, at the outset neither the Obreimow–Schubnikow method in its original form, nor other methods of growing single crystals, yielded crystals of the requisite purity demanded at Leiden, due to the inability of the methods to avoid built-in mechanical stresses. Schubnikow finally met with success by significantly modifying a method devised by Kapitza in his own efforts on the magnetoresistance of bismuth (in fields up to 30 tesla); Kapitza's starting point was, in fact, a variation on the original Obreimow–Schubnikow procedure.[35] By using the purest Kahlbaum or Hartman–Braun bismuth samples and repeatedly recrystallizing them, Schubnikow was able to grow exceedingly pure monocrystalline rods of predetermined crystallographic orientation.[36] With these, he and de Haas, assisted by, among others, Trapeznikova, undertook an important series of low-temperature magnetoresistance studies on bismuth; an upshot of this investigation was their discovery of the so-called Schubnikow–de Haas effect: the periodic oscillations in the resistivity of bismuth as a function of the magnetic field.[37]

On completion of his Leiden stipend in mid-1930, Schubnikow, at the suggestion of A. F. Ioffe, rejoined his former mentor Obreimow in the newly established cryogenic laboratory at Kharkov. In 1931 Schubnikow became scientific director of the laboratory. With great zeal he undertook to import and implement the requisite cryogenic technology, thereby creating a fine cryogenic laboratory in Kharkov. Liquid helium became available there in 1933, with the installation of a Simon liquefier. The timing could not have been better. A program on magnetization studies of superconductors—the hot topic of the moment—was soon under way under Schubnikow's supervision.

J. N. Rjabinin and Schubnikow reported perhaps the clearest confirmation of the Meissner effect in a letter submitted to *Nature* on 3 July.[38] Their preliminary results were actually submitted in April to the 2-year-old *Physikalische Zeitschrift der Sowjetunion* (published in Kharkov), but these findings hardly amounted to more than a "general survey" of the problem.[39] Their sample was a rod of polycrystalline lead, 5 mm in diameter and 50 mm long, mounted

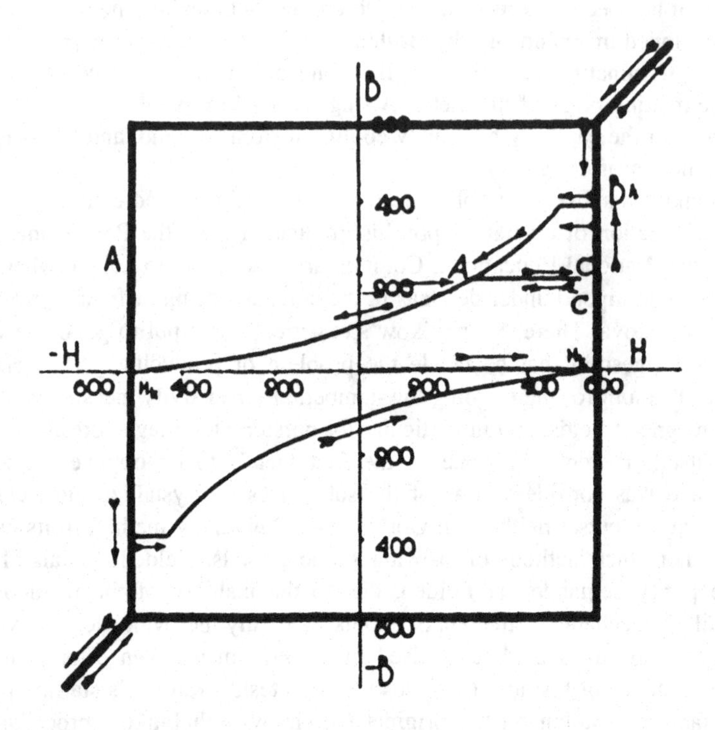

Figure 10-4. Magnetization curve, or induction versus external field, for polycrystalline lead cylinder (Rjabinin and Schubnikow). Heavy lines show classically expected magnetization behavior. Rjabinin and Schubnikow, NA, **134** *(1934), 286.*

coaxially within a long liquid nitrogen-cooled solenoid with compensated ends. All measurements were performed in boiling liquid helium at atmospheric pressure. The magnetic moment of the sample was measured in a constant field by observing the throw of the ballistic galvanometer, connected over an amplifier to the surrounding coil, when the sample was suddenly introduced to or removed from the coil.

The magnetization curve obtained, as shown in Figure 10-4, is the undisputed prototype of all subsequent curves of this genre, from those of Mendelssohn onward. In Figure 10-4, the classically expected magnetization behavior is represented by the heavy lines, providing a succinct description of the behavior depicted earlier in Figure 9-1, but from a new perspective. Recapitulating that behavior, let us assume first that the specimen, presumed already to be superconducting, is exposed to a rising field. Eddy currents are induced, preventing the field from entering (curve from $H = 0$, $B = 0$ to $H = H_k$, $B = 0$, where $H_k = H_c$). At $H = H_k$ the eddy currents are quenched, and $B = H_k$, since $\mu = 1$. Higher fields lead to a corresponding increase along the indicated diagonal. Upon decreasing the field from above H_k, the diagonal is traced backwards to $H = H_k$. Below H_k the induction remains constant until $H = -H_k$, where the induction drops abruptly to the lower diagonal. Increasing the field again leads to symmetric traversal along the diagonal and rectangle as shown, producing the expected hysteresis and frozen-in induction $B = +H_k$.

The actual behavior found by Rjabinin and Schubnikow is shown by the lighter curves. The initial process, when the sample is first magnetized, is as before: $H < H_k$, $B = 0$, and at $H = H_k$, B leaps to its value at the start of the upper diagonal. For higher fields, H and B behave as in an ordinary metal ($\mu \approx 1$). However, upon decreasing the field, the behavior is at first completely reversible. A further decrease in H causes a rapid fall in B, but only to a value B_h (a value completely nonreproducible in different experiments with the same sample). Proceeding along light curve A, at $H = 0$ a small residual induction is retained (≈18% of the maximum value of B at the moment of transition), which vanishes only at $H = -H_k$. Further changes in the field lead to a completely symmetrical curve.

The experiment thus confirmed that in the vicinity of H_k there occurs a sudden change in B with increasing field (expected classically) as well as in decreasing field (not expected). "The ... fact that a jump takes place in the induction in falling field strength we are inclined to ascribe to the formation of a new phase with $B = 0$. ... At $H = 0$, the persistent currents give rise to a residuary magnetic moment in a supraconductor. Mendelssohn and Babbitt, apart from ourselves, observed this phenomenon in a sphere of tin . ..."[40]

10.2 The Reichsanstalt Experiments

While the various teams abroad were absorbed in corroborating the tantalizing results of Meissner and Ochsenfeld, the measurements at the Reichsanstalt had barely begun. For a while, however, Meissner's program appeared to be in serious jeopardy with Ochsenfeld's sudden and unexpected departure from the laboratory in November for a promising teaching position at Potsdam. Not only that, but an academic position opened up at the Technische Hochschule in Munich, available for occupancy in the spring of 1934—a position Meissner could ill afford to turn down. (One reason for his leaving the Reichsanstalt just then was an ongoing disagreement with Johannes Stark, ardent Nazi and president of the Reichsanstalt since Hitler's ascendancy the previous year.) With a search feverishly under way for a successor to Ochsenfeld under the sponsorship of the Notgemeinschaft,[41] a cumbersome tentative arrangement was worked out that allowed Meissner's experiments to continue at the PTR. They would be carried out by a new assistant while Meissner himself settled in Munich. The new man, F. Heidenreich, came on board in February of 1934, and much of the remaining time before Meissner's departure was given to laying out a program of systematic measurements for him to undertake.

These were trying times at the PTR, with administrative upheavals going on as in everywhere else. Simon's guarded tone in his congratulatory letter from Oxford in January says much.

> I read that you have been appointed successor to [Hermann] Knoblauch; my heartfelt congratulations! Have you already relocated? You will of course have read that the Breslau affair developed differently. I venture, however, that Munich is better than Breslau. ... Your investigation of the magnetic lines of force on cooling below the transition point has excited the greatest of interest everywhere here; it is surely a beautiful piece of work. What a pity that your low-temperature program must come to a halt when you move to Munich, or will you be able to carry on?[42]

The red tape accompanying the new appointment was finally overcome, although prospects for subsequent laboratory facilities in Munich remained cloudy, as we gather from Walther Gerlach's letter written in early March.

> I am especially pleased that this stupid game has come to an end and your appointment [settled]. You may rest assured that all Munich physicists bid you a cordial welcome ... Concerning plans for the cryogenic laboratory, I have spoken with the Department of Buildings, although the matter has not yet progressed much further. In the first place, formalities stand in the way, namely an obscurely constituted cooperation between T. H. and the University. In the second place, the question has not been resolved to what extent resources for purely scientific purposes will be available ...[43]

In the event, the interim arrangement saved the day, at least for the time being. The new measurements at the Reichsanstalt seemed to be off to a good start under Heidenreich. Meissner presented a progress report in a lecture before the general meeting of the Deutschen Kälte-Vereins (German Cold-Association) in Berlin on 25 May,[44] and again amidst the lively discussions on superconductivity at the tenth Deutschen Physikertag at Bad Pyrmont in September.[45] These reports provide both the details lacking in the Meissner–Ochsenfeld letter, and a good account of the new series of measurements with Heidenreich; however, the chronological sequence of the measurements conveyed by them—especially in regard to the earliest measurements—has remained obscure until recently—and was only clarified with the release of the Meissner papers, among them his laboratory notebooks. The most extensive chronicle of the full investigation is found in Meissner and Heidenreich's final report submitted to the *Physikalische Zeitschrift* in mid-1936.[46] The account that follows is primarily a recapitulation of Meissner's lecture in May of 1934, but it relies on the 1936 report for filling in various details.

As Meissner emphasizes time and again, the measurements were originally prompted by his attempt to confirm the "superficiality" of the current distribution in the superconducting state. The first evidence for a change in the current distribution during the transition had come from the tell-tale hysteresis effects in Leiden's resistive transition curves at constant field. For polycrystalline tin with commutated currents, the hysteresis was relatively minor. The effect tended to be more noticeable without commutating the current and without compensating for the Earth's field. For a single tin crystal without commutation or compensation the effect was larger yet. This tendency made sense to Meissner only if the current underwent a sharp redistribution from a bulk to

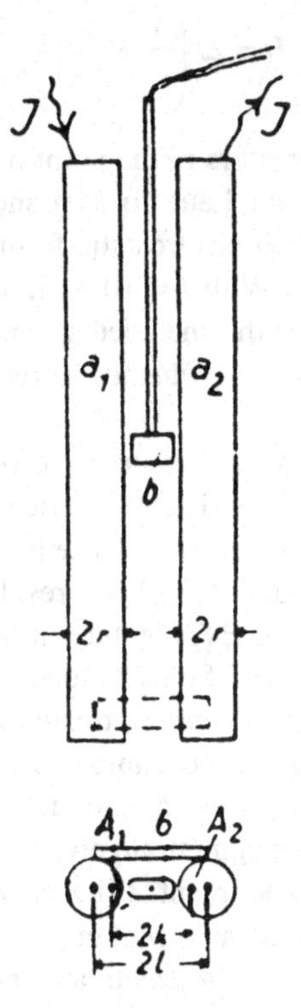

Figure 10-5. Meissner and Ochsenfeld's first experiment. The search coil b measures the magnetic field between two parallel monocrystalline tin cylinders connected in series and supplied with a current. J. Meissner and Heidenreich, PZ, **37** *(1936), 451.*

essentially a surface current during the transition to the superconducting state.

To settle the question, Meissner and Ochsenfeld resorted to the clever procedure—apparently conceived by Laue—of mapping the field distribution between a *pair* of parallel monocrystalline tin cylinders. A *single* cylinder would not have sufficed, because in that case the external field distribution, by the symmetry, is independent of whether the current is uniformly distributed over the cylinder cross section or confined to the surface.

The experimental arrangement is depicted by Meissner in Figure 10-5.[47] The two parallel cylinders, a_1 and a_2, were 3 mm in diameter, 150 mm long, and spaced approximately 5 mm between centers.[48] They could be connected in series and supplied with a current *J*. The magnetic field, whether self-field or applied externally, was measured by flipping the search coil *b*. The coil was10 mm long, 1.5 mm wide, and 1 mm thick. The current was purposely limited to about 5 A, to stay below the critical current in liquid helium. This made it necessary to employ a highly sensitive armored Du Bois–Rubins galvanometer.

For a uniform current distribution in wires of circular cross section, the field strength in the plane of the two cylinders may be expressed formally by:

$$H = 2J\left(\frac{1}{r_1} + \frac{1}{r_2}\right)$$

where r_1 and r_2 are the radius vectors to the point of observation. This equation is the subject of a postcard from Laue to Meissner dated mid-April 1933,[49] where Laue gives it in the explicit electrostatic form $H = 2I\,(1/r_1 + 1/r_2)/c$. ($I = J$ and c is the velocity of light.) With $I = 10$ A, it assured Laue that ≈10 gauss should be readily obtainable for the intended geometry (r_1 and r_2 some millimeters each); even so, he questioned whether larger dimensions might be wise, permitting stronger currents.

To ensure himself on this point, Laue took pains to solve the problem rigorously with the aid of his assistant Frierich Möglich (whom students dubbed Unmöglich) while Meissner and Ochsenfeld's experiment was further into the planning stage; Laue and Möglich's resulting paper[50] went to press scarcely a month after Meissner's ETH lecture in late June of 1933. Here they derive analytic expressions for the fields around superconducting samples of particular geometry (a thin loop or tourus), treated of course as perfect ($R = 0$) conductors. The paper, which bears the stamp of Laue's propensity for classical electromagnetic analyses, is clearly a product of his consultations with Meissner in this period even though, strangely, the paper contains no explicit reference to Meissner or his experiment. (The only explicit reference, besides one to Lippmann, is to the Canadians, McLennan, J. F. Allen, and J. C. Wilhelm.) As such, it is a purely analytical treatment, devoid of numerical examples. "The following treatment," state the authors simply, "will demonstrate the possibility of investigating the current distribution experimentally. … "[51] After some introductory generalities, Laue and Möglich focus on the specific problem of Meissner and Ochsenfeld's series-connected cylinders. They analyze the potential function of the field distribution in terms of the real and imaginary parts of a complex function $\chi = \varphi + i\psi$, where the field lines satisfy the condition $\psi = \text{const}$. Now (they explain), according to Lippmann and Maxwell, the normal component of the magnetic field at the boundary of a superconductor that is cooled in a field-free region invariably remains zero. Since the interior in this case is impervious to an external magnetic field, both circles in the cross-sectional representation of the two superconducting cylinders are automatically lines of force satisfying $\psi = \text{const}$. Moreover, from the symmetry of the problem, they show that, not only is the field distribution the same as that produced by a pair of cylindrical surface current distributions, but it is equivalent to the field distribution produced by a pair of line currents located at A_1 and A_2, respectively as in Figure 10-5, or at B_1 and B_2 in Figure 10-6, which is also due to Meissner.[52] (Neither of these figures are shown in the Laue–Möglich paper.) In Meissner's earlier notation, the field distribution may be expressed in terms of a set of transformed coordinates as:

$$H = 2J\left(\frac{1}{\rho_1} + \frac{1}{\rho_2}\right)$$

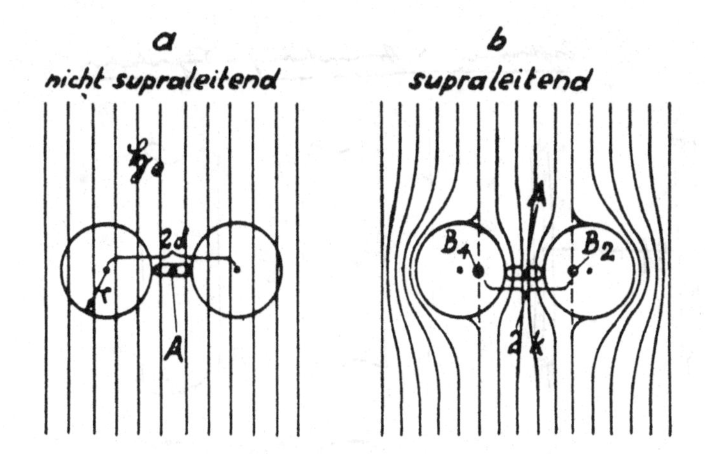

Figure 10-6. Qualitative portrayal of field distributions outside tin cylinders above and below the transition point. Point A is the field measuring point. In (b) the field distribution is as would be expected from two parallel linear current elements parallel to the cylinders and located at B_1 and B_2. Meissner, Ochsenfeld, and Heidenreich, ZK, **41** *(1934), 127.*

where ρ_1 and ρ_2 are the local radius vectors and the separation A_1A_2 is given by $k^2 = d^2 - r^2$. Here $2d$ is the separation between cylinder centers and $2r$ is the cylinder diameter.

Several sketches depicting the geometry in Meissner's hand, Figure 10-7, are found on page 63 of his laboratory notebook number VIII;[53] the essentials of the lower right-hand sketch also appear (and is the only figure) in Laue and Möglich's paper. The sketches are accompanied by various calculations based on Laue's formula.

The essential results from the first round of measurements are contained in two tables, designated here as Table 10-1 and Table 10-2. The first is adapted from Meissner and Heidenreich's 1934 report. The second appears, curiously, only in the much later and final report of 1936. In fact, what amounts to the two tables can be found in the end of notebook VIII, just before the last entries, which are dated 28 September 1933. Referring first to Table 10-1, the following measurements were performed. First, with the indicated constant current flowing, the ballistic deflection b_1 was measured with the cylinders at room temperature. The cylinders were then cooled below the transition point and the deflection was again measured (b_2). The deflection with current flowing above T_c was also compared with the deflection produced by switching on the same current only after the transition to the superconducting state (b_3). Invariably a larger ballistic deflection was seen below the transition point. To be sure of a quantitative test of the von Laue–Möglich prediction, one more measurement was performed: that of the deflection b_4 produced by replacing the tin crystals with thin copper wires located at A_1 and A_2. The result, b_4/b_1, is considerably smaller than b_2/b_1 or b_3/b_1; however, it agrees quite well with the calculated

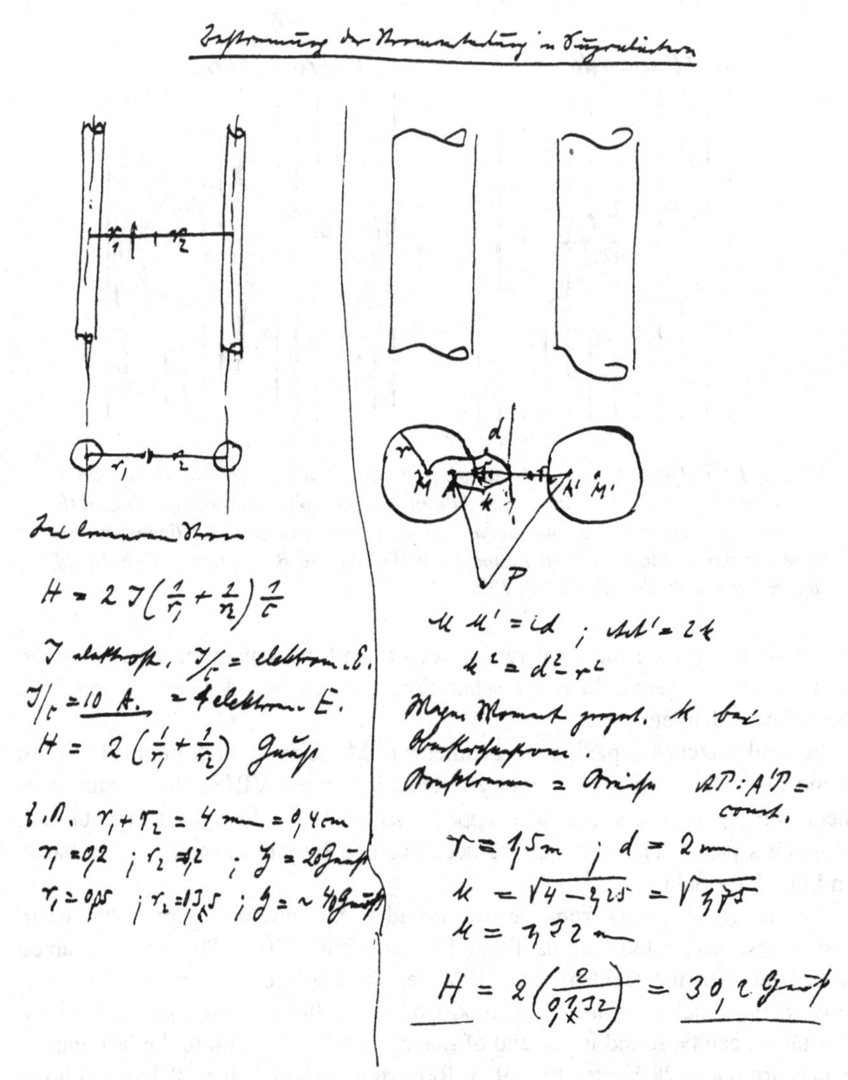

Figure 10-7. Notebook sketches in Meissner's hand. Courtesy Deutsches Museum, Munich.

ratio: b_4/b_1 (calc.) = 1.15. (Note that, in the table in Meissner and Heidenreich's 1934 *Kalte-Industrie* paper, b_4: b_1 is erroneously entered as b_3: b_1.)

In all but one of these series of measurements, the Earth's field was compensated for by means of a Helmholtz coil. However, as noted earlier, this precaution was omitted in series I. Meissner doesn't say whether this omission was on purpose or by oversight, and the actual sequence of measurements (Table 10-2) renders it a moot point. In this particular run, a noticeable dependence of ballistic deflection on the direction of current flow was observed. Meissner concluded that the flux through the search coil from the Earth's field

TABLE 10-1

Measurement series	Terrestrial field	Current strength	$\frac{b_2*}{b_1}$	$\frac{b_3**}{b_1}$
I (Sn crystal)	Not compensated	4.68 A	1.050	1.212
II (Sn crystal)	Compensated	4.90 A	1.233	1.208
III (Sn crystal)	Compensated	4.91 A	1.221	1.134
IV (Cu wire)	Compensated	4.86 A	$b_4***/b_1 = 1.10$	

b_1 = Deflection in normal state.
$*b_2$ = Deflection in superconducting state.
$**b_3$ = Deflection on switching on current at transition to superconducting state.
$***b_4$ = Deflection corresponding to theoretical surface current.

TABLE 10-2

Measurement series	Magnetic field	$\frac{a_2*}{a_1}$
I	Terrestrial	1.77
II	4.1 gauss	1.64
III	4.1 gauss	1.69

a_1 = ballistic deflection above T_c.
$*a_2$ = ballistic deflection below T_c.

had to be greater in the superconducting state *even in the absence of a transport current.*[54] Consequently, Meissner writes that a new round of measurements was performed *without* a current; first with the Earth's field uncompensated (but the crystals suspended vertically so that the field was mainly parallel to the wire axis), and then in the presence of a somewhat stronger field (≈4 gauss, still well below H_c) from a Helmholtz coil surrounding the dewar, directed perpendicular to the plane of the crystals.[55] The results are shown in Table 10-2. In either case, the ratio of the ballistic deflection below and above T_c was, on the average, 1.70—the value given in the Meissner–Ochsenfeld letter.

The actual sequence of measurements in the initial experiment was, as noted earlier, deliberately obscured in the spring report to the *Kalte-Vereins* in the stated interest of "simplicity and clarity."[56] It can, however, be reconstructed from notebook VIII—something of interest since it seems to lend credence to our earlier speculation that omitting the field compensation was done inadvertently. Measurements I of Tables 10-1 *and* 10-2, that is, all measurements made without compensating for the Earth's field, were performed in a single run on

14 July 1933. Galvanometer readings under the various conditions were taken in the following sequence: a_1, b_1 (based on averaging the results of 3 runs), b_2 (average of 2 runs), a_2 and a_3. The second set of measurements (II) was obtained on 28 July, followed by a longer hiatus while the results were analyzed. The third and fourth sets, (III) and (IV), were obtained on or about 8 September, including b_4 which is denoted by b_2' in Meissner's notes. The final data reduction was performed in late September.

A further series of measurements was made with the single tin crystals replaced by a pair of wires of very pure polycrystalline lead arranged geometrically as before. With the same applied field, 4.1 gauss, the ratio of ballistic deflections now was $a_2/a_1 = 1.77$. This value, in fact, was in perfect agreement with one deduced from a numerical calculation of the field between the adjacent cylinders for $\mu = 0$. The calculation involved the expansion of a slightly modified version of the expression for the scalar potential derived by von Laue and Möglich in the cited work, modified because their treatment was applicable for the two cylinders connected in a *closed circuit*, not disconnected as in the latest experiment. Qualitatively, the results of the calculation are as depicted in Figure 10-6. On entering the superconducting state, the same flux passes between the cylinders as flowed previously between the two parallel line currents located at B_1 and B_2.

Next, Meissner and Ochsenfeld chose a cylindrical tube of polycrystalline lead—a poor choice in hindsight. The small search coil could be operated outside or inside the tube, nearly filling the hollow. The external field could be directed vertically or horizontally, in either case perpendicular to the axis of the tube. For the vertical orientation, the observed deflection ratio ranged from 1.27 to 1.30, compared with 1.49 calculated for $\mu = 0$. For the horizontal orientation, the ratio was worse: only 0.66 to 0.89. If the originally present homogeneous background field was now switched off, and if the vertical component of the Earth's field was compensated for as well, a residual external field of 5 to 15% of the homogeneous field remained, as had been reported originally by Meissner and Ochsenfeld. Definitely more surprising were the measurements of the field in the *hollow* of the tube. In cooling the tube below T_c the internal field actually *rose* by about 5%. On switching off the external field, the interior field remained basically unchanged despite the fact that the external field essentially vanished.

It was at this critical, or confusing, stage in the experiments that Ochsenfeld made his untimely departure, more or less leaving Meissner in the lurch. Waiting for Heidenreich, Meissner took the opportunity of revamping the experiment. Satisfied that the question of the current distribution had been settled, he decided that further pursual of the magnetic behavior called for a test sample of simpler geometry and more accurate measurements. The essential elements of the improved apparatus, which was actually in the planning stage when Ochsenfeld left, are shown in Figure 10-8.[57] Basically, it allowed moving a search coil *S* on fixed circles concentric with the axis of a hollow cylindrical conductor *K*. *D* is the cover of the dewar vessel, and the tube *R* encloses a centrifugal liquid stirrer.

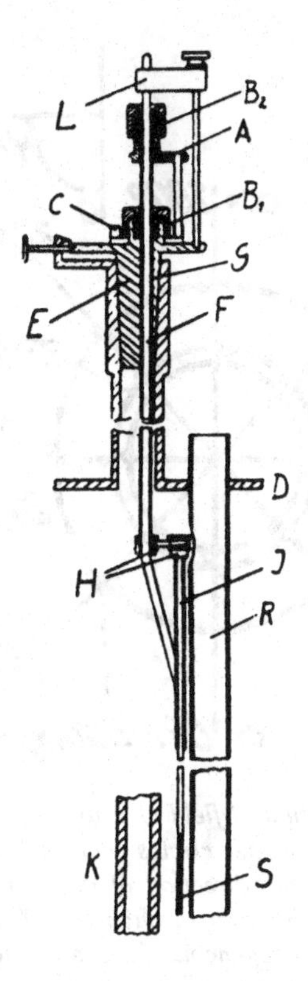

Figure 10-8. Meissner's new experimental arrangement in the spring of 1934, utilizing a single solid or hollow cylindrical crystal. Meissner and Heidenreich, PZ, **37** *(1936) 455.*

The first sample in the new round of measurements with Heidenreich was a solid (not hollow, as depicted in Figure 10-8) single cylindrical crystal of tin, 10 mm in diameter and 130 mm long. A transverse magnetic field of about 5 gauss, approximately homogeneous without the sample present, could be applied with a Helmholtz coil of 130 mm radius and 130 mm between coil pairs. The same search coil that was used earlier could be moved on two concentric circles: one 12 mm in diameter and thus close to the surface of the cylinder, and one 20 mm in diameter. For eight fixed points on either circle, the direction of the field was determined by flipping the search coil 180° for various initial angular positions and hunting for the initial angle that gave maximum ballistic throw—a tedious procedure at best. From the direction, the magnitude of the field could also be calculated.

In Figure 10-9,[58] the small tick marks indicate the direction of the plane of the search coil for maximum ballistic throw, and thus the direction perpendicular to the field direction, at each of the sixteen measuring points. The line through $\alpha = 0°$ and 180° corresponds to the direction of the external initially homogeneous field. In Figure 10-10,[59] these directions and their associated magnitudes are superimposed on a field pattern calculated for the case $\mu = 0$;

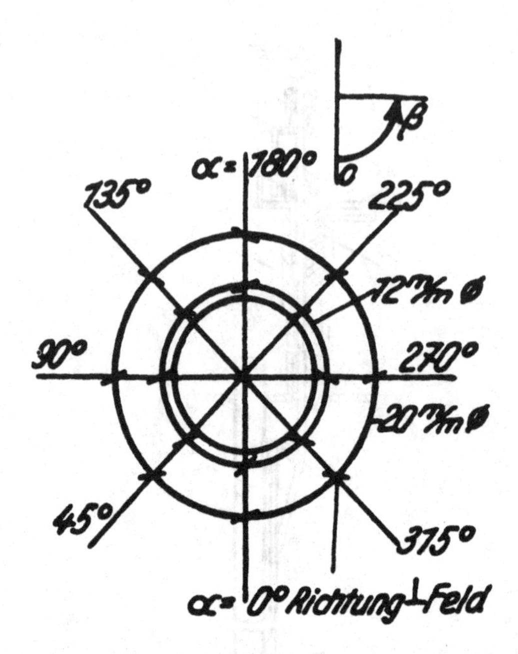

Figure 10-9. Qualitative depiction of field measurements at eight fixed points on 12 mm diameter and 20 mm diameter circles concentric to a cylindrical tin crystal. The small marks indicate the direction of the plane of the search coil for maximum ballistic throw. The diameter through α = 0° and 180° corresponds to the direction of the original homogeneous field. Meissner and Heidenreich, PZ, **37** *(1936), 458.*

here the local numerical magnitudes of field intensity designate, in arbitrary units, the density of field lines intercepted by the search coil. Actually plotted in Figure 10-10 is the pattern of streamlines in an incompressible, frictionless fluid flow around a circular cylinder—the magnetostatic analog encountered in Section 9.1. The field lines are given by the contours ψ = const., where

$$\psi = H_0 \left(R - \frac{R_0^2}{R} \right) \sin \theta .$$

(Meissner uses ρ for R, r for R_0 and α for θ.) Here H_0 is the strength of the homogeneous field, R_0 the radius of the cylinder, R the radius vector from the point of observation to the center of the cylinder, θ the angle between R and the direction of H_0. The flux intercepted by the search coil between two points with radius vectors R_1 and R_2 is simply $\psi(R_1) - \psi(R_2)$.

In Figure 10-11,[60] the magnitudes of the ballistic deflections on the two circles are plotted against angular position; the magnitude of the homogeneous field is also shown. "All in all one would probably agree that the distribution of lines of force after the onset of superconductivity corresponds closely to the condition $\mu = 0$."[61] The correspondence, in fact, is far from perfect. Nor was it to be expected, countered Meissner and Heidenreich.[62] Thus, the fact that for

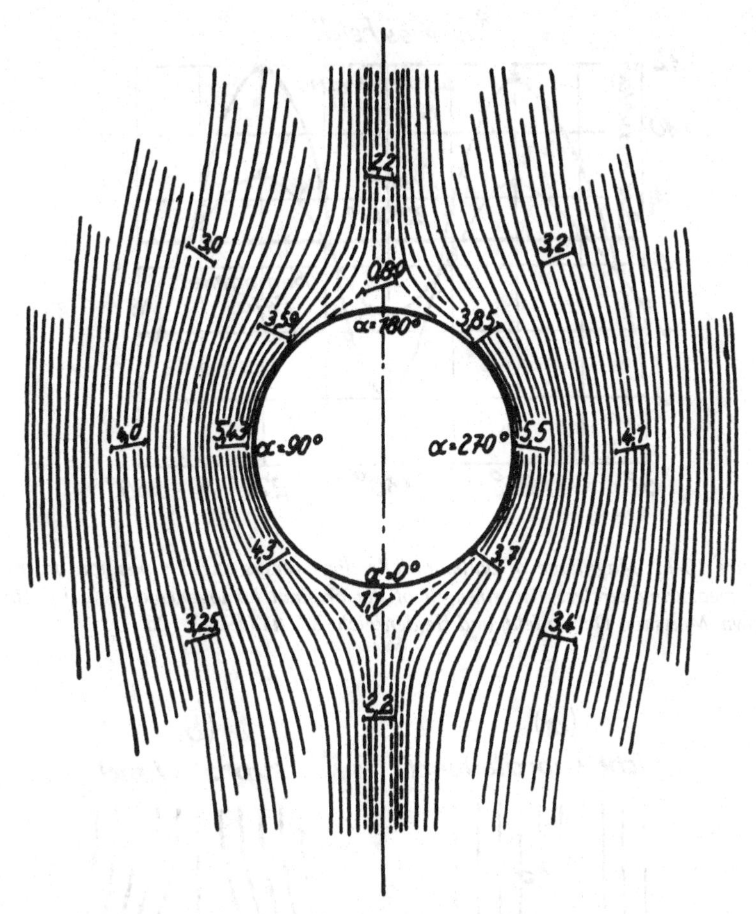

Figure 10-10. Calculated field pattern around a cylindrical conductor of infinite permeability, with the measured directional indicators of Figure 10-9 superimposed. The numerical magnitudes indicate density of field lines. Meissner, PZ, **35** *(1934), 934.*

$\alpha = 0^{\circ}$ and $\alpha = 180^{\circ}$ on the inner circle the ballistic deflection (Figure 10-11) was not zero was simply due to the fact that the field was measured ~1 mm away from the cylinder. More generally, the fact that the directions for $\alpha = 0^{\circ}$ and 90°, or for 180° and 270° (Figure 10-9) were not parallel and perpendicular to H_0, respectively, came about because the two circles were not quite concentric with the crystal axis—which in turn was not necessarily the cylinder axis. Moreover, small crystal defects, or small deviations from homogeneity in H_0, could also affect the results.

Finally, switching on the external field *after* the transition to the superconducting state again produced the same field pattern as in Figure 10-10, but this was to be expected classically.

With this more accurate arrangement, the somewhat disquieting results with the hollow lead tube could also be checked. The single cylindrical tin crystal

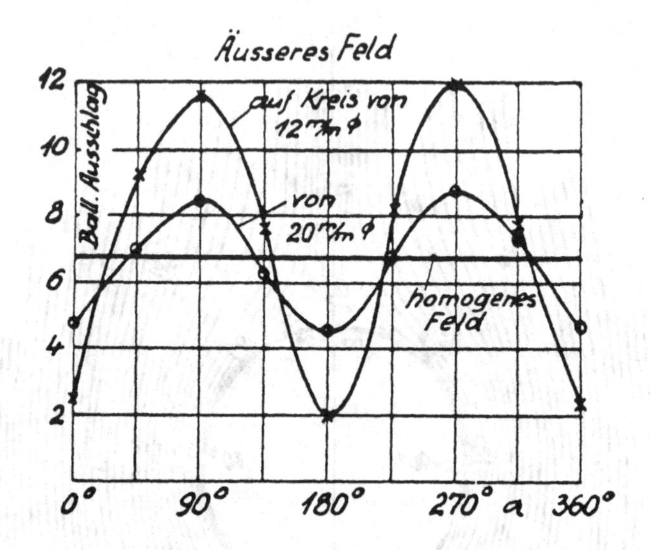

Figure 10-11. Magnitude of ballistic deflections versus angular position on the two measuring circles. The strength of the original homogeneous field is also shown. Meissner, Ochsenfeld, and Heidenreich, ZK, **41** *(1934), 126.*

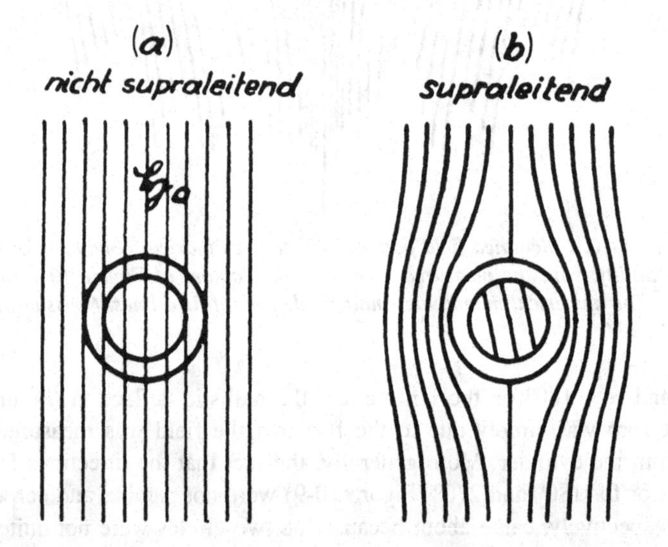

Figure 10-12. Effect of superconducting transition on field pattern in the vicinity of and in hollow of lead tube. Meissner, Ochsenfeld, and Heidenreich, ZK, **41** *(1934), 127.*

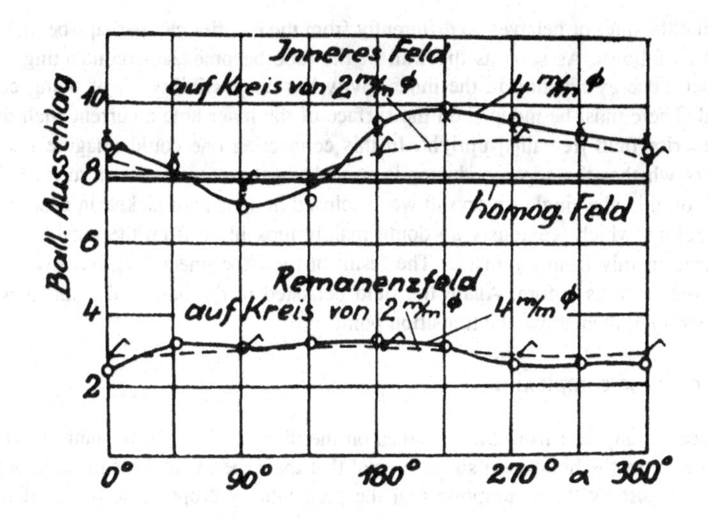

Figure 10-13. Field strength (ballistic deflection) inside the cylindrical hollow versus angular positions on two measuring circles, before and after removing applied field (top and bottom curves, respectively). Note that the vertical scale cannot be compared directly with that in Figure 10-9, because of differences in galvanometer sensitivity in the two experiments. Meissner, Ochsenfeld, and Heidenreich, ZK, **41** *(1934), 127.*

was bored carefully with a hole, 6 mm in diameter and concentric with the axis of the cylinder. This hole was large enough to allow discrete measurements with the 1.5-mm-wide search coil on concentric circles 2 mm and 4 mm in diameter, rather like the external field measurements. The Earth's field was compensated for with a Helmholtz coil. The results from cooling in a homogeneous applied field are shown qualitatively in Figure 10-12[63] and plotted in Figure 10-13.[64]

> The homogeneous field that was present in the interior above the transition point [Figure 10-12a] changed its direction a little (a few degrees) at the decrease below the transition point, but remained in actual existence. The rotation is shown greatly exaggerated in [Figure 10-12b]. The field strength rose by about 10% on the average. This can be seen more accurately in [Figure 10-13], ... [65]
>
> If the external field was now switched off, the magnetic field in the external vicinity of the tin tube vanished completely. In the interior, however, there remained the remanent field represented in the lower part of [Figure 10-13]. Also this did not change in the course of time, but maintained the same magnitude after two hours. If the external field was now switched on again, there again resulted on the outside the field distribution (found earlier for the solid crystal) in [Figure 10-11]. In the interior, however, the state represented in the upper part of [Figure 10-13] did not reappear, but rather the field strength increased only by 3 to 4% above the value of the remanent [*sic*] field given in [Figure 10-13]. On switching off again, there appeared again the same remanent field. ...

> That the interior behaves so differently from the exterior can perhaps be understood as follows. As soon as the wall of the tube becomes superconducting, the magnetic energy residing in the inner cavity has no possibility ... of being cancelled. There must be induced on the surface of the inner hole a current such that the interior field [remains] upright. In this connection one could imagine that it matters whether the superconductor is cooled via the outer or inner surface. To check on this, the single tin crystal was enclosed in a vacuum jacket, in order that the cooling, which previously no doubt mainly took place from the outside, now occurred mainly from the inside. The result of the experiment was, however, exactly the same as before. Again the field persisted in the bore and again it was even stronger than above the transition point.

In spite of these complexities,

> One sees at any rate from these studies on the distribution of a magnetic field in the interior of the hollow tin single crystal that the experimental results cannot be explained just by the assumption that the permeability drops to zero. For if this were generally the case, then also no field could terminate on the inner surface of the hole in the superconductor, as is obviously the case according to the experimental findings.[66]

At Bad Pyrmont later in the year, Meissner stuck to his capital point in all of this, namely, that something more is involved than simply vanishing permeability below T_c. Two years later, he was just as adamant. "As far as explaining these phenomena in the interior of lead and tin tubes is concerned," he declared, "I can simply note that, essentially due to the presence of a superconducting layer, the magnetic energy prevailing in the interior cannot without further ado be expelled."[67] In other words, though a superconductor is perfectly diamagnetic, flux can exist within a hollow in a superconducting body. If so, paramagnetic currents must be generated in the interior surface around the periphery of the hollow. Moreover, by Lippmann's rule, which still holds, since the flux threading the hollow cannot change, the flux and the circulating current associated with it will persist even if the applied field is changed or even reduced to zero.[68]

10.3 Taking Stock at Bad Pyrmont

Low-temperature physics was a principal subject of the meeting of the German Physical Society at Bad Pyrmont in September of 1934, chaired by Peter Debye. It was to be the last important meeting on superconductivity in Germany, with minimal international participation, for a very long time. The task of pulling together the ambitious program, no small undertaking in the best of times and surely not in the charged circumstances of Germany in 1934, fell mainly on Meissner and Eduard Grüneisen.

The original agenda for the meeting included an invited talk on condensed helium by Keesom and a major review of superconductivity by de Haas, with a shorter report by Meissner thrown in for good measure.[69] On Keesom's request, however, his own presentation was revised, with Grüneisen's concurrence, to cover the latest caloric results instead, including the atomic heats of tin and thallium, since he had covered helium at the *Vortragswoche* in Zürich the previous summer. Besides, the new caloric results "were likely to provoke more heated discussion."[70] On Grüneisen's urging Meissner's talk was expanded into a major progress report on the ongoing experimental research, concentrating on Berlin since "the co-operation with the Reichsanstalt ... must be strongly emphasized,"[71] but also touching on work at Leiden, Toronto, and Oxford.

Despite the political pressure to laud German progress, de Haas's contribution was considered important for lending scientific weight to the program. It proved problematic, however. For reasons of his own, de Haas balked, citing uncertainty in his own commitments that far in advance of the meeting. He first suggested H. B. G. Casimir as a replacement "in an emergency"—a suggestion that caused raised eyebrows among the organizers in view of the worsening political situation in Germany. "Do you know this individual personally?" wrote Grüneisen to Meissner on 24 June. "It strikes me that we may be dealing with someone of Russian or half-Russian, even Jewish, background. If so, his invitation at present is certain to excite the greatest offense."[72] A frank letter on the "extremely delicate matter" went immediately to de Haas, who saved the day by proposing his former pupil Gorter instead.[73] After some hesitation on Gorter's part about the wisdom of attending a meeting in Germany just then,[74] he agreed to give a paper on his and Casimir's work on the thermodynamics of the superconducting state.

One more shuffle of speakers became necessary when Arnold Eucken withdrew his name from the program. The difficult but influential Eucken, whom Simon had succeeded at Breslau and who now headed physical chemistry at Göttingen, was to have given a major review of caloric measurements, including von Laue's recent work on phase transitions of the third order. Irritated at his copping out, having "assumed that the arrangement with Eucken was as good as finalized" and fearing that "finding a replacement would be difficult," Grüneisen suggested to Meissner that Eduard Justi take his place, since Justi had worked with Laue in the area of phase transitions. (Justi succeeded Meissner as head of the low-temperature laboratory at the Reichsanstalt.) Apropos of Laue, he added, "how is it with him? ... with the attacks he has undergone, I would dearly like to see a demonstration of sympathy for him."[75]

Indeed, Laue had not had an easy time in recent months. On 16 December 1933, the ever troublesome Johannes Stark, installed as president of the Reichsanstalt the previous May (replacing the admittedly ineffectual Friedrich Paschen who was also a Jew) as a reward for backing Hitler, had fired Laue from his consultantship at the PTR. Stark, the archetypal Aryan physicist (with Philipp Lenard) had been an academic outcast for years, and Laue was his most vocal opponent in scientific circles. "The date (12/16) is rather amusing," confided Laue guardedly in a letter to Otto Stern the following year, complaining about the constant troublemaking by "you know who." "Two days earlier there took place a discussion about the very person [i.e., Stark] in a confidential meeting."[76] In a tit-for-tat that very month, Laue, with Otto Hahn and Wilhelm Schlenk, had raised such emphatic objections to the admission of Stark to the Prussian Academy that Stark's sponsors (friends of the new regime) were forced to withdraw the nomination.

As an aside, Stark's investiture at the Reichsanstalt gave rise to a bit of scientific trivia touching on superconductivity—a far cry from his attempt to explain the phenomenon in terms of *Schussflächen* two decades earlier (Section 5.3). No sooner had Stark been installed at the Reichsanstalt than he set out to cleanse Aryan physics by, among other things, substituting the Jewish concept of the electron with his own ring-shaped electronic structure—a construct acting like a resistance-free superconducting current. This nonsensical concept, Laue complained to Meissner, contradicts classical electrodynamics plain and simple—something Meissner frankly explained to Stark.[77] Nevertheless, it is the centerpiece of a 1935 issue of the *Physikalische Zeitschrift* devoted largely to superconducting and other investigations at the Reichsanstalt "on the suggestion of Herrn Präsident Stark"; it is repeated in Steiner and Grassmann's wartime monograph on superconductivity.[78]

Returning to Laue, his clash with Stark continued into 1934. At mid-year, Laue lost his chairmanship of the physics committee of the influential Notgemeinschaft that had served German science so well in a critical period—it was a position he had originally inherited after the organization defeated a power play by Stark involving a breakaway rival to the Physical Society.[79] Shortly afterward, the president of the Deutsche Forschungsgemeinschaft (German Research Association, which succeeded the Notgemeinschaft), Friedrich Schmidt-Ott, was dismissed and replaced by Stark as well.

The final agenda for the meeting at Bad Pyrmont featured a keynote address by Debye on magnetic cooling,[80] an outline of experiments demonstrating basic phenomena with liquid helium by K. Clusius,[81] Meissner's expanded superconducting review,[82] and Keesom's caloric report.[83] Laue, in place of Eucken and on the insistence of Meissner, wound up speaking on phase transitions of the third order in a paper coauthored with Justi,[84] followed by Gorter's thermodynamic paper delicately coauthored with the problematic Casimir.[85] In retrospect, the meeting proved to be a thorough success from a technical standpoint, among other things for the heated discussions accompanying the invited papers, most of which are still regarded as classic. Judging from the printed discussions, however, the participants of stature, except for Keesom and Gorter, were almost exclusively German scientists. From the standpoint of international participation and exchange, the Royal Society discussions on superconductivity held in London 8 months later would prove to be much more successful.

Increasingly, the focus of attention was shifting; it was no longer centered on the rule $B = 0$ but on the apparent exceptions to the rule. The biggest skeptic of Gorter's advocacy of zero induction as a general condition turned out to be Meissner himself who, in the informal discussions at Bad Pyrmont, offered the opinion that the induction in pure superconductors was zero "at Gorter's responsibility." "Perhaps," he allowed, "the internal remanent field in an ideal single crystal disappears completely,"[86] since the remaining field seemed stronger for the polycrystalline lead tube than for the single crystal tin tube. However, this had seemed less likely when it was found that reorienting the crystal axis by as much as 20° with respect to the original homogeneous field had no observable effect. Crystal faults had also seemed possible candidates for spots terminating the lines of force in the interior wall of the tin tube that exhibited no detectable field on the outside. This was subsequently ruled out while the Bad Pyrmont proceedings were in the proof stage; a footnote to Meissner's paper reports Heidenreich's finding that not only the lead tube, but also the hollow tin crystal, exhibited a weak remanent external field.

Meissner had a frustrating time coaxing these last results from Heidenreich who, he learned to his surprise, had accepted a job offer at Telefunken. With prospects for more experiments at the Reichsanstalt thus running out, and under pressure to sort out the results in time for the approaching deadline for the Bad Pyrmont proceedings, Meissner implored Heidenreich in early November to dispatch to Munich all manuscripts, notebooks, and calculations—most urgently the results for the hollow tin cylinder, as soon as Heidenreich had analyzed them.[87] By mid-December he still had no further word from Berlin;[88] the footnote to the Bad Pyrmont paper must have relied on what little he surmised from Heidenreich's bombshell letter confiding his Telefunken plans in September. Eventually Heidenreich did apparently forward most of the records, but experimental details would bedevil Meissner once more when he prepared for another major presentation half a year later.

In his own presentation at Bad Pyrmont, Laue attacked Ehrenfest's second order transition with gusto, naturally provoking vigorous objections from Keesom and Gorter that are only hinted at in the published proceedings. His point was that the free energies of the normal and superconducting phases would tend to osculate at the transition point (the curves of their temperature functions would not cross but they would exhibit the maximum degree of contact commensurate with the order of the transition), allowing the existence of the same phase above and below T_c. This possibility was ruled out by Gorter in his own presentation. The published version of Gorter's paper is important for introducing in print, if not in coining the term of, the "two-fluid" model of superconductivity.

Since a treatment of superconductivity from first principles seemed a remote possibility, Gorter and Casimir decided to attack the problem "from the phenomenological side, without making detailed electron-theoretical assumptions." After all, they had already used such a phenomenological approach with good results in treating the destruction of superconductivity in a magnetic field, arguing that the destruction is a consequence of zero induction in the superconductor. As soon as the sum of the free energy of the superconductor and the magnetic term $H^2V/8\pi$ exceeds the free energy of the normal phase at the same temperature, the latter will become thermodynamically stable and superconductivity is disturbed. On the other hand, while a comprehensive theory able to account for the vanishing of electrical resistance at a definite temperature was not in the offing, there was ample evidence, particularly thermodynamic, that the superconducting state is a distinct phase bounded by the $H_c - T$ equilibrium curve in the $H - T$ diagram; "the condition $B = 0$ represented a new and rigorous criterion for distinguishing this phase."[89] Moreover, the evidence—particularly crystallographic and mechanical—suggested that the superconducting phase is associated with the conduction electrons, not a feature of the crystal lattice.

They took as their starting point the relationship between threshold field and difference in entropies between the normal and superconducting phases that results from differentiating with respect to T the expression for the difference in Gibbs free energy, $G_n - G_s = VH_c^2/8\pi$, namely:

$$\Delta S = S_n - S_c = -\frac{VH_c}{4\pi}\frac{dH_c}{dT}.$$

Plotting ΔS against T for Hg, and using Leiden $H_c - T$ data, they obtained a curve that passed through a maximum with decreasing temperature. Extrapolated to absolute zero, it suggested that ΔS is proportional to T at the lowest temperatures of observation and reaches zero at $T = 0$. The simplest explanation offered was that the entropy of the electrons in the superconducting state approaches zero at these low temperatures while in the normal phase it is proportional to T. As discussed in Section 12.1, Kok had independently reached essentially the same conclusion—this was not surprising since the linear dependence on T for the normal phase was also a feature of Sommerfeld's latest electron theory of metals.[90] In fact, ΔS estimated from the linear portion of the $\Delta S - T$ curve agreed remarkably well with calculations based on a formula by Sommerfeld (making plausible assumptions regarding the number of free electrons per atom). Equally good agreement was obtained for tin and lead.

In light of all of this, Gorter and Casimir suggested that the superconducting state could be viewed as a one component, two-phase system akin to one first introduced by Kronig.[91] At very low temperatures, all the electrons are condensed into a "crystal phase," whose energy was tentatively assumed to be independent of temperature. At higher temperatures, a fraction x of the electrons is in the normal or "gas" phase similar to Sommerfeld's electron gas; x is an internal parameter indicating a degree of order or of superconductivity. At any given temperature x will adjust itself according to the minimum of the total free energy and consequently, x will, with rising temperature, gradually increase from zero to one.[92]

The internal parameter was the trick and the device that negated Laue and Justi's objection, since the number of superconducting electrons, $1 - x$, could not become negative.[93] In the discussion following Gorter's presentation, Laue offered no further opinion on the matter (at least in print), but Meissner doggedly inquired "whether the speaker, based on the conception of the electron gas precipitating into an electron crystal at the transition point, [was] able to deduce why the induction should always be zero in the supraconductive state."[94] Replied Gorter:

> Without additional assumptions about the nature of the "electron crystal" one can only expound on the thermal, not the electric or magnetic properties of metals. To explain the fact that even very few electron crystals suffice for the metal as a whole to exhibit the characteristics of a supraconductor, one must probably, with Kronig, assume that the crystal can form very long chains. The gross electron picture can, as Frenkel has also emphasized, lead to a very large negative susceptibility. A really satisfactory explanation of the fact that in a supraconductor $E = 0$ and $B = 0$ must await a definitive quantum-theoretical treatment of the problem.[95]

In fact, the presentation following Gorter's at Bad Pyrmont was an attempt at just that by R. Schachenmeier of the University of Berlin. On Meissner's urging, he summarized his recent attempts to employ quantum-mechanical electron–electron Coulomb interactions to explain superconductivity.[96] These attempts, first made in 1931,[97] were based on a resonant exchange between conduction electrons and the electrons bound to the ions, for frequencies above the maximum thermal lattice vibration frequency. Schachenmeier's prowess in superconductivity theory was far from universally admired, it might be added. Casimir had by chance attended a lecture of his on superconductivity in the famous seminar run by Schrödinger and Becker, during his stay in Berlin-Dahlem in the summer of 1932; he recalled the lecture as "nonsense."[98] Hans Bethe was more constrained, pointing out at the time that Schachenmeier "neglect[ed] fundamental facts of wave mechanics"[99]—an objection to which Schachenmeier did not take kindly.

Frenkel, alluded to by Gorter, was in some respects closer to the mark. Yakov Ilyich Frenkel was a member of the Physico-Technical Institute in Leningrad, and not a newcomer to the electron theory of metals.[100] His attempt to treat superconductivity in this period took account of the mutual electromagnetic (inductive) forces between electrons "moving as an organized crowd."[101] These forces, Frenkel argued, stabilize the motion against the perturbations due to the thermal agitations of the crystal lattice. A noteworthy aspect of Frenkel's paper is that it anticipated the Meissner effect by about a year: "A metal in the superconducting state must behave like a *diamagnetic* body with a large negative susceptibility. ... This is quite equivalent to saying that the interior of such a body will be screened from external magnetic fields by the system of surface currents induced by the latter."[102] Alas, Frenkel too was seriously taken to task by Bethe (and by Herbert Fröhlich), who pointed out that the magnetic interactions lead to only a very small correction of the electron effective mass.[103]

In fairness to both Schachenmeier and Frenkel, Hoddeson, Baym, and Eckert note that Schachenmeier's papers "hint at modern ideas of mixed valence" and that the Meissner effect "would verify [Frenkel's] intuition."[104] And both were in good company with Bohr, Kronig, Brillouin, Elasser, and others. In fact, quantum-mechanical electron theories proved to be of marginal assistance in the elucidation of superconductivity in this period and for a good while to come. A variety of complications intervened, among them too many ad hoc assumptions, the inability of proffered models to allow quantitative comparisons with experiments, declining interest in solid state problems by many (though not all) of the principal theorists as cosmic rays and nuclear physics took center stage, and disruptions caused by the political upheavals in German scientific circles. Rather, the phenomenological and thermodynamic approaches turned out to be more fruitful in promoting understanding, particularly that of Gorter and Casimir, and above all Fritz London's electrodynamics that went to press a month after Bad Pyrmont.[105]

CHAPTER 11

Mostly Alloys

11.1 The Anomalies of Alloys

The frustrating efforts to confirm the Meissner effect had at least one positive result: They revived interest in a class of superconductors that had suffered an ignominious setback in the high expectation for practical applications only months before. At that time, attempts to exploit the extremely high critical fields of superconducting alloys in high-field magnets had been thwarted by their inexplicably poor effective current density. But the unpredictable alloys suddenly took on new importance in the fall of 1934 with the suspicion that they held the key to more basic questions. That November the Oxford team of T. C. Keeley, Kurt Mendelssohn, and Judith Moore noted the extreme sensitivity of frozen-in flux (incomplete Meissner effect or remanent flux of induction on switching off the external field) to the addition of even minute amounts of another component to a pure superconducting metal.[1] (The addition of a mere 4% Bi to pure Pb trapped the entire flux.) This occurred despite the fact that the samples were carefully prepared in the prescribed form of long cylindrical rods, cycled in a uniform parallel external field—the prerequisite conditions for ensuring the maximum Meissner effect according to Gorter. In its letter to *Nature*, the British team drew attention to two other notable aspects of the alloy behavior—neither of which were by any means new but nevertheless they were striking—without presenting actual data. The first aspect, the change in induction at constant temperature, extended over a broad field interval, typically 10 to 20% of the threshold value. The other aspect of note was suggested by the extreme steepness of the curve of threshold field versus temperature (Figure 7-9). Since Rutgers's equation predicts a quadratic dependence on dH_c/dT of the difference between the specific heats of the superconducting and normal states, the alloys should exhibit an enormous jump in specific heat at the transition point. Here, then, was a way of testing whether the high threshold values found at Leiden for alloys did or did not have the same significance as the low values characteristic of pure metals, and without the need for cumbersome high-field apparatus.[2] Apparently they did not, the authors claim, because no sign of any discontinuity whatever—certainly not one of the predicted order of magnitude—was evident in a specific heat curve for $PbTl_2$ obtained recently by Mendelssohn and Moore but unfortunately not published.

In Leiden, de Haas and J. M. Casimir lost no time in adapting their bismuth wire technique to the new problem. They mounted fine bismuth wires inside axial channels drilled in cylindrical rods of $PbTl_2$ and Bi_5Tl_3, and measured the change of probe resistance as a function of transverse magnetic field. (Actually, because of the breadth of the step in the magnetic transition curve for alloys, the difference between the transverse and longitudinal field effects is much less pronounced for alloys than for pure metals.) At sufficiently low fields, the probes showed no change in resistance, they reported in a letter submitted to *Nature* on 7 December.[3] However, above a certain critical value of the field (much lower than the field necessary to restore the first trace of resistance) the probes indicated a gradual penetration of the field, even though the alloy itself remained superconducting. Only above a considerably higher threshold field was the resistive state gradually restored.

Here we have the first report to make a qualitative distinction between the "critical value of the field, that is, the value at which the field starts to penetrate into the alloy" and the "threshold field at which the resistance is coming back."[4]

At Kharkov, Schubnikow, and colleagues were also in hot pursuit of both the specific heats of the alloys and their magnetic behavior, and this neophyte team proved to be the first to publish quantitative results in both areas. In the second week of December, Schubnikow and W. J. Chotkewitsch led off with a preliminary report in their new journal on the specific heat of a PbBi alloy.[5] They confirmed that though its specific heat tends to be slightly higher than that of pure lead over the temperature range from 5.5 to 12 °K, the specific heat curve for this alloy, like $PbTl_2$, shows not the slightest hint of a jump at the transition point. A second paper, this time by J. N. Rjabinin and Schubnikow, followed in late January of 1935;[6] it reported on both magnetic and critical current measurements with single crystals of $PbTl_2$. (Schubnikow was an authority on the production of single crystals from a molten metal, as noted in the previous chapter. At Leiden, during his lengthy tenure with de Haas, it was he who had been largely responsible for producing the near-perfect crystals of bismuth that figured so prominently in all of the measurements obtained by his host and J. M. Casimir.) The magnetic moment of the superconductor was measured in a constant field by Schubnikow's standard technique of rapidly withdrawing the sample from the coil surrounding it and observing the resultant throw of the ballistic galvanometer. Figure 11-1 shows their B–H curve at T = 2.11 °K. Clearly, they reported, two critical fields are involved:

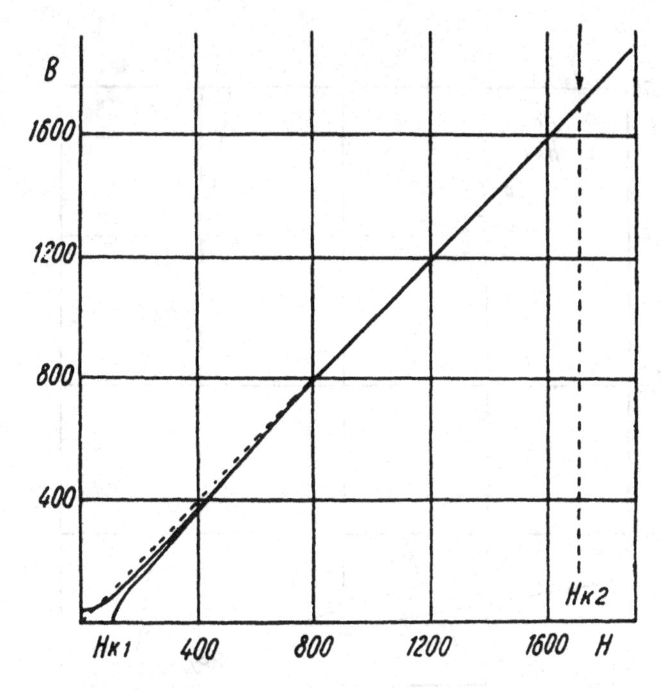

Figure 11-1. Rjabinin and Schubnikow's magnetization curve for $PbTl_2$, showing two critical fields for alloys. Rjabinin and Schubnikow, PZS, 7 (1935), 123.

1. Up to a definite critical field-strength H_{k1}, which depends on the temperature, B remains nearly zero. In this region the behavior of alloys and pure metals is the same.

2. In the field interval from H_{k1} to H_{k2} the induction increases with the field-strength, gradually approaching the value characteristic for the non-superconducting metal. Electrical measurements carried out on wires from the same melt showed that the potential difference remained equal to zero right up to the field-strength H_{k2}. It was found that the induction does not depend on time, which means that the currents excited on the surface of the specimen do not die out in the course of time. ...

3. At the second critical field-strength H_{k2}, the alloy loses its superconductivity. The magnetic properties undergo no change at this point. ...

4. When hereupon the field-strength is lowered a light hysteresis is observed and at zero field-strength a small residual magnetization remains, which does not depend on time.[7]

The two critical fields established here, and essentially Schubnikow's notation for them as well, would remain a permanent feature of the superconductivity of alloys (properly speaking, of type II superconductors) in the years ahead. The lower critical field, where penetration starts (and for a while aptly termed

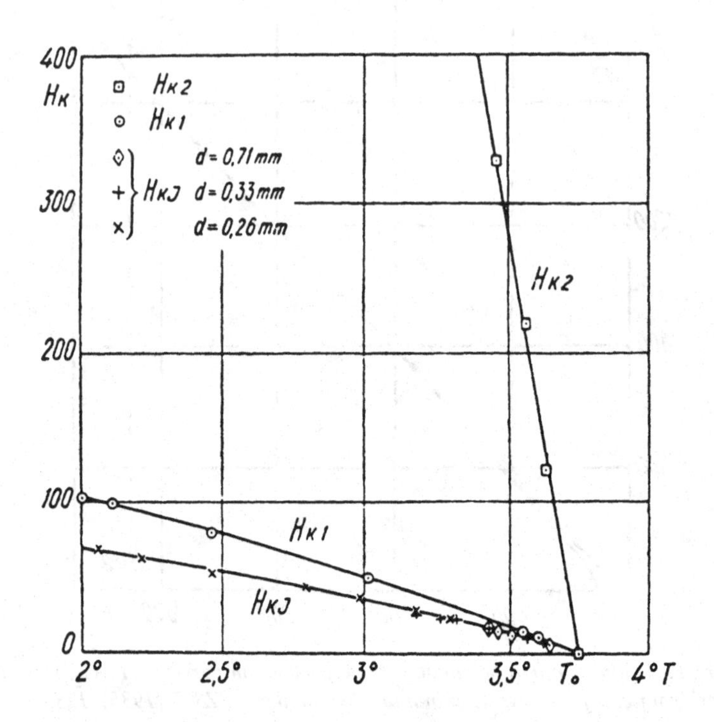

Figure 11-2. Temperature variation of two, and possibly three, critical fields for $PbTl_2$. Rjabinin and Schubnikow, PZS, **7** *(1935), 124.*

the penetration field), remained in the parlance of post-World War II superconductivity simply as "H_{c1}." The upper critical field, signifying complete penetration, accordingly, is H_{c2}.

At any temperature below the transition temperature for $PbTl_2$ (3.75 °K), the *B–H* relationship of Figure 11-1 was found to apply. The results of Rjabinin and Schubnikow's measurements at various temperatures are shown in Figure 11-2, where the two critical field strengths are plotted as a function of temperature. (A qualitatively similar plot appears in de Haas and J. M. Casimir's aforesaid letter to *Nature.*) Jointly, these results for the magnetic properties and the specific heats left little doubt that Gorter's general condition $B = 0$, and with it Rutgers's equation whose derivation was predicated on the condition, was inapplicable or at least not entirely fulfilled in the case of alloys. However, the tendency exhibited in Figure 11-2 (ignoring for the moment the curve marked H_{kj}) hints at several possible avenues out of the dilemma. First, the critical field to be used in Rutgers's equation is the lower, not the upper critical field H_{c2}. Second, the suggestion seems quite compelling that several coexisting phases with markedly different superconducting properties are involved in superconducting alloys, perhaps with one characteristic of the bulk of the sample and the other of impurities, defects, or local concentrations. At the time, Rjabinin and Schubnikow refrained from speculating, observing merely that

> ... possibly one may be able to keep up the concept of two phases in some form, if it appears that this unusual behavior of the alloys is caused by their inhomogeneity, which may be due to the decomposition of the solid solution and the formation of a new very dispersed phase.[8]

One more relationship was jeopardized by the latest Ukrainian experiments. The lower curve marked H_{kj} in Figure 11-2 represents critical field strengths calculated from $H_{kj} = 2\ I/R$, where I is the critical current at a given temperature and R the wire radius. Points thus calculated from measurements on wires of varying thickness are seen to "lie admirably on one curve for all three wires."[9] That is, the self-field at the surface of the alloy wire produced by the current, which caused the first trace of resistance, was independent of the diameter of the wire and of the same order of magnitude as the longitudinal field, which first penetrated the wire but much lower than the field that restores the resistance. As with Rutgers's equation, the apparent failure of Silsbee's hypothesis for alloys could be circumvented if the lower critical field H_{cl} was used. It has remained a moot point whether a third, even lower critical field is actually involved in these experiments as Figure 11-2 would have, or whether the field at which the first penetration took place was too weak to be detected by Rjabinin and Schubnikow's experimental arrangement.[10]

Whether this was so or not, Schubnikow's short-lived but productive team would remain intact for several more years, long enough to have the last word in the experimental showdown between two competing viewpoints engendered by the experiments of 1934–1935.

11.2 Mendelssohn's Sponge, Gorter's Razor Blade, and More

If Schubnikow was reticent about speculating on the alloy structure, Mendelssohn was not. His celebrated "sponge model" took form in relatively short time in the early spring of 1935. This tenacious, albeit in the long run counterproductive model was destined to exert an authoritative influence out of all proportion to its worth for over two decades. So pervasive was its influence in the face of strong counterarguments of more theoretical substance, that it held sway well into the 1950s, distracting investigators from the final formulation of the theoretical underpinnings of high-field superconductivity and discouraging concomitant progress in realizing the potentially enormous current densities that were then within sight.[11]

The failure of Rutgers's equation in particular convinced Mendelssohn that the high threshold value of alloys was not a property of the whole alloy but confined to local regions in it, whereas the bulk of the alloy has approximately the same threshold value as pure metals. The meshes of the sponge, which become superconducting first, cause all lines of induction penetrating the specimen to be caught by persistent currents. The retention of flux and hysteresis showed that these regions are multiply connected ("rings" or "annular regions"). Since the bulk of the alloy has a low threshold value, it will cause only a very small discontinuity in specific heat.

With Judith Moore, Mendelssohn soon put the model to a simple test. Their results were communicated by letter to *Nature* in early April.[12] The test consisted of watching the change in induction in Pb–Bi rods as they were warmed slowly in a constant field by switching off and on a small additional field. When passing the penetration (H_{c1}) curve, flux entered the alloy when switching on, but no flux left when switching off; that is, the meshes of the sponge (as they described the scheme) were filling up with flux. This continued until near the threshold (H_{c2}) curve the changes became reversible, and finally the sample was entirely normal after passing the curve. However, on cooling in a field, flux neither entered nor left below the reversible region; that is, now the meshes were already saturated with frozen-in flux and no flux entered or left the sponge until the threshold curve was again passed.

Mendelssohn reported further confirmation of the sponge model at the end of May, this time based on calorimetric data presented at the one-day discussion on superconductivity and "other low temperature phenomena" held at the Royal Society in London on 30 May. These measurements, carried out by Moore,[13] showed that after a sufficiently high magnetic field was applied to and removed from a superconducting alloy (thereby producing a state of frozen-in flux) the bulk of the sample was left in the normal state. Moore first measured the temperature dependence of the specific heat of a tin sample alloyed with 4% Bi as it was cooled in zero magnetic field, and then when a high field was switched on and off at the lowest temperature. In the first measurement a specific heat discontinuity was seen clearly at the transition, but in the second the specific heat curve followed that of the normal state without appreciable discontinuity. This would suggest that superconductivity inside the meshes of the sponge had been suppressed by the frozen-in flux. The method was somewhat similar to Keesom and Kok's method for estimating the proportion of a thallium block that became superconducting during the transition in a field, and was thus a way of determining the fraction of the alloy occupied by the superconducting meshes.

The discussion at the Royal Society that May provided the opportunity for perhaps the last important European forum, with the clouds of war gathering, for the exchange of results and ideas on superconductivity. It was chaired by the footloose McLennan, who had been residing in Surrey near Virginia Water since his retirement from his chair at Toronto and was now strongly involved in Royal Society affairs. Characteristically, he "undertook to provide hospitality" for several of those invited from abroad, including Ralph Kronig as well as Meissner.[14] He also gave the keynote address—an address that was up to his usual high standards, succinctly summarizing the status of the many facets of superconductivity at that time. He can be forgiven for overstressing the contributions of his associates in Toronto.

The gathering in London proved to be a last chance for McLennan to interact with many of his farflung colleagues and friends. One more honor awaited him, when in July he was made a Knight Commander of the British Empire for "fundamental discoveries in physics and scientific work in peace and war."[15] (His career, in addition to creating a physical laboratory of world class stature from what was originally hardly more than a physical cabinet of instruments, had spanned radioactivity, spectroscopy, and low-temperature physics. His work "in peace and war" had ranged from antisubmarine warfare in the first world war to a near-abiding preoccupation with promoting radium therapy since his retirement.) In the two months that he had left after the bestowal, superconductivity remained important to him despite his newfound devotion to more pressing causes, and he stayed in touch with Kronig and others. Thus, on 18 September, he wrote Kronig: "I saw London recently at Oxford and he is [working] on the mathematical formulation of superconductivity he brought forward last spring at the Royal Society. I am living a quiet yet busy life over here, but try to keep up my readings on modern physics. ... "[16] Three weeks later, 8 October, he wrote again, this time from Paris:

> I am in Paris as a member of the [Committee of the International Bureau of Weights and Measures]. You will recall ... the quantization of space as it applies to the Zeeman effect. I have been wondering if this concept was [*sic*] involved in the process of a metal becoming a superconductor. At the lowest temperatures ... the energy differences might become comparable and of considerable relative importance.[17]

The next day, 9 October, with the meeting over, McLennan boarded the boat-train for Boulonge. He began to feel ill during the journey, and was found dead when the train made an unscheduled stop for aid at Abbeville near the Channel coast.

Among those attending the Royal Society discussions in May were J. D. Cockroft, D. Shoenberg, Keesom, Meissner, Kronig, Brillouin, Kurti, Simon, the Londons, Mendelssohn, J. D. Bernal, and N. F. Mott. A good portion of the agenda for the meeting was given to the Meissner effect. In his own report, Meissner emphasized once more that cooling a hollow cylinder below T_c in a field produces phenomena more complicated than those characteristic of a solid sample.[18] These phenomena, he added, were apparently explained by Gorter and Casimir in terms of normal regions into which the field lines are contracted locally to such an extent that the critical field is exceeded. Heidenreich's latest results with the bored-out tin cylinder had, if anything, clouded the phenomena even more. This time the culprit was the remaining field around the *outer* surface when the homogeneous external field was switched off; it too refused to vanish completely. Adding to the confusion, the experimental results on this very point were found to be in disarray when Meissner once again troubled Heidenreich (who was now located at Berlin-Friedenau) for them in early May while preparing his London report. In particular, a certain diagrammatic record of the declination of the external remaining field was nowhere to be found—a "downright catastrophic" development.[19] Fortunately, close scrutiny of the data at hand allowed reconstructing the behavior to Meissner's satisfaction.

> ... The field near the outside surface of the crystal tube before taking off the outer field is not exactly that to be expected for zero permeability. The contraction of the lines of magnetic induction is less than for a solid crystalline cylinder (without interior holes), and the variation of the field strength from point to point near the surface is more complicated. When the outer homogeneous field is taken off, there remains a field in the interior bored-out region as well as near the outer surface. ...
>
> These experiments of Dr. Heidenreich confirm the statement of our first publication, namely that the hypothesis of a vanishing permeability does not enable one to describe the results of the experiments with hollow cylindrical tubes. The recent experiments also give rise, I believe, to doubts as to whether the hypothesis of Gorter and Casimir is valid for all experiments. Especially the diagrams, taken from the observations, lead, so far as can be seen now, to the conclusion that the normal component of the remaining field at almost all points of the inner and outer surface of the crystal tube differs from zero, a fact which seems to me not to be favorable to the *general* validity of the hypothesis of Gorter and Casimir and therefore also of the theory given by the Londons.[20]

In his report, Mendelssohn also dwelt on the empirical evidence for zero induction in light of varying amounts of permanent frozen-in flux and irreversibility accompanying the observations.[21] Both effects were strongly dependent on sample size, shape, crystal arrangement and—above all—purity. He confined himself mainly to alloys where "this dependence ... seems to have a deeper significance," barely touching on his parallel, ongoing effort with Babbitt on solid and hollow tin spheres. In so doing, he ignored a new paper of theirs, following up on the preliminary letter to *Nature* the year before, and submitted for publication only two months before the Royal Society meeting.[22] The new results, in fact, shed little further light on the matter besides confirming that deviations from the Meissner effect were more pronounced in the hollow sphere. Possibly Mendelssohn was becoming uneasy about the purity of the tin that was utilized, which indeed was soon revealed to be no better than commercial grade.[23]

A certain point in Mendelssohn's new paper with Babbitt stirred Meissner to confide in Fritz London the following spring, in an argumentative letter that was soon brought to Mendelssohn's attention secondhand.[24] Upset, Mendelssohn wrote to Meissner at once.[25] Meissner's pique had been raised, not by the uncovering of further deviations from $B = 0$—deviations that Meissner was all too aware of—but by Mendelssohn and Babbitt's "control experiment" (cooling in zero field and then applying the field). Such experiments had confirmed that "under special circumstances a superconducting sphere can behave as if it were perfectly diamagnetic."[26] Meissner regarded this conclusion as self-evident, with the experiments proving nothing vis-à-vis the new (Meissner–Ochsenfeld) effect. Mendelssohn countered that since Maxwell's electrodynamics for perfect conductivity appeared to hold for tin, mercury, and lead but not for very pure tantalum, it had seemed worth the trouble to put Maxwell's prediction to an experimental test. Meissner immediately replied that if the results of such spurious control experiments disagreed with Maxwell it could only signify that regions in the sample must have been nonsuperconducting. "As interesting as the tantalum experiment may have been, it cannot have been of fundamental importance."[27]

Meissner ends his admonishment with an interesting remark that seemingly places the credit for the Meissner–Ochsenfeld effect in perspective. Noting that a comprehensive paper on all the experiments with Ochsenfeld and Heidenreich is intended but for the time being is thwarted by "sundry external circumstances," he adds: "Ochsenfeld, by the way, remained in the laboratory but few weeks, and participated in the discovery of the new effect only by chance."[28]

A newcomer to the scene attended the discussions in London. David Shoenberg reported on the first experiments on superconductivity from Cambridge's Mond Laboratory. Kapitza's novel helium liquefier had been operational at the Mond for some time. (By coincidence, Kapitza's note announcing the first liquefaction of helium at Cambridge appeared in an issue of *Nature* containing a note by Lindemann proclaiming the same achievement at Oxford.) Schoenberg's experiments involved measuring the force on a superconducting lead sphere in a low inhomogeneous field with the aid of a balance; the force divided by the field gradient gave the magnetic moment. The magnetization curve thus obtained behaved as expected from the Meissner effect and demagnetizing coefficient appropriate for a solid sphere; the slight frozen-in flux suggested that the lead was purer than that used by Keeley, Mendelssohn, and Moore.

Shoenberg, formerly a student of Peter Kapitza, was then an "Exhibition of 1851 Senior Student" at Cambridge. The Mond Laboratory, adjacent to the Cavendish, had been created with funds from the Mond Nickel Company by Kapitza circa 1930 under the auspices of the Royal Society and under the general supervision of Rutherford, Kapitza's mentor.[29] In 1935, the laboratory was still in a state of abject shock following Kapitza's detention in the USSR while he was vacationing there the previous fall. Rutherford himself took over responsibility for the Mond, but day-to-day administrative affairs were looked after by Cockroft, who had personally collaborated with Kapitza both on the high-field project and on liquefiers. The protracted negotiations with the Soviet authorities over the transfer to Moscow of Kapitza's high-field apparatus, especially the famous generator, were in full swing, and no doubt sorely tried Cockroft's patience. Eventually, Rutherford acquiesced, believing the apparatus was in fact something of a "white elephant" and opting instead to concentrate on transforming the Mond into a first-rate cryogenic laboratory with high-field electromagnets such as the large one near Paris. Meanwhile, to get Kapitza started properly, Rutherford dispatched Shoenberg, Kapitza's laboratory assistant, Emil Laurmann, and Henry Pearson, chief mechanic at the Mond, to Moscow; all three stayed in the Soviet Union for 3 years.[30] (As an aside, it was Rutherford who had been mainly responsible for Lindemann's election to the chair of physics at Oxford, and thus he played an unsung role in establishing both of the two premier British cryogenic laboratories, the Clarendon and the Mond. Rutherford and Lindemann were two strong personalities, and the relationship between them was never happy. Gradually it chilled to the point where they were no longer on speaking terms.)

The physical raison d'être for Mendelssohn's sponge model was inhomogeneities in the alloys, exemplified by lead–bismuth which is a eutectic mixture of two distinct phases. As Mendelssohn remarked at the close of his presentation at the Royal Society discussions:

> We think that all experimental results so far obtained on impure metals and on alloys can be explained by their inhomogeneity which causes the formation of a "sponge" of higher threshold value. Such a model would substitute the *ad hoc* assumption of a fine partition in supraconductive and normal regions in Gorter's recent theoretical treatment and thereby avoid a new assumption for which otherwise no simple reason could be given.[31]

In fact, Gorter had not remained quiet during the mounting debate on alloys; Mendelssohn's reference is to a short but prescient paper on the subject submitted by Gorter to *Physica* a month prior to the Royal Society meeting.[32] In it Gorter suggested that the difference between the behavior of pure single-crystal superconductors and that of alloys could be explained by postulating (ad hoc, to be sure) that the alloys subdivide into extremely thin superconducting regions ("razor blades") parallel to the external field, and separated by normal regions. Splitting into regions thinner than, or of the order of, the penetration depth lowers the free energy of the alloy, allowing superconductivity to persist up to higher fields than possible in ordinary superconductors. Gorter's paper is remarkably prophetic in anticipating several concepts in the mature formulation of type II superconductivity. One is Gorter's notion of a "minimum size k for the supraconductor" embodying the modern concept of coherence length. If l is the penetration depth of Becker, Heller, Sauter, and Braunbek, and H_0 the thermodynamic critical field, then if $k < l$ superconductivity will be possible at fields up to a maximum critical field $H_0(l/k)$. This too hints strongly of (Abrikosov's) eventual upper critical field H_{c2} for type II materials. If on the other hand, $k > l$, then "as soon as $H > H_0$, every trace of supraconductivity will vanish. This appears to be on the whole in agreement with the behavior of single crystals of very pure metals"—that is, of the type I superconductors. Gorter added that in "some respects however rather pure metals behave like an 'alloy' "; perhaps here $k \approx l$.

> The behavior of supraconductive alloys is certainly explained far from completely by these remarks [he cautioned], which do not even offer a suggestion, why l should be especially large or k especially small for an alloy; but still, the consideration of the possibilities connected with the limit[s of] k and l seems to offer a useful point of view for the discussion of results and for the instigation of new experiments.[33]

Or, as Gorter put it much later, "the problem remained why this behavior [of alloys] did not occur in common superconductors too."[34] The answer to the problem was eventually supplied by A. Brian Pippard of Cambridge, 15 years later. In the short run, however, important insight was provided within a few weeks from a more quantitative elucidation of Gorter's concepts by another member of the Clarendon circle. The new analysis was published in a paper submitted to the Royal Society in June of 1935,[35] although it was actually written *before* the appearance of Gorter's note.[36] To deal with it, it is necessary to digress and take note of two new arrivals at Oxford: the London brothers.

11.3 Fritz and Heinz London

The older of the two London brothers was introduced to superconductivity by a circuitous route. Fritz London had preceded his brother Heinz by about a year in the exodus of Jewish scientists from Germany in 1933, settling temporarily in Oxford on one of the fellowships from the Imperial Chemical Industries at Lindemann's disposition. Fritz had received his doctorate in philosophy at the University of Munich in 1921, earning the degree entirely on his own initiative by petitioning the faculty with a philosophical paper written without any formal guidance.[37] After some further follow-up work in philosophy and a stint as a *Gymnasium* teacher, he became dissatisfied with philosophy as a result of sensing imminent breakthroughs in physics. (As well he might; he himself would contribute to the new enlightenment!) Fritz spent a brief period on his own in Göttingen, and then returned to Munich where in 1925 he became assistant to Arnold Sommerfeld; subsequently he held appointments in theoretical physics with Paul Peter Ewald (an early student of Sommerfeld) at Stuttgart and then with Erwin Schrödinger, first in Zürich and then in Berlin. He soon built a strong reputation in molecular physics, especially on the quantum-mechanical theory of the chemical bond (London and Walter Heitler wrote their famous paper on chemical bonding in 1927 during London's stay in Zürich). However, his forced departure for Oxford coincided with a further shift in his scientific propensity, this time to an abiding preoccupation with a concept of long-range order. This concept would become a central and recurring theme underlying his work on superconductivity, and in his later broader scientific interests. More specifically at the center of his immediately forthcoming work in superconductivity would figure certain concepts supplied by his younger brother Heinz. Heinz, in fact, would be responsible for interesting Fritz in the problem of superconductivity in the first place.

Heinz London began his undergraduate studies at the University of Bonn in 1926. After a brief period with the chemical firm of Heraeus, he continued his studies in Berlin and Munich, finishing in 1931.[38] Next, he came under the tutelage of Simon at Breslau (where the elder Londons had first resided and where Fritz was born). From Simon Heinz acquired a passion for thermodynamics,[39] something that would soon serve him well. No longer perceiving himself as treading on Meissner's territory, Simon suggested that London look into superconductivity for a thesis topic. London elected to follow up a suggestion of Walter Schottky of Siemens and take a crack at a largely unexplored superconducting domain: to investigate the possibility of Joule heating in superconductors from high-frequency alternating transport currents. His attempt to measure such ac effects was frustrated for the time being, mainly because the highest frequency available to him, 40 MHz from a radio source, proved to be

Figure 11-3. Fritz London, circa 1954. Courtesy Physics Today Collection, AIP Niels Bohr Library.

too low for the sensitivity of the detector at hand. A certain lack of manual dexterity—a characteristic that was affectionately noted by colleagues—probably contributed to the failure. The upshot was that he lost patience and decided to attack the problem theoretically instead. It was just as well, because in mulling over the problem he developed some interesting notions about the superconducting state in general and high-frequency ac superconductivity, in particular. Had it not been for his characteristic reticence to publish, his ideas would have earned him several firsts in superconductivity; as it was, they earned him a doctorate despite the lack of experimental data.

The first notion was that of a finite penetration depth in superconductors due to the inertia of the electrons, similar in concept to that of Becker, Heller, and Sauter. London described the influence of the inertia by the equation

$$\Lambda \dot{\mathbf{J}} = \mathbf{E} \tag{11-1}$$

which is simply Eq. (9-4) with the constant $\Lambda = m/ne^2$, nowadays known as the London order parameter. Combining this equation with those of Maxwell leads to the penetration depth $\lambda = c\sqrt{\Lambda/4\pi}$, as noted in Section 9.3. By the time Lon-

Figure 11-4. Heinz London. Courtesy Francis Simon Collection, AIP Niels Bohr Library.

don's thesis was completed in 1934, the paper by Becker, Heller, and Sauter was out, and the tendency has been for Becker's team to be credited with this important precursor to the subsequent macroscopic theory of Fritz and Heinz London. The equation actually seems to have been in the air in 1933–1934; Braunbek, for one, can also lay claim to its derivation.

Heinz's second notion was conceived as a mechanism for ac superconductivity based on a two-fluid model of the superconducting state. In it, normal electrons furnish a resistive path parallel to the inductive path of the superconducting electrons. Direct currents are carried entirely by the inductive, nonresistive path, while alternating currents divide between the two. As the frequency increases, the supercurrent contribution diminishes and the resistive losses increase rapidly. As promising a notion as it was, London was upstaged again, this time by Gorter and the publication of Gorter and Casimir's model in 1934.

11.4 The Londons at Oxford

When Simon left Breslau for Oxford with Kurti and Mendelssohn in 1933, Heinz London stayed on to complete the formalities for his thesis. His doctorate was issued late in 1933—one of the last awarded to a Jewish candidate in Nazi Germany. London forthwith joined the circle of Breslau émigrés in Oxford in 1934, which included his brother. His 2 years at the Clarendon are best remembered for his collaboration with Fritz on their famous phenomenological theory of superconductivity, but also for his own theory of phase-equilibrium of superconductors, the latter based on a thermodynamic analysis of the equations constituting the former theory. Heinz also resumed his experimental search for ac losses, but still without much success even though the frequency was pushed to 150 MHz. (Only after his move to Bristol in 1936 did the effort finally pay off, after the frequency was raised a further factor of ten—and just in time to be interrupted by the war.) More immediately successful was an experiment that disproved a suggestion originating in a collaboration with Fritz: that an electric field might penetrate to the same depth as a magnetic field in a metal upon undergoing a thermal transition to the superconducting state.

In spite of the close professional rapport between Fritz and Heinz London, the contrasts between the two were legion. Fritz was a superb organizer; Heinz was "shockingly untidy."[40] Fritz was a deep conceptual thinker; Heinz was highly pragmatic. The latter has been described as excelling in interpreting experimental results, "but having to be kept away from the apparatus."[41] Despite ribbing from colleagues about his clumsiness in the laboratory, Heinz seems to have regarded himself primarily as an experimenter. Nevertheless, this characterization was apparently far too simple. Perhaps Fritz put it best. He is said to have been fond of remarking that he would have become a mathematician if his father (a professor at Bonn) had not been one, and that Heinz would have become a theoretician if Fritz had not been one.[42]

In what follows, just enough of the phenomenological theory will be discussed to bring out some of its main features as it relates to the Meissner effect and bears on Heinz London's equilibrium theory.

Fritz enthusiastically pounced on the ideas Heinz had nurtured at Breslau, in particular the acceleration Eq. (11-1). The problem, begins their famous joint paper of 1935, was "that actually this equation ... implies more than is verified by experiment; moreover, presupposing an acceleration without any friction it implies a premature theory, the development of which has presented a hopelessly insoluble problem to the mathematical physicist."[43] The insoluble problem was the old one of Bloch, namely, that the most stable state of a superconductor was invariably one with zero current (Section 8.1). The way out, they reasoned, was to: 1) essentially abandon Eq. (11-1), which was too general, and 2) abandon the attempt to devise a model of a superconductor with a stable persistent current *without* assistance of an external magnetic field.

To be sure, Eq. (11-1) could not be dismissed outright; it had to be retained provisionally to show "where it must be corrected." The first step was to combine the equation with Maxwell's equations for curl **E** and curl **J**, deriving in the process Eq. (9-6) discussed earlier in connection with Becker, Heller, and Sauter. Integration with respect to time again gave nothing very new: a nonhomogeneous differential equation for **H**. The general solution to this equation was a solution to a homogeneous equation for **H** (representing cooling and *then* applying a field) plus a particular solution given by an arbitrary constant of integration $\mathbf{H}_0$ representing the "frozen-in" field existing in the body of the metal when it last lost its resistance *in* a field. However, in light of Meissner's results (unknown earlier to Becker et al. and Braunbek), the Londons took the ad hoc liberty of setting the constant of integration equal to zero. The remaining solution to the homogeneous equation was a phenomenological equation connecting the current with the *magnetic* field, rather than the electric field:

$$c\Lambda \operatorname{curl} \mathbf{J} = -\mathbf{H} . \tag{11-2}$$

This equation is the centerpiece of their joint paper submitted to the Royal Society in October of 1934. It leads to a consistent description of the state of magnetization, irrespective of the path by which that state is produced; it replaces, then, Ohm's law in the case of superconductors. (Λ is the analog of specific resistance—i.e., a new characteristic constant depending on the material.)

In their seminal paper, the Londons note that Eqs. (11-1) and (11-2) possess the same degree of generality. Eq. (11-2) embraces the Meissner effect but in some respects embraces less since Eq. (11-1) cannot be deduced from it, only the weaker statement curl $(\Lambda\dot{\mathbf{J}} - \mathbf{E}) = 0$. Integration of this relationship allows $\Lambda\dot{\mathbf{J}} - \mathbf{E}$ to be represented as the gradient of an arbitrary scalar constant μ, or equivalently

$$E = \Lambda\,(\dot{\mathbf{J}} + c^2 \operatorname{grad} \rho) , \tag{11-3}$$

where $\rho = \operatorname{div} \mathbf{E} = \mu/\Lambda$ is a charge density. In their joint paper, Eq. (11-3) was proposed as a second fundamental equation connecting current and electric field. A feature of this formulation, they added, was the possibility it raised of an electrostatic field penetrating to a depth of order λ in a superconductor (in sharp contrast to a normal conductor). As noted earlier, this question was eventually put to an experimental test, by Heinz London himself at the Clarendon in 1936; he looked for a decrease in the capacity of a condenser with superconducting plates (two half-cylinders of mercury separated by mica foil) when it was cooled below the transition temperature. No decrease was found. "It follows from these measurements," concluded London's report on the experiment "that no electrostatic fields exist in a pure supraconductor, not even in a thin surface layer. ... In all stationary cases the surface of a supraconductor is therefore exactly an equipotential surface."[44] Hence it could also be concluded that grad $\rho = 0$.

Aside from the tantalizing question of the electrostatic field, which was still open to experiment in 1934, the new significance accorded a magnetic field by their equation was to the Londons the essential fruit of their happy collaboration that year.

> In contrast to the customary conception that in a supraconductor a current may persist without being maintained by an electric or magnetic field [ends their paper], the current is characterized as a kind of diamagnetic volume current, the existence of which is necessarily dependent upon the presence of a magnetic field. That field itself may be produced reciprocally by the current.[45]

Just as Heinz London's theoretical ruminations were the inspiration for the macroscopic London theory, so the macroscopic theory provided the apparatus for constructing Heinz's own equilibrium theory shortly afterward. The equilibrium theory furnished the hitherto lacking quantitative distinction between alloys and pure metals in terms of a new thermodynamic surface parameter, and a theoretical construct that quantified Gorter's ad hoc but equivalent scheme based on the concept of domain size. Abandoning Gorter's discontinuous transition from a state with $\mathbf{B} = \mathbf{H}$ to one of $\mathbf{B} = 0$, Heinz, in his paper of 1935, uses the two new Eqs. (11-1) and (11-2) to derive the connection between current density and the magnetic field that sustains it.[46] From Eq. (11-2) with $\mathbf{B} = \mathbf{H}$ and using c curl $\mathbf{H} = \mathbf{J}$, he shows, first, that $\mathbf{H}$ and $\mathbf{J}$ decrease exponentially inward from the surface of the specimen. If the electron density n is approximately equal to the number of valence electrons per cubic centimeter, then the penetration depth D (in London's notation) is given by $D = c\sqrt{\Lambda} \approx 10^{-6}$ cm. London next treats the equilibrium conditions for a mobile boundary between a normal and a superconducting phase in a body exposed to an applied field (which makes possible the coexistence of the two phases). (The space is assumed to be subdivided into volume elements sufficiently small that they encompass only a single phase.) The field and screening currents decay exponentially from the interface surface into the superconducting phase; in the (assumed) absence of an electric field the current density is discontinuous at the interface and zero outside. In considering the energy balance during a virtual displacement of the boundary surface, London equates the divergence of the Poynting vector (energy flow) to the time rate of change of the local energy density. Integrating over the metal, applying Gauss's theorem, and taking account of a discontinuity in the energy density at the moving phase boundary given by $1/2\Lambda J_s^2$ (where J_s is the current density on the superconducting side of the boundary), he shows that when the superconducting region expands there is unaccounted additional energy that must disappear; this is the energy due to the phase transformation occurring at the boundary. For an isothermal process one must use the free energy, which decreases by the phase transformation. Equating the energy due to the phase transformation to the energy deficit finally produces the desired equilibrium condition:

$$\Delta F/V = F_n - F_s = 1/2\Lambda J_c^2 . \tag{11-4}$$

That is, whereas for superconductors of large dimensions the difference in free energy per unit volume is given by Gorter's relation

$$\Delta F/V = F_n - F_s = 1/2H_c^2 ,$$

for the superconductors consisting of superconducting phases of small dimensions the equilibrium condition is dictated, *not* by a magnetic threshold, but by a *threshold for the current density* at the boundary of the two phases.

London then considers in greater detail the stationary conditions for **J** and **H** in a plane superconducting lamina of thickness d comparable to the penetration depth, exposed to a longitudinal field. Calculating the contribution to the thermodynamic potential of the field, using the new equations and in view of the equilibrium condition (11-4) introducing the current density $\sqrt{\Lambda}J_c = (D/c)J_c$, he obtains the threshold field for a lamina of thickness d:

$$H_T = H_c ctgh(d/D) \,. \tag{11-5}$$

This reveals that the laminar superconductor should have a much higher magnetic threshold than a bulk superconductor. From (11-5), "one is tempted to conclude that every supraconductor in a magnetic field [above H_c] should split up into a great number of thin supraconducting laminae or fibres separated from each other by thin, normal conducting regions [analogous to Gorter's ad hoc razor blades], as then supraconductivity could persist at these higher fields."[47] London attributes the fact that this is *not* true for undeformed single crystals of pure metals to a normal-superconductor interface surface energy (or surface tension). The surface energy must be positive for pure metals, but negative for alloys. In alloys, the state of lowest energy is one of splitting as much as possible into two phases. For $H < H_c$ the metal is a superconducting matrix subdivided by normal laminae; for $H > H_c$ the metal is a normal matrix subdivided by superconducting laminae. Presumably the inhomogeneous structure of the alloys facilitate the formation of such microscopic regions. The lower limit for the interface surface energy γ sufficient to restore the homogeneous phase could be expressed by the inequality

$$\gamma > DH_c^2/c \,.$$

States of negative surface energy might also account for the hysteresis observed with commutated currents in the descending magnetic transition curve in a transverse field, as suggested by London—that is, for the fact that the onset of the superconducting state is delayed in an unreproducible manner as the field sinks below the threshold value. Analogous to the supercooling of a vapor, the hysteresis occurs because of the difficulty in initiating the growth of a nucleus of the stable phase. Another effect eventually attributed to the model has been alluded to earlier: the "0.58 mystery" of Section 9.2, which seems to be accounted for by the size dependence of the characteristic length D.

11.5 Schubnikow's Final Contribution

By mid-1935 a pattern was emerging from the spectrum of magnetization behavior that was gradually being pieced together at Oxford, Leiden, and Kharkov. At first it suggested simply a clear distinction between elemental and alloy superconductors, provided the samples were extremely pure and free from strain. Elements, under ideal conditions never quite attained, would exhibit complete flux exclusion, a reversible transition at a well-defined critical field, and a final state independent of the magnetization history. Alloys invariably exhibited gradual flux penetration, starting at a critical field somewhat lower than H_c typical of a pure metal and up to a second, considerably higher threshold field; in decreasing fields alloys often exhibited some degree of hysteresis leading to trapped flux on reaching zero field, even for pure specimens. The last observation, however, hinted at a further subdivision of the alloys—some exhibiting distinct irreversible magnetic and thermal properties and some apparently not.

It was his attempt to account for the observed magnetic hysteresis and trapping of flux (reduced Meissner effect) in two-phase alloy systems of the former type, such as Pb–Bi, that led Mendelssohn to introduce his sponge model with multiply connected filaments.[48] The exact role in the model of lattice imperfections "had to be left open, but the problem was fully realized at the time," insisted Mendelssohn in later recalling the early years.[49] It was particularly so in 1936 with the discovery that not only alloys but also "hard" elemental superconductors, such as tantalum, showed evidence of this behavior.[50] In view of this, Mendelssohn and Judith Moore concluded that "only further experiments on pure Ta specimens with undistorted lattice (if possible single crystals) can decide whether the splitting up in small supraconductive regions is spontaneous or caused by disturbances of the lattice";[51] however, they left no doubt that it could be explained by inhomogeneities, whether impurities or strains.

The competing model advocated independently by Gorter and Heinz London sought to account for high-field materials exhibiting reversible behavior, and in constructing it they laid the foundation for a concept of *two ideal* types of superconductors distinguishable by a bulk parameter—in London's version the sign of an interphase surface energy. To be sure, both Gorter and Heinz London recognized that inhomogeneities, such as polycrystalline structure and internal stresses, would probably have to be invoked to account for any hysteresis and trapped flux.[52] However, their basic scheme of a finely divided mixture of superconducting and normal laminae represented an *intrinsic* thermodynamic feature of the bulk alloy—one not predicated on inhomogeneities.

The stage was thus set for a decisive experimental showdown invoking a homogeneous alloy free of defects; according to Mendelssohn such a structure should not exhibit sponge behavior but rather show behavior typical of pure elements; according to Gorter and Heinz London it should be capable of remaining superconducting to high magnetic fields. The definitive test was not to be John Daunt and Mendelssohn's follow-up experiment with much purer but nevertheless polycrystalline rods of tantalum and niobium.[53] Instead, the distinction between the two models was dramatically underscored in a series of experiments reported in mid-1936 by Schubnikow, Chotkewitsch, Schepelew, and Rjabinin—albeit experiments not explicitly designed, it would seem, to force an experimental showdown between the two theoretical schemes.[54] Unfortunately, the importance of the Russian work was unappreciated for a very long time.

The new measurements performed under Schubnikow's ongoing magnetization program not only covered polycrystalline and monocrystalline metals, but also covered measurements on single-crystal samples of Pb–Tl, Pb–Bi, Pb–In, and Hg–Cd. In particular, systematic magnetization measurements on Pb–Tl alloys with 50, 30, 15, 5, 2.5, and 0.8 by weight % Tl in Pb now revealed a critical alloy composition no more than 0.8-2.5% Tl "sufficient to bring out a dependence of induction on field which is characteristic of the alloy."

> The result of the investigation of different alloys which for the most part form solid solutions, reveals in all cases a quite similar dependence of induction on field. Up to a certain limit of the field H_{k1}, the permeability remains almost equal to zero; on increasing the field, it increases and gradually it approaches the value of one. The interval of field H_{k1} - H_{k2} increases with the amount of admixture in the superconducting metal. Such unusual properties of the superconductor *cannot be explained by hysteresis phenomena, since it is precisely at high increasing and decreasing fields that the phenomenon is sufficiently reversible and the hysteresis quite small* [italics added].[55]

Moreover, despite the disparity between the upper threshold field H_{k2} for alloys and the critical field H_c for pure metals, calculations had showed that the difference in free energy of the superconducting and normal material, obtained by integrating under the magnetization curve, "is of the same order and dependent on the temperature in the same way."[56] Therefore, the jump in the specific heat of an alloy at zero field should also be comparable to that of a pure superconducting element. Unfortunately the calculated jump, 7%, was "not very accurate" and Mendelssohn and Moore's corresponding specific heat measurements on $PbTl_2$ were too inaccurate at just these temperatures to be reliably compared with the calculations.

In spite of these uncertainties, it is generally agreed today that Schubnikow and colleagues rank among, and probably were, the first to appreciate the thermodynamic character of alloy superconductivity that would constitute the basis for type II superconductivity elucidated by their countrymen 20 years later. They attributed ideal alloy behavior to the bulk homogeneous alloy, not to inhomogeneities—a point stressed by Berlincourt in his analysis of factors leading to the Ginzburg–Landau–Abrikosov–Gorkov theory.[57] Curiously, Berlincourt notes, the 1936 paper of Schubnikow et al. makes no reference to Heinz London's paper *or* to the Gorter–H. London theory in general—a theory nicely corroborated by the latest Kharkov experiments! Nor does it refer to Mendelssohn's sponge, even though it acknowledges Mendelssohn and Moore's letter of May 1935 that introduces the sponge; thus Schubnikow's own theoretical leanings at the time remain a matter of speculation. Nor, for their part, did Gorter, Heinz London, or Mendelssohn take notice of the definitive Kharkov paper, even though reprints were exchanged on a regular basis between Schubnikow and his colleagues in the West.

In fact, nothing more was heard on the subject from Schubnikow and his team. A few more papers on superconductivity generally appeared under his signature in the Soviet journals during 1936–1937, and it is known that he continued to work with another group of young scientists at the Kharkov Institute (including his wife) at least into early 1936 on the magnetic transformation of paramagnetic salts and allied subjects. That year, however, Schubnikow and his colleagues failed to appear at the Sixth International Low Temperature Conference at the Hague, despite having indicated to the organizers their intention to attend. We now know from a eulogy given in 1966 by O. I. Balabekyan that Schubnikow was arrested in the purges of 1937 and sentenced to 10 years imprisonment.[58] He died in 1945. In April of 1957, he was exonerated posthumously by the Military Board of the Supreme Court of the USSR. That year marked an All-Union Low Temperature Conference in Moscow, attended by a small number of physicists from Oxford and Cambridge who met personally for the first time fellow Soviet scientists previously known only from the Soviet literature of the 1930s. Foremost among the long list of names toasted to great applause at the conference banquet was Schubnikow's, who was honored in the presence of his widow and former colleague Olga Trapeznikova. At the same conference, A. A. Abrikosov read his seminal paper on type II superconductivity. With Schubnikow's public exoneration, Abrikosov could at long last acknowledge Schubnikow's paper of 1936—not hitherto possible "since up to then Soviet etiquette required that anyone who had disappeared in the purges had never lived."[59]

CHAPTER 12

Theoretical and Practical Breakthroughs

12.1 Winding Down, circa 1940

Several novel experimental lines of attack bore fruit on the eve of the second world war. One, on magnetic penetration effects in superconductors, provided the earliest quantitative confirmation of London's phenomenological theory, and was the precursor to a whole new experimental subfield in superconducting research that blossomed after the war. Another, on the temperature dependence of specific heats, provided perhaps the first inkling of the existence of an energy gap in the excitation spectrum of a superconductor—a watershed discovery on the long road to a microscopic theory of superconductivity.

Heinz London's prediction of enhanced threshold fields in very small superconducting specimens—that is, Eq. (11-5)—provided the incentive for investigating the field penetration in fine superconducting wires and thin films. At the same time, all of the accumulating sundry evidence for the importance of surface currents and the concomitant implications of a finite penetration depth, not to speak of the renewed focus on alloy behavior, collectively hinted to a rich field of experimentation ready to be exploited on finely divided superconducting specimens. The hint was actually first taken by the Canadians. Experiments in the McLennan Laboratory at Toronto by A. D. Misener, H. G. Smith, and J. O. Wilhelm during 1934–1935 gave sketchy indications of a size dependence of H_c in thin films.[1] As luck would have it, their films were deposited on a substrate of normal metal, and consequently the possibility of alloying at the interface could not be discounted. But in 1937 Rex Pontius, an American Rhodes scholar working under Mendelssohn at the Clarendon, succeeded in preparing lead wires as thin as 6 microns by the "Taylor process": a method in which wires are drawn down in pyrex capillaries in an oxygen flame, and the pyrex is subsequently dissolved away with hydrofluoric acid. Resistance measurements on these wires in a longitudinal field clearly showed an inverse dependence of critical field on wire size below a certain critical diameter.[2] Despite great care in preparation and handling, $\approx 10^{-4}$ cm was essentially the smallest wire size possible, however. Nor was the effect very large, amounting to a maximum increase in critical field of about 4%.

Significantly finer samples, say a factor of ten smaller, was only possible by vacuum evaporation of thin films—not a routine laboratory procedure in 1937. Curiously, in this the Russians now seized the initiative, the second time for that matter and despite the upheavals in all walks of Soviet life, including scientific and technical circles. In a letter to *Nature* in mid-1938, Alexander Shalnikov of the Institute for Physical Problems in Moscow reported resistance measurements with thin (10^{-5}–10^{-7} cm) films of lead and tin evaporated on glass backings. They confirmed that though "the value of I_c at a given temperature decreases with the thickness of the film," on the other hand, "the magnetic fields required to restore the resistance were not lower, but much higher than for the bulk metal."[3]

The Moscow Institute, formally the S. I. Vavlov Institute of Physical Problems but affectionally known as Kapitza's Institute, had been founded by decree in 1934 as compensation to Kapitza for retaining him against his wish in the Soviet Union that year. Shalnikov, formerly of Leningrad, became the first of Kapitza's new colleagues, and the two personally roamed the streets of Moscow searching for a suitable site to place the institute. In 1937, Lev Landau, who until then headed the Theoretical Division of the institute in Kharkov, joined the new institute on Kapitza's invitation as head of its own Theoretical Division.

For reasons undoubtedly connected with deteriorating Soviet political conditions, it took Shalnikov's letter over a year to appear in print in the West. Its date of submission is actually 27 April 1937, or a month ahead of Pontius's. Meanwhile, in 1938 the first of a new series of similar measurements on mercury films as thin as 400 Å (4×10^{-6} cm) were reported by a joint collaboration between E. T. S. Appleyard, J. R. Bristow, and H. London at the H. H. Wills Laboratory of the University of Bristol and Misener, who had by then joined the Mond Laboratory.[4] The apparatus was constructed and calibrated at Bristol and laboriously transported to Cambridge for the actual cryogenic measurements that were to be supervised by Misener.[5] Heinz London had moved to Bristol in 1936, the same year his brother left Oxford to spend 2 years at the Institut Henri Poincare in Paris (before settling permanently at Duke University in North Carolina). Appleyard died shortly after the final paper of the Bristol–Cambridge collaboration was submitted to the Royal Society.

The latest measurements showed that the critical fields for thin films were indeed much higher than the bulk critical field H_c. They confirmed that the critical field increases rapidly with smaller film thickness; for a given thickness it also increases rapidly with temperature near the transition point, and may even become infinite there. From these results they concluded "from purely dimensional arguments" that the penetration depth λ, assuming it to be independent of film thickness, rises rapidly as the temperature approaches the transition point. "If we accept the theory of the London type," wrote Appleyard and colleagues, "this result can be most simply explained by assuming that the number of electrons responsible for superconductivity is small or zero at the transition point and increases rapidly as the temperature is lowered."[6] The actual functional relationship between critical field, film thickness, and penetration depth required postulating more definite assumptions about the equilibrium between the normal and superconducting phases in a magnetic field. Two such relationships were soon proposed by Isaac Pomerantchuk,[7] a former student of Landau, and by von Laue,[8] but neither gave very good agreement with experiment.

Although the resistance measurements gave convincing empirical evidence for a gross size dependence of the critical field, they shed less light on the details of the field penetration itself. Far more revealing were the magnetic measurements by Shoenberg, who was back at the Mond Laboratory where he would remain permanently ensconced, following a 3-year stint at Kapitza's Institute in Moscow. His classic susceptibility measurements on small mercury colloidal spheres suspended in chalk or in emulsions revealed that for radii comparable to λ the susceptibility exceeds the bulk susceptibility value expected for perfect diamagnetism (e.g., $\chi_0 = -3/8\pi$ for a sphere).[9] To be sure, the magnetic moment is proportional to sample volume and thus becomes exceedingly small if the sample size is reduced to the point where penetration effects appear ($\leq 10^{-5}$ cm). Shoenberg's trick, however, lay in measuring the *total* magnetic moment of a colloid or emulsion, in which case the small volume of each particle is compensated by the large numbers of particles present. From the temperature variation of χ/χ_0 he was able to deduce the variation of the reduced quantity λ/λ_0 with temperature (λ_0 is the value of λ near absolute zero) by an application of a particular solution of the London equation $\nabla^2 H = H/\lambda^2$ for the case of a sphere of radius r in a uniform field.[10] Shoenberg, first of all, quotes Fritz London for the size dependence of the magnetic moment per unit volume of a small sphere:

$$\chi/\chi_0 = 1 - \frac{3\lambda}{r}\coth\frac{r}{\lambda} + \frac{3\lambda^2}{r^2}$$

which reduces to $\chi/\chi_0 = 1/15\ r^2/\lambda^2$ for $r \ll \lambda$. (Pomerantchuck's and Laue's functions were simplified solutions to the London equation for a plane slab in a parallel field.) Assuming this formula, Shoenberg was able to calculate λ/λ_0 as a function of temperature. In retrospect, the resultant curve for the relative variation of λ/λ_0 with temperature, Figure 12-1, agrees remarkably well with the law that became so well known to later students of the subject:

$$\left(\frac{\lambda}{\lambda_0}\right)^2 = \frac{1}{[1-(T/T_c)^4]}\,. \qquad (12\text{–}1)$$

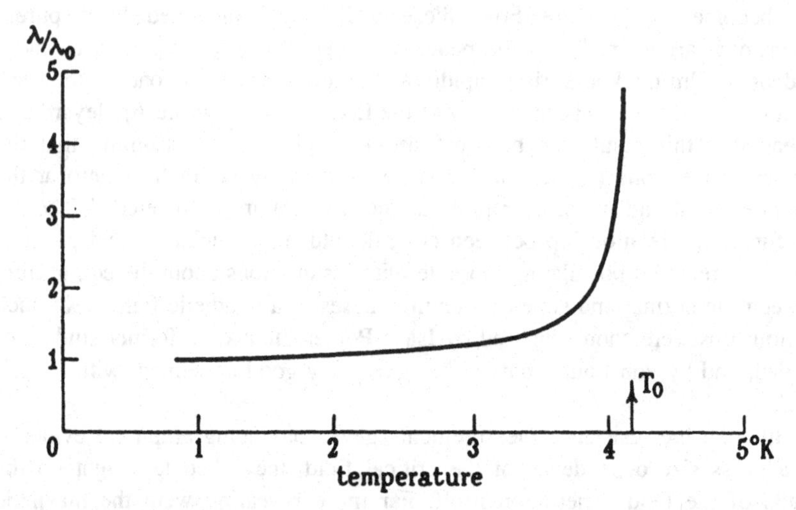

Figure 12-1. Variation of penetration depth, λ/λ_0, *with temperature. Shoenberg, PRS,* **A175** *(1940), 66.*

This expression echoes Gorter and H. B. G. Casimir's two-fluid model, if it is combined with the London theory, since the model[11] has the proportion of superconducting electrons going as $1 - (T/T_c)^4$ and the expression for λ deduced by the Londons (and originally by G. L. de Haas) is $\lambda^2 = mc^2/4\pi ne^2$; that is, λ^2 is inversely proportional to the number of superconducting electrons. (The *absolute* variation of λ with temperature could not be determined by Shoenberg's experiments, since the size distribution of the particles was not known.)

Shoenberg's attention was first drawn to Eq. (12-1) by John Daunt in a private conversation in 1946,[12] by which time the equation had been long considered "an article of faith" and accorded in some circles "a status of almost holy writ."[13] Daunt had subsequently championed the expression and its "theoretical interpretation" in a letter to the *Physical Review* in 1947.[14] Unfortunately, owing to a long undetected misprint,[15] an exponent of 3 was substituted for 4 in the right-hand term.

Soon after the publication of Shoenberg's results, Casimir himself made a comparably insightful experimental observation. He realized that, in view of the rapid variation of λ with temperature, it should not be necessary to confine studies of λ to samples comparable in size to λ. Instead, useful information should be possible with macroscopic samples: the inductance of a coil wound tightly around a cylindrical superconducting sample should exhibit a slight change with temperature due to the change in penetration. Casimir actually tried the experiment in 1940, but his experimental prowess was not up to the task. A slight swelling of the quartz tube containing the mercury sample as the helium bath temperature was reduced caused an increase in the mutual induction; this compensated for the decrease caused by the change in λ and yielded a negative result.[16] After the war the experiment was repeated by more experienced experimentalists with predictably better results. Both low frequency measurements (at 70 Hz) by E. Laurmann and Shoenberg and measurements at microwave frequencies by Pippard gave absolute values of the difference $\lambda(T) - \lambda(0)$ in good agreement with values obtained by various other methods.

A highlight among the many findings covered by Keesom at Bad Pyrmont had been the telltale discontinuous jump in the specific heats of tin and thallium. Less was known about the detailed temperature dependence of the specific heats. About all one could say was that after the initial jump at T_c, the specific heat of the superconductor appeared to decrease more rapidly with decreasing temperature than the specific heat of a normal metal. The temperature variation of the specific heat of normal metals was, however, known to be well represented by

$$C_n = \beta' \left(\frac{T}{\theta}\right)^3 + \gamma T = \beta T^3 + \gamma T . \qquad (12\text{–}2)$$

Here the first term is contributed by the ions of the crystal lattice (θ is the Debye temperature). The second, linear term comes from the free electrons; according to its expositor, Arnold Sommerfeld, $\gamma \propto V^{2/3} n^{1/3}$,[17] where V is the atomic volume and n the number of free electrons per atom. As Sommerfeld himself explained in the general discussion following Keesom's report at Bad Pyrmont, "the γT term is the simplest inference of the elementary theory of metallic electrons and is independent of assumptions about the nature of the crystal lattice."[18] This was so presumed despite the dubious reputation of the free electron gas concept by 1928, the year of publication of Sommerfeld's definitive paper on the subject.

Keesom, at Bad Pyrmont, dwelt mainly on experimental evidence for the electronic term of normal metals at the lowest temperatures. However, appended to the printed version of his report is the abstract[19] of a paper[20] by his colleague J. A. Kok on the less certain electronic contribution, if any, in the superconducting state. Newly available data on tin and thallium suggested a fairly pure T^3 variation below the transition point. Since the lattice contribution also goes as T^3, and is presumably the same in the normal and superconducting states, the electronic term in the superconducting state can be estimated by subtracting the lattice term from the observed specific heat. In addition, however, Kok made a shrewder observation. Assume, with him, that the specific heat in the normal state varies as in Eq. (12-2) and assume purely a T^3 variation in the superconducting state:

$$C_s = \beta T^3 + \alpha T^3 . \qquad (12\text{-}3)$$

Now, differentiation with respect to T of the expression for the difference in Gibbs free energy of the normal and superconducting phases gives, as noted in Section 10.3, the difference in entropies between the two phases

$$\Delta S = S_n - S_s = -\frac{VH_c}{4\pi}\frac{dH_c}{dT}, \tag{12–4}$$

and a second differentiation and multiplying by T gives

$$\Delta C = C_s - C_n = \frac{VT}{4\pi}\left[\left(\frac{dH_c}{dT}\right)^2 + H_c\frac{d^2H_c}{dT^2}\right]. \tag{12–5}$$

Thus far, nothing new is involved. Kok cites Keesom's derivation of both expressions in his undergraduate lectures in 1933, as does Gorter,[21] who repeated it on several occasions. In fact, Keesom's old latent heat formula from 1924, Eq. (6-6), results from multiplying by T the difference $(S_n - S_s)$ as given by Eq. (12-4). In the absence of a magnetic field (i.e., for $T = T_c$), Eq. (12-5) reduces to Rutgers's equation. However, if we now substitute $C_s - C_n = \alpha T^3 - \gamma T$ in Eq. (12-5) and integrate twice (also supposing the specific heats are not to be functions of field) we obtain, after some manipulation, the parabolic threshold field curve that is expressed today in the familiar form

$$H_c = H_0\left[1 - \left(\frac{T}{T_c}\right)^2\right] \tag{12–6}$$

with

$$T_c^2 = \frac{H_0^2 V}{2\pi\gamma}$$

where H_0 is a constant of integration (the threshold field for $T = 0$). Somewhat surprisingly, threshold field data can be used for determining the specific heat of the *normal* phase, assuming the $H_c - T$ curve is indeed parabolic. The derivation sketched here is actually a slightly modified version first given in Shoenberg's monograph.[22]

No sooner had Kok addressed the problem, than Mendelssohn and his student John Daunt also turned to the temperature variation of specific heats and entropies. Specific heat investigations, an ongoing priority at Leiden under Keesom's direction, had been pursued at the Clarendon ever since Simon's team settled in. They were pursued by Mendelssohn (with Judith Moore on $PbTl_2$) as well as by Simon himself and Kurti (on paramagnetic salts), thus keeping Nernst's old tradition alive. This time, Mendelssohn and Daunt's expressed purpose was to derive Sommerfeld's γ term without invoking Kok's ad hoc assumption of a T^3 dependence. Instead, they simply calculated the entropy difference from Eq. (12-4), utilizing the latest measured H_c values for Pb, Hg, Sn, Ta, and Nb. Their results were published in 1937.[23] Plots of ΔS versus temperature suggested that ΔS decreases linearly to zero at the lowest temperatures, from which they inferred "that the electrons pass into an arrangement of higher order than that corresponding to the so-called 'Fermi gas'."[24] Assuming the entropy of the superconducting phase is negligible at these low temperatures (roughly below 1 °K), the slope of the $\Delta S - T$ curve gives γ, if the entropy of the normal phase is truly a linear function of T. That is,

$$\Delta S = S_n - S_s \approx S_n = \gamma T,$$

the general validity of which Sommerfeld had reiterated the same year.[25] To be sure, Gorter and Casimir had concluded as much from calculations for mercury, tin, and lead (Section 10.3); after all, the low-temperature behavior of ΔS with T constituted the essential basis for the two-fluid model. But by 1936–1937 considerably newer threshold data were available from purer samples.

From the curves of ΔS versus T the entropy differences of the normal state for mercury, tin, and lead could be read off at some particular low temperature, for example 1.5 °K where the H_c data stopped. These values, in turn, could be compared with the total electronic entropies calculated from Sommerfeld's formula

$$C = 3.26 \times 10^{-5} \times n^{1/3} \times \left(\frac{A}{d}\right)^{2/3} \times T.$$

In fact, Daunt and Mendelssohn observed

> ... it is interesting to note that in all three cases the entropy difference is found to be higher than the total entropy calculated from Sommerfeld's formula. Whereas the discrepancy is small ... in the case of tin, it is strongly marked in mercury and lead. In the latter case our value is twice the theoretical one, in other words, even if we assume the entropy of the supraconductive state to be already zero at 1.5 °K the entropy of the normal must be about twice as great as the value derived from Sommerfeld's formula.[26]

We know from Mendelssohn's correspondence with Sommerfeld on these matters that the Oxford measurements continued through 1937. Thus, in September Mendelssohn wrote him, among other things:

> We are presently engaged in our measurements of the temperature dependence, extended below 1 °K, to determine γ-values ... for additional metals. We also hope to be able to more accurately determine the specific heat of supraconducting electrons for which we employ the second differential quotient of the threshold value curve. It appears to me justified, in contrast to [an assertion] in your paper (Ann. Phys. *28*, page 6, lines 15–17, 1937), to speak of the specific heat of the supraconducting electrons. Of course, the possibility exists that the additional specific heat in the supraconducting state (which is not, as claimed by Kok, strictly proportional to T^3, see pp. 134–135 of our paper) means that with increasing temperature more and more supraconducting electrons pass into the normal state.[27]

In the same letter Mendelssohn takes Kok to task for uncritically accepting the "widely assumed" validity of a strictly parabolic H_c–T relationship—a prerequisite for Kok's methodology. (The offending statement plucked from Sommerfeld's paper reads: "In the supraconducting state there are no free electrons; the specific heat here is only determined by the Debye lattice oscillations.")[28]

The weak evidence for excess specific heat in the superconducting state over what could be accounted for by the lattice vibrations underscored the need for a more direct determination of the specific heat of the superconducting electrons. One possibility was measuring the temperature distribution in an unevenly heated superconducting wire carrying a persistent current; any exchange of thermal energy between the superconducting electrons and the lattice should reveal itself as a change in the temperature distribution on reversing the direction of current flow. The experiment amounted to a determination of the Thomson heat—that is, of the heat generated when a temperature gradient is applied to a current-carrying conductor. The difference between the Thomson heats of two metals is related to the so-called thermoelectric power e_{12} between them by

$$\sigma_1 - \sigma_2 = T\frac{de_{12}}{dT}\ .$$

Since de_{12}/dT between superconductors must vanish, this relation implies that the Thomson coefficients of all superconductors are equal. Moreover, Nernst's theorem requires that $\sigma = 0$ at 0 °K. Consequently, the Thomson heat was expected to vanish over the full superconducting temperature range.

The experiment was performed by Daunt and Mendelssohn in 1938, with time running out as the clouds of war drew closer. "With the conviction that war was approaching fast, and with the most interesting problems of superfluidity and superconductivity on [their] hands,"[29] they found time for only a brief letter to *Nature*.[30] The full paper waited until 1946.[31] The hastily contrived experiment involved a lead ring attached to two vacuum-jacketed copper vessels containing liquid helium. A temperature gradient was maintained in the ring by allowing the helium in the two vessels to boil under different pressures. The entire apparatus was mounted in a vacuum chamber. The temperature distribution in the ring, read with a gas thermometer, was first measured with a persistent current established in it and then when the direction of the current was reversed. No change was detected. With 3×10^{-4} degrees the minimum detectable change in temperature, the upper limit on the Thomson heat coefficient was $4 \times 10^{-9} V / °K$. Within the error of measurement, this showed that σ in the superconductor was zero or (what amounts to the same thing), that the Thomson heat of a persistent current is zero. Daunt and Mendelssohn concluded "that the electrons taking part in [a persistent current in a superconducting ring that is gradually warmed] remain in the same thermal state they had at absolute zero, since they have had no opportunity of taking up thermal energy." Not only that, the result suggested

> ... that a superconductor may be characterized by a gap [10^{-4} eV wide] in the

> energy spectrum which coincides with the upper limit of the Fermi distribution and that the existence of such a gap leaves no alternative for freely mobile particles when subjected to external forces but to be lifted into metastable states in which no dissipation of energy is feasible.[32]

The concept of an energy gap underlying the superconducting state was actually not entirely novel in 1946. The possibility had been raised before the war, and it resurfaced elsewhere in 1946, with the concept being revived as a purely theoretical construct by V. L. Ginzburg. It seems to have been explicitly floated first by Fritz London at the Royal Society discussions in 1935. On that occasion Fritz London suggested that, from the microscopic point of view, it ought to be possible to explain his phenomenological equations by supposing "the electrons to be coupled by some form of interaction, in such a way that the lowest state may be separated by a finite interval from the excited ones."[33] The concept was further developed in an unsuccessful attempt to explain the Meissner effect by Heinrich Welker of the University of Munich, who discussed his theory at the Baden-Baden *Physikertag* in 1938.[34] Welker's model incorporated a gap of the order of magnitude width kT_c above the Fermi level in the energy spectrum of single electron levels; however, the gap was not temperature dependent. Pressed by Schachenmeier following the presentation on whether "the perception of such a gap in the energy spectrum [was] based on analytical considerations of the mutual exchange of electrons, or merely an assumption," Welker had to admit that it was "purely a hypothetical concept."[35]

An upper limit for the value of a possible energy gap can in fact be construed from an experiment conducted under Mendelssohn's direction as far back as in 1936. Daunt, Keeley, and Mendelssohn had pursued Heinz London's early notions on high-frequency heating effects in superconductors, electing to approach the problem from the short wavelength end. In the event, infrared absorption measurements down to 10^{14} Hz revealed no change in absorption when a superconducting lead or tin sample was quenched by application of a magnetic field.[36] (Moreover, this was actually a repetition of inconclusive measurements performed by Mendelssohn and R. Suhrmann as early as 1932—the first measurements of the optical properties of superconductors undertaken since those of Hagen and Rubens.) The lack of change in electron–lattice interaction implied by this result was interpreted to signify "that all electrons pass into the supraconductive state but are by no means entirely free."[37] Four years later, Heinz London's own effort to measure ac losses in tin by calorimetric means finally met with success. Significant absorption well below the ordinary transition point for tin was observed at 1500 MHz,[38] providing as well a lower limit to a possible gap. As noted elsewhere, London took this as evidence for the simultaneous presence of normal and superconducting electrons; it was not seen as conflicting with the measurements of Daunt et al., because in the infrared region "the wave-length is reached below which the absorption of energy is due mainly to optical transitions and not to the conductivity."[39]

In the course of this work, London's loss measurements on tin in the *normal* state brought an unexpected dividend: discovery of the "anomalous skin effect" in normal metals, which London tentatively attributed to the electronic mean-free path exceeding the classical skin depth under the conditions of high conductivity and high frequencies.[40] The onset of World War II halted further work on all of this. The problem was tackled again after the war, particularly by Pippard, who used microwave techniques developed during the war. We will have something to say about Pippard's experiments in the next section. Much of the experimental material resisted plausible interpretation for years because, in Serin's words, "the problem becomes one of grafting a two-fluid model on to an already complicated theory of conductivity."[41] For now it suffices to note that Pippard's "coherence length" that emerged from this work would, like the energy gap, become a cornerstone for the microscopic theory of superconductivity.

Unambiguous evidence for the energy gap ultimately came from dogged pursual of the controversial T^3 variation of the specific heats. The most careful work carried out on the temperature dependence prior to the war had been Keesom's on tin with P. H. van Laer[42] and on tantalum with M. Desirant.[43] In fact, systematic deviations from a T^3-dependence are quite evident in these data,[44] and this was Mendelssohn's point. The T^3 law was inexorably linked to the parabolic H_c–T relationship; one presupposed the other and both were approximations only. Not until 1952 could the problem be examined anew by A. Brown, M. W. Zemansky, and H. A. Boorse at Columbia University with more precise measuring techniques at hand. Specific heat measurements on niobium (where, like vanadium and tantalum, the lattice contribution is small compared to the electronic contribution) revealed a decrease "more rapidly than a T^3 law would allow."[45] However, the exact functional dependence on temperature remained unresolved.[46] Credit for that step goes to W. S. Corak, B. B. Goodman, C. B. Satterthwaite, and A. Wexler of Westinghouse. In 1954 they fitted electronic specific heats for vanadium[47] with an *exponential* expression of the form

$$C_{es}/\gamma T_c = ae^{-bT_c/T}, \tag{12-7}$$

where γT_c is the usual electronic specific heat in the normal state. Such a dependence suggests itself quite naturally on the basis of a two-fluid model and is the hallmark of a true energy gap, not simply a reduced density of states as might be construed by the T^3 variation.[48] Or, in the words of the Westinghouse team,

> On any single-electron model of a superconductor with a gap in the energy level spectrum, the expression for the specific heat would be expected to be dominated, at sufficiently low temperatures, by the term $\exp(-\varepsilon/kT)$. The experimental evidence for vanadium, niobium, and tin ... supports the concept of such an energy gap; the magnitude of this energy gap, deduced from the experiments, is of the order of kT_c.[49]

12.2 Postwar Highlights: A Microscopic Theory

The resurgence of experimental investigations after 1945, investigations interrupted by the war but revitalized by the legacy of wartime technology and unprecedented funding levels, was accompanied by a profusion of new superconducting theories—by Born and Cheng, Heisenberg and Koppe, Schafroth and Blatt, Bohm, Ginzburg, and others. "Some theories," according to the recent ruminations of Brian Pippard—a brash newcomer to the field in 1945—"have perished as though they had never been, while others merged into the soil from which the great growths sprang."[50] The "great growths" date from 1950, a memorable year marking theoretical as well as experimental milestones. That year saw the nearly simultaneous appearance of the mature phenomenological theory of Ginzburg and Landau and of Frölich and Bardeen's landmark papers, cornerstones for a fledgling microscopic theory. It is also remembered as the year of the isotope effect—the experimental key, with the Meissner effect, to the BCS (Bardeen–Cooper–Schrieffer) theory.

The genesis of the microscopic theory, according to the architect of its final synthesis,[51] can be traced to the singularly productive Royal Society discussions in 1935, where Fritz London suggested a way in which the phenomenological equations might follow from quantum theory. On the premise that the diamagnetic aspects are the most basic feature of superconductivity, London argued for an energy gap interposed between the ground state and the lowest excited states, as noted earlier. He also suggested that if the ground state is "rigid" (unperturbed by a magnetic field) the current density will be proportional to the vector potential:

$$\Lambda C \mathbf{J} = -\mathbf{A} . \tag{12-8}$$

This equation is equivalent to Eq. (11-2), since taking curl of the former leads to the latter with the aid of $\mathbf{H}$ = curl $\mathbf{A}$. In later years, London grappled with radically new concepts of the nature of the superconducting state. In his monograph published in 1950 (another achievement in superconductivity that year!), he replaced the velocity field of the superconducting electrons with the local mean value of their momentum vector.[52] By expressing the equations of superconductivity in terms of momentum, he showed that "it is *not an alignment of the current elements* $\mathbf{v}_s$ which is characteristic of the superconducting state. It is the local mean value of the *momentum vector* $\mathbf{p}_s$ which ... tends to establish a long range order, as widespread as possible within the superconductor."

> ... the long range order of the mean momentum vector would be due to the *wide extension in space* of the wave functions representing the *same* momentum distribution throughout the whole metal in the presence as well as in the absence of a magnetic field. Hence this would *not* be a mobile electronic lattice, but rather a quantum structure on a macroscopic scale.[53]

A corollary is that the magnetic flux threading a multiply connected superconductor, such as a ring, should be *quantized*; for example, there must be an integral number of waves around the ring.[54] London's "universal unit for the fluxoid" was $hc/e \approx 4 \times 10^{-7}$ gauss · cm^2, or *twice* the unit of flux predicted by the BCS microscopic theory of superconductivity and eventually demonstrated experimentally. The factor of two comes from the effective charge of $2e$ of an electron *pair* in the BCS theory.

A generalization of the London equation in the version of Eq. (12-8) is due primarily to A. Brian Pippard, whom we have already met. Newly arrived in Cambridge in 1945 for graduate studies after wartime service, Pippard soon took up superconductivity under Shoenberg's tutelage. With 4 years of experience in microwave physics at the Great Malvern Radar Research and Development Establishment, he elected to concentrate on measurements of the surface resistance of superconductors and normal metals at microwave frequencies. This involved determining the bandwidth of tin or mercury resonators excited to a variable frequency; the amplitude of excitation was obtained by extracting from the resonator a portion of the energy that was rectified to give a dc reading on a galvanometer.[55] The resistance measurements per se "created more puzzles than they solved."[56] However, the resonant frequency, ν, was found to change slightly on passing from the superconducting to normal state, with $\Delta\nu$ proportional to $\delta - \lambda'$. (Here δ is the classical skin depth for the normal metal[57] and λ' the corresponding depth in the superconducting state, not necessarily the same as the penetration depth at low frequencies or in a dc field.) Because for tin δ is independent of temperature, the measurements allowed determining $[\lambda'(T) - \lambda(T_0)]$, as already noted. The results for tin at 1200 Mc, plotted against the quantity $[1 - (T/T_c)^4]^{1/2}$, agreed well with the corresponding data for λ at the lowest temperature, but deviated as T_c is approached. This was as expected from the two-fluid model, being simply the effect of the radio-frequency field on the normal electrons whose proportion grows as the temperature rises.[58]

By 1950, Pippard's ambitious program had been extended to measurements of the variation of the penetration depth with magnetic field. That λ might possibly depend on field as well as temperature was not a new suggestion. Heinz London had convinced Pippard that the temperature variation of λ implies, as a thermodynamic consequence, a variation of entropy with field, corresponding to a modification of the electronic configuration on application of a field.[59] In 1947 Ginzburg suggested a field dependence of λ, citing a similar argument, as did Koppe and von Laue in 1949—albeit on different grounds. In the event, Pippard's measurements, hastily performed on his return to Cambridge from the 1949 Conference on Low Temperature Physics at MIT,[60] revealed a surprisingly small variation: at most a few percent in fields near the threshold value.[61] In light of this finding, coupled with London's argument, Pippard examined the theoretical consequences of the thermodynamic effects being truly confined to a surface depth λ.

> Near the transition temperature ... the change in entropy density in the surface layer [would amount] to 1/4 of the entropy difference between the two phases, yet the corresponding change in λ is only $\approx 1\%$. It seems incredible that any theory should be able to account for this result, and it is much more likely that the

> experimental results should be interpreted as showing that there is no great change in entropy density on magnetization. ... Since the change in total entropy is determined by thermodynamic arguments we are led to the conclusion that the entropy change is distributed, not in the penetration layer, but in a considerably thicker layer at the surface, so that the change in entropy density is correspondingly less. This hypothesis is consistent with the idea that the superconducting phase is one having the property of long-range order, as has been suggested previously. ... [62]

Pippard estimated that, to obtain qualitative agreement between the measurements and a simple two-fluid model in which the layer thickness depends on the surface field as well as the temperature, the long-range order would have to extend over a range of about $20\lambda_0$, or $\approx 10^{-4}$ cm.

> This distance is of course very much greater than the smallest specimen which may exhibit the superconducting properties of the bulk material; for example Shoenberg's mercury colloids, though smaller in diameter than λ, nevertheless had almost the same transition temperature as bulk mercury. The range of order must therefore not be regarded as a minimum range necessary for the setting up of an ordered state, but rather as the range to which order will extend in the bulk material.[63]

Later in the year, unaware of Gorter's early ideas on the subject,[64] Pippard proposed that the interphase surface energy was a manifestation of an extended transition layer between the phases, and that the long-range order preventing λ from varying rapidly implies that the density of superconducting electrons, and with it the associated wave functions, must also vary slowly over the coherence length.[65]

The concept of coherence also suggested a natural explanation of superconducting alloy behavior, in terms of a reduction of the extended coherence length to less than λ by impurity scattering. The obvious next step was an experimental inquiry into the influence of small amounts of impurities, and Pippard lost no time in mounting one. Preliminary results on tin samples alloyed with indium were reported at the conference on low temperature physics organized at Oxford in 1951.[66] Sure enough, a much greater variation of λ was observed in impure tin than in pure samples on application of a magnetic field. However, a much more striking and quite unexpected phenomenon was revealed as well: The penetration depth itself was increased by alloying in the absence of a magnetic field, but without producing a corresponding change in the thermodynamic properties (T_c and H_c). This discovery "led to a shelving of the original purpose of the investigation,"[67] set in motion a new systematic investigation of the dependence of λ on purity, and quickly gave rise to a new phenomenological theory.

These decisive measurements showed that the maximum indium concentration (3%) increased λ by more than a factor of 2 while modifying T_c or H_c "to only a trivial extent."[68] Such behavior flatly contradicted the London theory. According to it, λ depends only on m/n_e. Yet, neither the effective mass of the superconducting electrons, nor their density, should be significantly affected by impurity, as attested to by the small observed change in the thermodynamic parameters. The clue to the breakdown of the theory lay in the fact that the rapid variation of λ with impurity, or with the mean-free path l, began where λ and l are comparable; this suggested to Pippard an analogy with a free path effect in normal conductors, namely, the anomalous skin effect that had been elucidated in 1948 by Reuter and Sondheimer and more recently treated by Maxwell, Marcus, and Slater.[69] The trouble, he reasoned, lay with Eq. (12-8), which is a "local" equation. By analogy with the theory of the anomalous skin effect, Pippard boldly introduced instead a nonlocal modification of the equation as follows.[70] Writing $\xi_0 \Lambda \mathbf{J} = -\xi \mathbf{A}$, where ξ_0 is the coherence length in a pure metal and ξ its value reduced by impurity scattering (analogous to l), he adopted an expression that is a generalization of the relation $\mathbf{J} = \sigma \mathbf{E}$, namely

$$\mathbf{J} = \frac{3\sigma}{4\pi l} \int \frac{\mathbf{r}(\mathbf{r} \cdot \mathbf{E}) e^{-r/l}}{r^4} d^3 r .$$

This formula, known as Chambers's formula, expresses the steady-state current density at a point in question (the origin) for electrons traveling on the Fermi surface in the presence of a spatially varying electric field $\mathbf{E}(\mathbf{r})$ as an integral of the field over a region around the point of dimension l. The formula has some claim to historic importance. According to Pippard, R. G. Chambers disclaims responsibility for it when in fact he taught Pippard how to perform the integration for a Fermi gas.[71] "By late 1952 Chambers's formula was available, and I simply resurrected in a tidied-up form what was in my thesis, slipping in a free-path dependent coherence length."[72] The resurrected formula, then, is

$$\mathbf{J} = \frac{3}{4\pi \xi_0 \Lambda} \int \frac{\mathbf{r}(\mathbf{r} \cdot \mathbf{A})\, e^{-r/\xi}}{r^4} d^3 r , \qquad (12\text{–}9)$$

which equates the superconducting current density, not to the vector potential, but to an integral of the potential.

Meanwhile, substantial progress was being made on the problem in another quarter as well, namely in Lev Landau's institute in Moscow. In 1950, Vitaly L. Ginzburg and Landau published their more specific nonlocal modification of the phenomenological theory.[73] In their approach, the "strength" of the superconducting state is described by an order parameter ψ, a complex quantity with some of the features of a quantum-mechanical wave function. The superfluid density is proportional to $|\psi|^2$ and the free-energy density in a magnetic field is assumed to be proportional to

$$\left| -i\hbar \operatorname{grad} \psi + \frac{eA}{c} \psi \right|^2 .$$

Thus, spatial variations in the concentration of superconducting electrons contribute to the free energy, as does the kinetic energy of the supercurrent. The equilibrium form of ψ is calculated by minimizing the free energy. This leads to a rather complicated nonlinear differential equation for ψ which, together with an electrodynamic equation relating ψ to the current density, forms the pair of simultaneous differential equations known as the Ginzburg–Landau equations.

The spatial gradient of ψ produces a tendency for long-range order, causing ψ to vary on the scale of a characteristic length which is basically Pippard's coherence length ξ. The other characteristic length, the penetration depth λ, also appears naturally in the theory, including its variation with field—which is found to be very small. Both lengths define a new parameter, the Ginzburg–Landau parameter kappa: $\kappa \approx \lambda_0/\xi_0$ where λ_0 is the penetration depth at $T = 0$ (Ginzburg and Landau used δ_0). This important parameter provides a succinct measure of the surface energy at the interface between the two phases. Superconductors with $\kappa < 1/\sqrt{2}$ are characterized by a positive interphase surface energy, representing superconductivity of type I in the parlance that came later. The case of $\kappa > 1/\sqrt{2}$ would correspond to a negative surface energy, or to type II superconductivity. However, in their seminal paper of 1950, Ginzburg and Landau inexplicably noted simply that for this case superconductivity would persist at fields above the thermodynamic critical field, concluding with the curious remark that

> It has not been necessary to investigate the nature of the state which occurs when $\kappa > 1/\sqrt{2}$, since from the experimental data ... it follows that $\kappa << 1$.[74]

This was curious indeed, in light of Pippard's explanation, just then, of alloy behavior by electron scattering reducing the range of coherence below λ and thereby causing the interphase energy to become negative! The old evidence for the novel thermodynamic characteristics of alloy superconductivity furnished by Schubnikow et al. was just as compelling. Here it is a matter of speculation whether the late Schubnikow's political status with the authorities in 1950 played a role in his snubbing by his scientific countrymen—something one suspects from Abrikosov's acknowledgement of Schubnikow on the subsequent occasion of the latter's exoneration. For his part, Pippard appears to have been equally disdainful toward Ginzburg and Landau:

> There is no need to repeat the excuses ... for why we in Cambridge were so lukewarm about G-L for several years, indeed until developments springing from BCS theory put it on a secure basis.[75]

Elsewhere, however, Pippard dwells on the excuses; to wit:

> Why in 1950 did it take such a long time for the Ginsburg–Landau theory to be recognized? Why, bearing in mind the extreme lucidity and clarity of their paper, didn't everybody accept it? Why, I ask in particular, did I not accept it, since I soon came to know it perfectly well? Well, there is quite a good reason for this. It wasn't merely the dislike of someone else's ideas, though we all suffer from that. It was that in the Ginsburg–Landau theory the parameter κ, whose size controls whether the superconductor is type I or type II, is determined by the penetration depth and by certain other parameters such as the transition temperature. The

penetration depth in London theory, which the Ginsburg–Landau theory incorporates, is fixed by the number of superconducting electrons and their mass. In other words, the penetration depth is a fundamental parameter according to London. What made me skeptical, at that time, was that in the early 1950's we knew that the penetration depth was changed by scattering. When the mean-free path was made shorter, the penetration depth increased, as could be explained easily by a non-local equation. One didn't have to infer that the number of superconducting electrons changes because of scattering. But Ginsburg and Landau implied that when you alloy a superconductor, making the mean-free-path shorter, the penetration depth increases and κ changes because *the fundamental parameters which go into the theory change*. I found that quite unacceptable.[76]

The new developments alluded to by Pippard were not long in coming; by chance both the microscopic theory and the formalism providing a macroscopic basis for type II superconductivity made their formal debut in the year 1957. The first step in elucidating the phenomenological theory after Ginzburg and Landau was prompted by an ongoing investigation of the critical fields of thin films by N. V. Zavaritskii, a young student of Shalnikov's at Landau's Institute. In 1952, he reported that whereas the behavior of well-annealed films of Sn and Tl crystals was in good agreement with the Ginzburg–Landau prediction for $\kappa < 1/\sqrt{2}$, that of amorphous films of the same pure metals deposited at low temperatures was not.[77] In the former case first order phase transitions in a magnetic field were observed for films down to a certain minimum thickness, below which the transitions reverted to those of second order; in contrast, those for amorphous films exhibited second order transitions regardless of thickness.

In discussing the possible origin of this discrepancy with his colleague Alexey A. Abrikosov (a student of Landau), the two were struck by the similarity of the amorphous film results to superconducting alloy behavior. Looking back on these discussions, Abrikosov recalls that they "came to the idea that the approximation $\kappa \ll 1$ based on the surface tension data ... could be incorrect for objects such as low-temperature films. Particularly one could suppose that $\kappa > 1/\sqrt{2}$."[78]

According to Ginzburg and Landau [Abrikosov continues], the surface energy should be negative under these conditions. Intuitively it was felt that in this case the phase transition in a magnetic field would always be of second order, and this was in fact what Zavaritskii observed.

When I calculated the dependence of the critical field on the effective thickness with $\kappa > 1/\sqrt{2}$, it appeared that the theory corresponded to the experimental data. This gave me the courage to state in my article of 1952 containing this calculation that apart from ordinary superconductors whose properties were familiar, there exist in nature superconducting substances of another type, which I proposed to call superconductors of the second group (now called type II superconductors).[79]

In the 1952 article referred to here, Abrikosov argued that the upper critical field in a bulk [type II] superconductor is given by $H_{c2} = \sqrt{2}\kappa H_c$.[80] Studying the phase transition in the neighborhood of H_{c2} in these superconductors, Abrikosov's concept of a "mixed state" began to take form. Landau encouraged him to publish without undue delay, but Abrikosov dallied, wishing to explore the mixed state over the full range of transition in a magnetic field.

> At this time I became ill and had to stay in bed for almost three months. One day Landau visited me. The conversation, as in most cases, concerned everything but physics, and Landau sipped with great pleasure from a glass of glühwein, which was not at all like him. And then suddenly I destroyed all this paradise by telling him what I had invented for the mixed state, namely the elementary vortices. As Landau's eyes fell on the London equation with a δ function on the right-hand side, he became furious. But then, remembering that an ill person should not be bothered, he took possession of himself and said, "When you recover we shall discuss it more thoroughly." Then he hastily bade farewell and disappeared.[81]

In the event, Landau, wary of "pseudoscience," continued to balk at the elementary vortex lattice that was emerging as the centerpiece of Abrikosov's model, and publication was put off for the duration. And yet, Landau himself grappled some time later with the problem of treating Kapitza's superfluid helium rotating in a vessel, an old interest of his.[82] Although Landau would receive the Nobel Prize for his work on superfluid helium,[83] this particular aspect proved to be problematic, namely, the critical flow velocity and energy dissipation of the fluid in the rotating vessel. That problem was solved by Richard Feynman in 1955, who suggested that single-quantum superfluid vortices, predicted in 1948 by Lars Onsager, were at work in the rotating vessel; dissipation of energy, whereby the superfluid becomes a normal fluid, occurs by interaction of the vortices with vibrational quanta (phonons) or with excitations of the vortex spectrum (rotons) above a critical velocity.[84] Abrikosov immediately recognized the analogy with his second group of superconductors; in retrospect, "He II could be considered an extreme case of a type II superconductor with a correlation length of the order of interatomic distances and infinite penetration depth."[85]

As he was putting the final touches on the vortex picture, Abrikosov remembered that somewhere he came across some familiar magnetization curves for alloys with two critical fields.

> Digging for the ... experimental data, I found the old work (1937) of Shubnikov, of Khotkevich, Shepeliov, and Riabinin on the magnetization curves of Pb-Tl alloys. The authors prepared their samples very carefully, annealing them for a long time close to the melting temperature. So their samples were probably sufficiently uniform, and this was also confirmed by a rather small hysteresis. But at that time and during the subsequent twenty-five years everybody explained this form of the magnetization curve in terms of the formation of a "Mendelssohn sponge," i.e., of a nonuniform structure with a distribution of critical parameters. It is worth mentioning that even many very good experimentalists finally believed in the mixed state only after they saw the powder figures of a vortex lattice obtained in 1966 by Essmann and Träuble.[86]

After some further argumentation with Landau, who eventually was persuaded of the analogy,[87] in 1957 Abrikosov published his celebrated paper.[88]

This landmark publication constituted the first theoretical explanation of the mixed state in type II superconductors based on a mathematical solution of the Ginzburg–Landau equations for the case $\kappa > 1/\sqrt{2}$, even though it had a shaky debut.

> In the same year I reported [my work] at a low-temperature conference in Moscow at which some physicists from Oxford and Cambridge took part. Nobody understood a single word. This could be explained, however, by the fact that I had a terrible cold with high temperature and had hardly any idea myself of what I talked about. The translation of the article was then published in the *Journal of Physics and Chemistry of Solids*, but with more than 100 errors in the formulas and text, and this of course did not improve the situation.[89]

The essential feature of Abrikosov's theory is the now firmly entrenched concept of a vortex lattice. As the applied field is increased above H_{c1} the flux penetrates the material in the form of small flux tubes or vortices parallel to the field. The core of each tube is normal, but the surrounding material remains superconducting; in it flows a vortex of persistent current that maintains the field in the normal core. As the applied field increases, the number of flux tubes increases. Eventually a field is reached where they completely overlap and their combined field is equal to the applied field; at this point the material is no longer superconducting.[90]

The correct approach to a microscopic theory was signaled by the experimental verification in early 1950 of the isotope effect; that is, the verification of a dependence of T_c on isotopic mass. To be sure, Kamerlingh Onnes and Tuyn had performed the experiment on lead with negative results in 1922; equally negative was a much later attempt by Eduard Justi in 1941. (Justi's experiment on the isotope effect in the midst of World War II was reminiscent of the suspicions raised in Vienna by Onnes's preparations for the same experiment in World War I.) However, the availability of substantial quantities of separated isotopes after World War II improved prospects for a definitive test. The test was duly performed, independently and virtually simultaneously, by Reynolds and Serin of Rutgers University (who reported their results at a conference sponsored by the Office of Naval Research in Atlanta on 20–21 March 1950) and by Maxwell at the U. S. Bureau of Standards. Their letters to the *Physical Review* are both dated 24 March.[91] From measurements of H_c versus T for mercury samples enriched in various isotopes they inferred an inverse dependence of transition temperature on isotopic mass. They did not, however, offer a more explicit statement of the precise functional form of the mass dependence, something that was predicted within weeks by Herbert Fröhlich of Liverpool.

Unaware of the ongoing experiments, Fröhlich was just then putting the finishing touches on a theory of superconductivity that he had constructed while on leave at Purdue University. His version was based on his own concept of *indirect* electron–phonon interactions, that is, one in which the electron–electron interaction relies on the intermediary of the lattice vibration, unlike Heisenberg's, which was predicated on direct Coulomb interaction between electrons. Fröhlich's indirect interaction was brought about through the polarization of the ionic lattice by individual electrons; this polarization reacts on the remaining electrons, and this gives rise to a small interaction energy between electrons. Although the detailed theory per se proved to be short lived, a notable triumph was its prediction of the isotope effect that arises naturally from the postulated interaction mechanism. (The theory's notable shortcoming was its failure to predict the Meissner effect.) Fröhlich learned of the experimental results of Reynolds et al. and Maxwell as he was preparing to submit his paper to the *Physical Review*.[92] A note added in proof draws attention to the inverse dependence of the normal-superconducting free energy difference on isotopic mass, from what source follows immediately the effect in question. To underscore the point, Fröhlich dispatched a letter to the British *Physical Society*[93] within days after submission of his main paper to the American journal.

Further, quantitative confirmation, provided by his British colleagues, was not long in coming. Pippard recalls the circumstances, beginning with Fröhlich's presentation of his theory at the summer provincial meeting of the Physical Society at Liverpool on 7–8 July of 1950.[94] Fröhlich's call for experiments with more isotopes prompted Pippard, in the ensuing discussion, to remark that the Cambridge team stood ready to undertake them, lacking only the few milligrams of separated isotopes needed. William D. Allen of the Atomic Energy Research Establishment, Harwell, rose at once and offered to provide the material; his newly commissioned mass spectrograph was ready for the task.

> It was vexatious to hear, some weeks later, that he had been persuaded by Mendelssohn that he had earlier made the same promise to Oxford, who therefore must have the first batch while we would get ours some two weeks later. Shoenberg and Lock spent the interim getting their apparatus into first-class order, and immediately the samples arrived set to work. Within a week or two they had excellent confirmation of Fröhlich's prediction about T_c, but also critical field curves which we analyzed to show that the electronic specific heat was virtually independent of isotopic mass. We wrote a short account for *Nature*, but Shoenberg sent a copy to Mendelssohn suggesting that if Oxford had some results we should be prepared to wait a week so that both laboratories could report together. I believe this letter arrived while the Oxford samples were still sitting on Mendelssohn's desk, and he sent at once for his students Jorgen Olsen and Marianne Bär and told them to waste no time; working round the clock they managed to put together just about enough for a letter to join ours.[95]

The British results were published as back-to-back letters from the Mond and Clarendon Laboratories (with Harwell also represented) under a common title, dated 25 November 1950.[96] They confirmed that T_c for tin isotopes is proportional to $1/M^{1/2}$, as predicted by Fröhlich [and Bardeen]. The conclusion was compelling in confirming that, while superconductivity is an electronic phenomenon, it depends on interactions between electrons and lattice vibrations, or phonons.

John Bardeen was inspired to resume work on superconductivity in May of 1950, when he was informed by Serin of the isotope effect.[97] Bardeen had unsuccessfully labored on a theory of superconductivity in the late 1930s, strongly influenced by Fritz London's thinking and, like Fröhlich, starting from the one-electron model of Felix Bloch, Heisenberg's first graduate student. That model, in which each electron moves independently in a self-contained field determined by the ions and other conduction electrons, was published as Bloch's thesis in 1928 and formed the basis for a successful quantum-theoretical explanation of normal metallic conduction—but not of superconductivity.[98] Bloch's subsequent attempt to treat that subject failed, as we noted in Section 8.1. Nor was Bardeen's attempt very successful and it was, in any case, interrupted by the war. In his renewed attack on the problem, he sought to develop a theory in which the effect of the electron–phonon interaction was such as to lower the energy of the electrons near the Fermi surface, but as a result of dynamic interactions with the zero-point vibrations of the lattice rather than by periodic lattice distortions (his earlier approach). Fröhlich soon paid him a visit at the Bell Telephone Laboratories, and the two reviewed the problem and their respective approaches.

> Although mathematical difficulties existed in both approaches, primarily because of a use of perturbation theory in a region where it is not justified, Frölich and I were both convinced that at last we were on the road to an explanation of superconductivity.[99]

Alas, the road proved far from smooth. Granted that experiments on thermal properties were providing evidence of an energy gap in the excitation of the electrons above the ground state. Moreover, it was shown that a "reasonable" model of the energy gap could lead to Pippard's nonlocal modification of the phenomenological theory and thence to the Meissner effect. Unfortunately, however, there seemed to be no way to derive an energy gap model from *microscopic* theory. To be sure, the Heisenberg–Koppe theory based on Coulomb interactions could be interpreted in terms of a gap, but it did not yield the isotope effect. Thus, the major problem perceived at the time was the inability of electron–phonon interactions to account for an energy gap.

> That electron–phonon interactions lead to an effective attractive interaction between electrons by exchange of virtual phonons was shown by Fröhlich using field-theoretic techniques. His analysis was extended by Pines and myself to include Coulomb interactions. In second order, there is an effective interaction between the quasi-particle excitations of the normal state which is the sum of the attractive phonon-induced interaction and a screened Coulomb interaction. In the *Handbuch* article, I suggested that one should take the complete interaction, not just the diagonal self-energy terms, and use it as the basis for a theory of superconductivity.[100]

The next significant step forward was made by Leon Cooper, who joined Bardeen at the University of Illinois in 1955. Bardeen had left Bell Laboratories for Illinois in 1951. In the spring of 1955 Bardeen came east looking for postdocs to work with him on superconductivity, and he ran into Cooper, a former graduate student of Robert Serber at Columbia and then a postdoctoral fellow at the Institute for Advanced Study in Princeton. Cooper accepted Bardeen's offer and joined his small group in Urbana early in the fall. Among Bardeen's students was Robert Schrieffer, a third-year graduate student, to whom Bardeen had assigned superconductivity as a thesis topic.

Cooper became convinced that the essence of the problem was an energy gap in the single particle excitation spectrum,[101] something almost demanded by the exponentially decreasing electronic specific heat near $T = 0$; had not Bardeen already shown that such a gap was likely to yield the Meissner effect? Moreover, superconductivity was thought to occur as a result of the interactions between electrons, and the work of Bardeen and Fröhlich had indicated that the interaction arose as a result of phonon exchange that produces an attractive interaction between electrons near the Fermi surface. However, there was no clue to how this behavior actually produced the superconducting state. "In total frustration," Cooper decided to "shelve diagrams and functional integrals and step through the problem from the beginning."[102]

The clue was revealed in the spring of 1956, when Cooper turned to certain submatrices of the Hamiltonian of the many-electron system, namely, those that came about due to transitions of zero-spin electron *pairs*. The energy of a surprisingly coherent state composed of such pair states would be proportional to $(\hbar\omega)^2$ and thus inversely proportional to the isotopic mass.

> It seemed clear that if somehow the entire ground state could be composed of such pairs, one would have a ground state with qualitatively different properties from the normal state, with the ground state probably separated by an energy gap from single-particle excited states and thus likely, following arguments that had already been given by Bardeen, to produce the qualitative properties of superconductors. In addition, since all of this could be accomplished as a variational solution of the many-electron Schrödinger equation with demonstrably lower energy than the state of independent particles from which the pairs had been formed, it seemed that this must be considered an interesting candidate.[103]

The conceptual framework for the theory was completed at the beginning of the new year, when Schrieffer incorporated Cooper's quasi-particle pairs in a wave function that satisfied the Pauli principle. This scheme occurred to him while he was in New York at the end of January, 1957; "... I returned to Urbana a few days later where John Bardeen quickly recognized what he believed to be the essential validity of the scheme, much to my pleasure and amazement."[104] These developments finally convinced Bardeen that they were on the right track, and he suggested that the time was ripe for the three of them to prepare a major paper on the theory of superconductivity. In February they dispatched a letter to the editor of the *Physical Review* announcing the first successful microscopic theory of superconductivity.[105] There followed 5 or 6 weeks of intensely concentrated and productive work as the final manuscript took form, punctuated by nervous lapses when the Meissner effect still seemed in doubt.[106] Their landmark manuscript was mailed to the *Physical Review* in July of 1957.[107]

12.3 A Technology Is Born

Despite the near-simultaneous breakthroughs on both the microscopic and phenomenological fronts, these developments contributed little toward the realization of Onnes's old dream of high field "magnetic coils without iron." Prospects for exploiting the novel and promising superconducting properties of alloys and compounds had been dashed once and for all when Keesom and Mendelssohn abandoned the Pb–Bi eutectic in the early 1930s with hardly a cursory trial. Ever since, the predominate belief held that the critical current densities in question are invariably too low for serious contemplation in solenoidal magnets. One reason for this defeatist attitude, in the opinion of several who would play a significant role in the eventual breakout from this quagmire, was failure to appreciate the distinctly nonlinear relationship between critical current density and field for superconducting alloys, and the concomitant failure to extend short sample measurements to fields relevant to solenoid application (several teslas).[108] Another, more serious hang up was Mendelssohn's widely accepted sponge; it too undermined the concept of high critical currents, on account of the small fraction of the bulk volume attributed to the hypothetical filamentary mesh. As late as the early 1950s Shoenberg's authoritative monograph[109] maintained, solely on the evidence of the prewar magnetization data, that superconducting solenoids held little promise even for fields no higher than a few thousand gauss.[110] By then, the burgeoning Russian theoretical school was pointing the way out of the stalemate, but this was unknown to colleagues in the West. The reason for this ignorance must be partly or wholly blamed on the ravaging cold war in those years, with but a trickle of translated Soviet publications available abroad.

And yet, while the 1950s was unquestionably first and foremost a watershed decade in the theoretical and phenomenological elucidation of superconductivity in Western circles, it also proved to be a time of unexpected practical progress in the laboratory. It was this low-key experimental progress, not the impressive theoretical strides, that was to be mainly responsible for propelling superconductivity from a scientific oddity to a fledgling technology in only a few years.

The genesis of the new experimental advances was largely the handiwork of two young scientists. In the early 1950s, John K. Hulm and Bernd T. Matthias laid the foundation for a new school of superconducting research based on what they termed the "materials approach" to experimental superconductivity, in contradistinction to the sacrosanct "single crystal approach" that had been the guiding principle until then. Hulm had been a student of Shoenberg at the Mond. On completion of his graduate work on ferroelectrics and superconductivity,[111] he joined the newly established Institute of Metals of the University of Chicago in 1949 as a postdoctoral fellow. Here he struck up with Matthias, who was somewhat his senior and then on leave of absence at the institute from Bell Laboratories. Matthias was a graduate of Paul Scherrer's school at the ETH in Zürich; in those years Scherrer's group vied with Shoenberg's for leadership in the blossoming field of material science. Hulm and Matthias collaborated for a while on ferroelectrics, but then, partly on the urging of Enrico Fermi, who was ensconced next door on Ellis Avenue in the Institute for Nuclear Studies, and felt that superconductivity held more promise as a frontier topic of physics, they instituted a systematic search for superconductors among alloys and compounds. Matthias, who was a novice in superconductivity research, supplied the expertise in materials chemistry and Hulm the necessary low-temperature physics know-how. They concentrated their search on materials of the transition metals, particularly those with unfilled d-shells—materials that had received little previous attention, with one exception. They were, in fact, following in the footsteps of the pioneering research on the "hard" transition metal carbides and nitrides by Eduard Justi and fellow members of Meissner's school in the early 1940s. A fruit of the early German program had been the discovery in the unlikely year of 1941 of superconductivity in NbN at 15 °K.[112] The discovery brought superconductivity within reach of pumped liquid hydrogen—a rather impressive scientific milestone in the context of wartime Germany.

A more particular reason for the renewed spate of interest in the transition elements was the flurry of attention caused by the uncovering of a departure from the "classical" isotope effect, that is, from the rule $T_c \approx 1/\sqrt{M}$ characterizing all of the non-transition elements. This departure, the demonstration of an exponent in the expression closer to zero than 1/2, was widely believed to be the harbinger of a departure from the electron–phonon interaction as the basis for superconductivity.[113]

The implications of Hulm and Matthias's work for the basic BCS theory were mainly limited to solidifying the experimental basis for the material parameter in the BCS expression for the energy gap. From a practical point of view, their collaborative venture that began at Chicago and continued off and on after they went their respective ways—Matthias returning to Murray Hill after 2 years and Hulm to Westinghouse in 1954—paid off handsomely, however. Their "cookbook" approach unearthed a succession of new superconductors, starting with the solid solution of NbN and NbC with a T_c of 17.86 °K (discovered by Matthias in 1953) and the intermetallic compound V_3Si with $T_c \approx 17$ °K (exposed by Hulm and George Hardy in the same year, still at Chicago). The telltale pattern for maximizing T_c became embodied in the empirical rule of Matthias, which correlates high T_c (and Sommerfeld's γ-term) with peaks in the average ratio of valence electrons to atoms. Hulm's V_3Si, a compound erroneously identified for some time with the "beta-tungsten" structure but actually classified as A15, would soon prove to be a singularly promising prototype superconductor for high-field superconducting magnet applications.

From a strictly materials standpoint, and from the vantage of the late 1980s, the monotonic progress made in maximizing T_c in these and earlier years (gaining perhaps a quarter of a degree Kelvin per year and topping off at 23.2 °K for Nb_3Ge in 1973) seems utterly pathetic in light of the high-T_c explosion that eventually shook the field. In fact, Hulm and Matthias's starting point was compounds of the perovskite structure, members of which would figure so prominently in the startling developments in the 1980s. Those developments were still some years away when Hulm, unaware of things to come, looked back on the early days.

> While working in Cambridge, I wondered if similar behavior [to barium titanate, a compound then also under study by Matthias at ETH] could occur in other Perovskite-structure compounds and I went as far as replacing barium by strontium ... [However,] my knowledge of structural and preparative chemistry was not then sufficiently adequate to encourage me to explore other substitutions in the Perovskite lattice.[114]

In the event, the first successful superconducting magnet proper—as opposed to small nonbifilar sample coils—was constructed entirely without the benefit of these important advances in material science; in fact, the magnet utilized neither superconducting compounds nor alloys. In 1954, George Yntema, while at the University of Illinois, saw a potential need for superconducting magnets operated in the persistent mode for cooling paramagnetic salts by adiabatic demagnetization at low temperatures. Blissfully ignorant of the traditional doomsday wisdom on achieving meaningful current densities in alloys, much less in elements, nor discouraged by Shoenberg's repeated pessimism, he skipped the relevant text in Shoenberg's monograph, and turned to its appendices instead. Here he was struck by a graph of H_c–T curves for various elemental superconductors (Figure 12-2). Niobium was clearly in a class by itself. With a properly shaped iron core and presupposing the effect of a transport current on the superconductor to be "small," he reasoned that "a magnet could produce perhaps 8 or 10 kilogauss without 2 kilogauss at the windings."[115] Without further ado, he procured several spools of Nb wire from Fansteel, the only supplier of the metal. Preliminary short sample measurements in a background field gave puzzling results as the current was ramped in a constant field: zero voltage drop up to a certain current value, followed by a linear rise and finally an abrupt quench. (Before long this behavior would become known as "flux creep" and it would be the bane of superconducting magnet builders.) Undaunted, since a wire sample nevertheless carried 5 amperes nondissipatively in boiling liquid in a 3-kG field, Yntema had the whole spool sent out to be Formvar-insulated and then retested it in one piece in a bifilar winding; sure enough, it still carried 5 amperes. Next, the wire was wound around the backleg of a small *C*-shaped soft iron core. No further insulation was added between winding layers; however, the Nb winding was covered by a shorted secondary copper winding that protected against induction of excessive voltages in the niobium. This modest magnet, augmented, to be sure, by the iron core, produced 7.1 kG (with 1.82 A) in a 7/64-inch gap without much difficulty.

One of the few who appears to have paid attention to Yntema's abstract, which was presented orally at the New York meeting of the American Physical Society the following January,[116] was John Hulm. Hulm, by then established at Westinghouse (where in time he rose to become director of corporate research and research and development planning), felt that even better magnet performance should be possible. Disdaining the use of iron, he opted instead for a

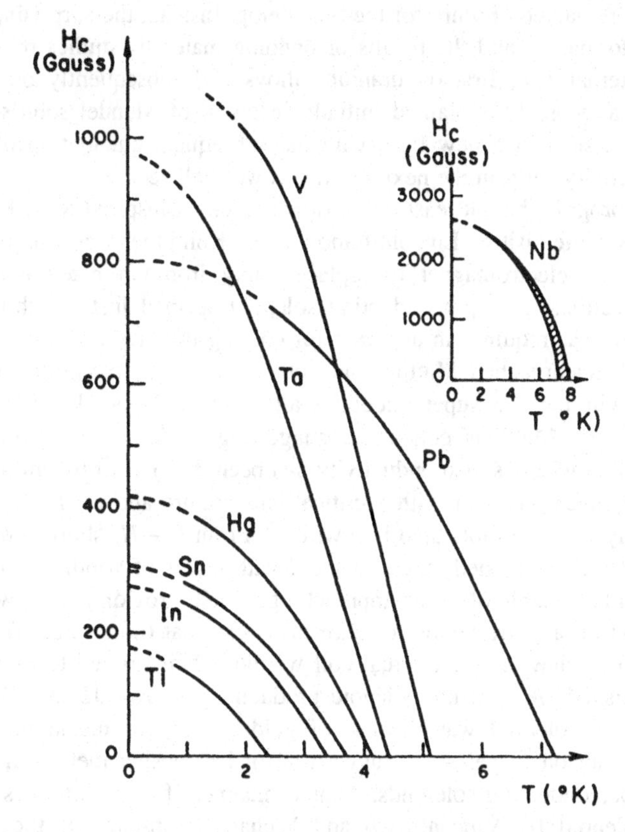

Figure 12-2. Yntema's clue: critical field versus temperature for certain elements, including niobium. Shoenberg, Superconductivity, 224.

simple solenoid wound from better cold-worked enameled Nb wire. It reached 6 kG in a 3/4-inch bore—even better performance, indeed, from the point of view of the conductor. This solenoid also exhibited a new irksome phenomenon that plagues virtually all high-performance superconducting magnets to some extent even today: "training."[117] That is, each time the magnet is energized (ramped slowly from zero current to a value high enough to drive the magnet "normal") it reaches a progressively higher quench current (presumably as built-up mechanical stresses are relieved) until a limiting "quench plateau current" is reached, either dictated by the short sample properties of the superconductor or by limitations of the coil support structure subjected to the high Lorentz (electromechanical) forces at work.

The second half of the 1950s saw a number of additional niobium magnets, among them a series of solenoids and iron-cored magnets constructed under Stan Autler's direction at Lincoln Laboratories;[118] Autler reached 25 kG at 4.2 °K with the aid of iron and nearly 10 kG at 1.5 °K in an air-core solenoid. Perhaps the most important result of all of this activity was that it focused on the effect of cold-work on the superconductivity of niobium, even though its significance remained obscure for the time being. Instead, the surprisingly good magnet performance, and the results of ongoing materials studies on alloys at Atomics International, first on uranium alloys and subsequently on titanium alloys,[119] was vaguely explained initially in terms of Mendelssohn's sponge, but was soon seen in a new light with the subsequent elucidation of type II superconductivity early in the next decade, as we shall see.

Autler's program had at least one positive result; it inspired Rudi Kompfner of Bell Labs while visiting Lincoln Laboratories. Kompfner was searching for a small, compact electromagnet to replace bulky iron-dominated magnets in maser applications. A superconducting solenoid seemed just the thing, since liquid helium was required in any case for cooling the maser. On his return to Murray Hill, he broached Matthias and others about the possibility of doing even better with the new superconducting alloys and compounds that were then coming to light. Matthias concurred, suggesting he look into Mo-Re for a starter, an alloy whose superconductivity had been discovered by Hulm and the only ductile superconductor with a critical temperature above 11 °K. An ingot was promptly cast and fabricated into wire.[120] From $I_c - H_c$ short sample measurements it was concluded that a solenoid with layers of winding about 1 cm thick should be capable of fields approaching 15 kG, providing, as always, that measurement on a short sample of wire in an external (transverse) field is indicative of its behavior in an actual coil winding. This proved to be the case, with the finished 1/8-inch i.d. solenoid producing just over 15 kG. One novel feature of this solenoid was the use of gold plating for insulation between turns—a variant on Onnes's old suggestion for interposing metal foil between turns of superconducting solenoids. Another magnet of this genre was a 10 kG niobium solenoid by Vincent Arp and Richard Kropschot of the National Bureau of Standards in Boulder.

While this performance was encouraging, there was reason to believe that even better performance should be possible. Matthias had suggested trying the ductile alloys Nb-Zr (known to be a superconductor since 1953) and Nb-Ti (since 1961), and particularly the brittle A15 compound Nb_3Sn. This compound was already on record at Bell Labs (since 1954) for having the highest known critical temperature (18 °K), and was expected to have the highest critical field as well. "It was apparent," recalls Eugene Kunzler who had involved himself in this problem as a diversion from other work, "that magnets capable of fields of 20 or maybe 25 kgauss were possible, if not with the ductile alloys, almost certainly with brittle Nb_3Sn, providing, of course, that a way could be devised for fabricating a solenoid with it."[121] As it turned out, the discovery of high field-high current superconductivity was made without the benefit of a superconducting solenoid, even though the Bell team wasted little time in devising a method of preparing Nb_3Sn wire by reacting elemental Nb and Sn packed in thin Nb tubing. More immediately important to the team was the availability of an in-house 88 kG copper solenoidal magnet. On 14 December 1960, a small rod of cast Nb_3Sn was found still to be superconducting at 88 kG, even with a substantial transport current flowing in the sample (Figure 12-3). More amazing yet, a sample of the Nb_3Sn "wire" proved just as good; in fact, it sustained a

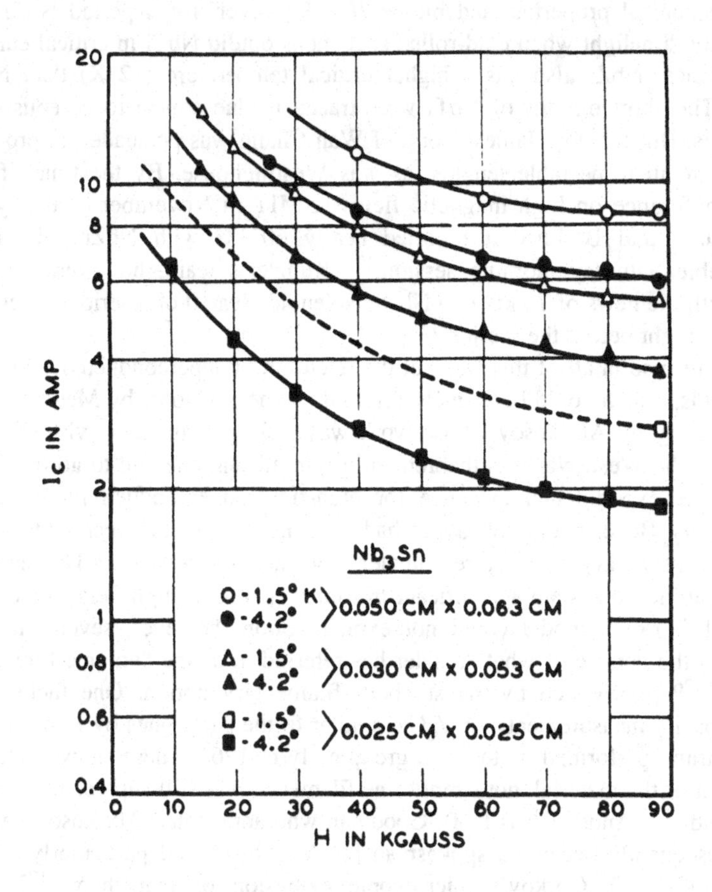

Figure 12-3. Critical current versus applied magnetic field for three sample sizes at 1.5 K and 4.2 K. Kunzler, RMP, **33** *(1961), 504. Courtesy J. E. Kunzler.*

current density of 10^5 A/cm^2 at 88 kG, or 50 times higher than the bulk sample.[122]

Superconducting solenoids exploiting the new materials were not far behind. Within weeks of the discovery at Bell Labs, a score of solenoids were being wound on laboratory benches, at Bell under Kunzler, at Westinghouse by Hulm and co-workers, at Atomics International under Berlincourt, and by Autler and his team at Lincoln Laboratories. The intricate art of fabricating solenoids from Nb_3Sn wire by the "wind and react" procedure was gradually mastered and refined. Despite variations, the procedure itself remained basically the same. A composite of Nb and Sn pellets packed in a Nb tube is drawn down. The drawn wire is wound onto a metallic former and this solenoid is then baked in a furnace at nearly 1000 °C for several hours, thereby activating the superconductor by a reaction at the Nb–Sn interface. Alternatively, it appeared that the whole problem could be circumvented with Matthias's two promising alloy candidates, NbZr and NbTi. Niobium-titanium was the first choice, with the best mechanical properties and higher H_{c2}. However, for a period NbZr occupied the limelight when cold-rolled specimens outdid NbTi in critical current performance. NbZr also has a higher critical temperature (12 K) than NbTi (9 K). The shortcomings of NbTi were traced in due course to gaseous contaminants. But for now James Wong of Wah Chang was persuaded to produce NbZr wire in respectable lengths, as was Westinghouse. By the time of the 1961 conference on high magnetic fields at MIT in November of that year, both Hulm and Berlincourt reported nearly 60 kG with NbZr coils in a memorable Saturday morning session.[123] Kunzler's team, however, usurped them with the news of 70 kG in a Nb_3Sn solenoid cleared of a serious electrical short the night before the session.[124]

The surprise in all of this was not the retention of superconductivity in such high fields, which could be vaguely foretold by the Londons, by Mendelssohn, and certainly by Abrikosov whose work was still, unfortunately, virtually unknown in the West. Nor was theoretical insight all that one had to go on; there was ample experimental evidence for high-field superconductivity by early 1961. Thus, Bozorth and colleagues had good indications of superconductivity in Nb_3Sn at 70 kG well before Kunzler's magnet was tested.[125] The surprise was the ability to sustain a significant *transport current* at high field, something even Abrikosov's model could not explain. Soon, however, several factors joined in the demise of what Kunzler has referred to as the "volume limitation hangup"[126] perpetuated by the stubborn filamentary notion. One factor was heat capacity measurements on V_3Ga, similar to the early ones by Keesom and M. Desirant, performed by the Bell group early in 1962. These showed plainly that most of the material, not simply fine filaments, was still superconducting at high fields.[127] Another was B. B. Goodman who anticipated Abrikosov's work and subsequently became a spokesman for Abrikosov, and particularly a proponent of L. P. Gorkov's microscopic extension of that theory;[128] with Gorkov's version, Goodman was able to derive accurate values for kappa and upper critical fields from experimental values of bulk normal state alloy parameters. It was gradually realized that H_{c2} and J_c are not inexorably linked parameters; rather, H_{c2}, like T_c, is an *intrinsic* property of the atomic structure and composition, while J_c is an *extrinsic* property governed by the metallurgical microstructure (especially grain size). The presence of an electric current creates a force on the superconducting fluxoids that consequently tends to move with the generation of voltage and attendant power dissipation; however, it is possible to "pin" the fluxoids and prevent them from moving by interactions with lattice dislocations and other defects.[129] These imperfections, associated up to this time with the proverbial superconducting filaments, are deliberately or unknowingly introduced—say by cold working as in Nb (the only elemental type II superconductor with the possible exceptions of V and T_c).

However, obtaining a satisfactory transport current was one thing; reliable high-field magnet performance proved to be an even more daunting proposition. First of all, there was the extreme brittleness of Nb_3Sn. Despite ingenious, if cumbersome, ways of winding and reacting coils, the problem remained sufficiently pervasive that the ductile alloys took center stage before long; their somewhat inferior superconducting properties were more than compensated for by their vastly superior mechanical properties, which allowed fabrication by conventional wire drawing procedures. NbZr, being more difficult to process and having the lowest H_{c2}, was gradually discarded in favor of NbTi, even though the transition from NbZr to NbTi was slow, requiring the collective effort of the several commercial vendors, including Atomics International, Wah Chang, Supercon, and Westinghouse, that were striving to capitalize on the potential technology. Before long a plethora of coils were in the making; initially they were wound mostly from multistrand cables of twisted superconductor and copper strands impregnated with indium. The first magnet utilizing NbTi to achieve 100 kG was a "hybrid" solenoid constructed by H. T. Coffey, Hulm et al. at Westinghouse in 1964; it used NbTi for the inner high-field winding and NbZr for the lower field region. With few exceptions, however, these pathbreaking coils exhibited disappointing performance, being prone to excessive training and generally failing to live up to short sample expectations. It required the better part of a decade to sort out the various mechanical and magnetic disturbances responsible for the poor performance and to devise methods to "stabilize" the conductor against them.

The most common disturbance plaguing the new generation of coils was "flux jumping"—sudden discontinuous field changes accompanied by heating as flux penetrates a type II conductor during magnet excitation. The remedy, basically, consisted of two kinds of stabilization, "dynamic" and "adiabatic."[130] In the first, the superconductor is fabricated in intimate contact with normal metal, usually very pure copper, which slows the propagation of the magnetic disturbance and helps to conduct heat to the coolant—usually liquid helium. Alternatively, the superconductor is subdivided into fine strands or filaments, which decreases the energy dissipated to the point where it can be absorbed by the conductor itself with a tolerable rise in temperature. Both stability criteria were met in the ubiquitous "composite" conductor, which consisted of a few or many NbTi filaments (typically of 30-μm diameter) embedded in a high-conductivity copper matrix.

An extreme form of dynamic stabilization, "cryogenic" or "full" stabilization, called for enough copper to carry the full transport current for the duration of the disturbance; that is, flux jumping was simply ignored and the Joule heating smothered with cooling. Pioneers in this approach were Charles Laverick of Argonne National Laboratory (himself a pioneer in the construction of large superconducting coils) and Z. J. J. Stekly of Avco Everett.[131] However, the penalty in this case, a very low overall current density, somewhat defeated the original purpose. More fruitful for some applications was a subsequent, elegant development that took the concept of adiabatic stabilization an additional step, introduced in the late 1960s in the form of "intrinsically stable" multifilamentary conductors. The mature form of intrinsic stabilization was largely the handiwork of the English team of P. F. Smith and colleagues at the Rutherford Laboratory. Now the composite (Figure 12-4) was *twisted* as well; in addition,

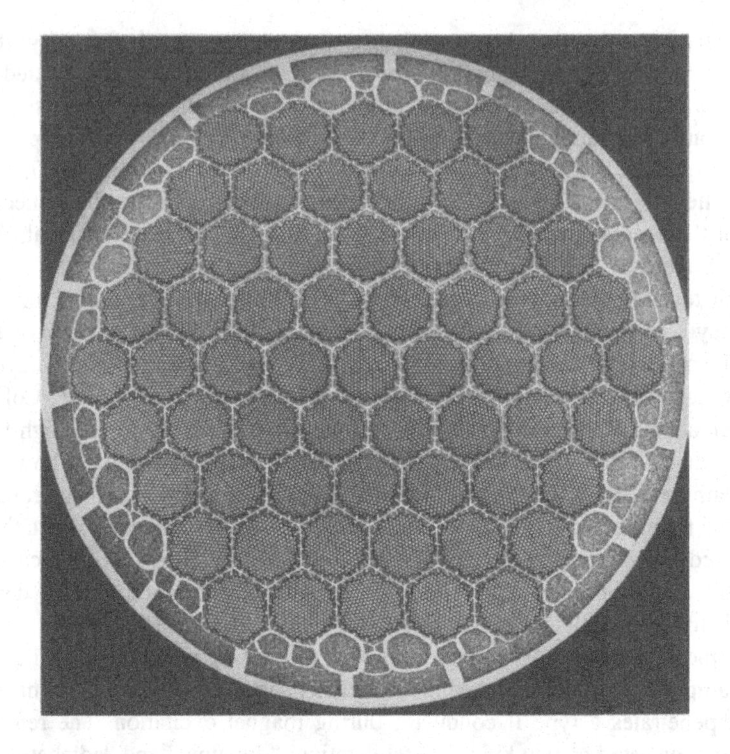

Figure 12-4. Cross section of a composite wire containing 14,701 filaments of niobium-titanium. Each wire is surrounded by copper and then cupronickel, and groups of 421 filaments assembled in a "spider can" of cupronickel with cupronickel radial spokes separating copper regions. Courtesy Imperial Metal Industries Ltd.

the filament diameter was chosen not simply to eliminate flux jumps but to eliminate power losses under ac conditions—a more stringent filament criterion[132] opening up the possibility of superconducting magnets operating in a pulsed mode, as in motors and generators.

CHAPTER 13

Large-Scale Vindication

13.1 Putting the Technology to Work: The National Laboratories Assume Leadership

In the midst of the elucidation and applying of the various stability criteria, there arose, not entirely by coincidence, a fruitful commonality of purpose between the fledgling field of superconducting technology and the one field whose very survival depended critically on technological breakthroughs in generating highly homogeneous, high-strength magnetic fields over ever-larger volumes: that of high-energy or elementary particle physics. The quest for fundamental particles and their mutual interactions has depended on particle accelerators of steadily increasing energy, from the earliest cyclotrons to the latest generation of behemoth colliders—the cathedrals of modern science. The technical centerpiece of all of these facilities are the focusing and bending electromagnets that guide the beams of protons, electrons, and their antiparticles on their circular paths. At the same time, larger and larger volumes of magnetic fields are required in mammoth particle detectors for resolving and analyzing the telltale trajectories of evanescent debris of secondary particles created in the flash of the primary particle interactions.

This is not to say that the exploitation of the new technology was confined to the esoteric reaches of high-energy physics. Far from it, for by the time high-field superconductivity was firmly entrenched as an indispensable design element in the National Laboratories, a rash of projects were being launched (in academia and in particular in industry) in the United States, Western Europe, the USSR, and Japan, to exploit the commercial promise of Onnes's dream. Spanning superconducting energy storage, power transmission and generation (both via ac and dc machinery), ship propulsion, levitated rail transportation,

MHD conversion, and fusion devices, much of the fervor of this multifaceted effort peaked in the nervous years of the energy crisis in the early 1970s and waned as the perceived crisis subsequently faded. With few exceptions none of these diverse projects got beyond the pilot or prototype stage. The one notable exception was superconducting magnetic resonance imaging (MRI) systems, which has blossomed into an industry serving roughly 1000 hospitals with capital expenditures exceeding 1 billion dollars. In any case, a chronicle of this far-ranging subject of applied superconductivity is beyond the scope of the present volume.[1] Instead, we have elected to trace, albeit briefly, the development of state-of-the-art superconducting magnet technology in a noncommercial setting, which has been characterized by unflagging persistency and drama of a high order.

By the early 1960s the centers of high-energy research, principally Brookhaven National Laboratory (BNL), Argonne National Laboratory (ANL), and Lawrence Berkeley Laboratory (LBL) in the United States and CERN (the European Laboratory for Particle Physics), had almost pushed to the limit the field strengths possible with conventional water-cooled copper-iron accelerator magnets. The energy usage, the cost, and the real estate associated with the huge facilities relying on those magnets had also become excessive. Sensing that conventional magnet technology would be capable of supporting at most one more generation of particle accelerators—specifically the expected near-term energy leap from Brookhaven's newly commissioned 30-GeV Alternating Gradient Synchrotron (and its sister machine at CERN) to the 100-GeV range—it was realized that the subsequent longer-term leap by a respectable factor of 10 (to 1000 GeV or 1 TeV) would have to rely on superconductivity. BNL, ANL, and LBL all launched superconducting magnet development programs at about the same time[2] or within 2 years of Kunzler's niobium-tin solenoid. Relying on the commercial superconductor materials technology in place by then, these R&D initiatives would spearhead the development of superconducting magnets on, ultimately, a gigantic scale in two directions that were seemingly at cross-purposes. In the relatively near term, they strove to exploit fully (cryogenically) stabilized NbTi conductors in the state-of-the-art particle detectors of the time, the ubiquitous bubble chambers in which the paths of electrically charged particles are revealed as photographable tracks of vapor bubbles in a thermodynamically unstable liquid. In the long run, these R&D teams would lay the foundation for the ultimate in high-field superconducting magnet applications: accelerator magnets eschewing full stabilization and predicated instead on intrinsic stability, culminating in those for the Supercollider of the 1990s.

Outstanding examples of large, megajoule magnets for high-energy physics were the magnets designed for the final round of bubble chambers of the 1970s, after which this style of particle detector faded into obsolescence. These formidable magnets, needed for introducing curvature in the particle tracks, represented the top of the line in solenoid design and engineering: some of them were air-core magnets and some incorporated an iron yoke for further field enhancement or shielding the laboratory environment from the hefty stray field.

All were wound with a fully stabilized monolithic (that is, not cabled from elementary strands) NbTi conductor with the superconductor encased in a stabilizing matrix of high-purity copper. The iron-supported magnet for the 12-ft diameter bubble chamber at Argonne was first energized in 1968 (a mere 7 years after Kunzler's 4-inch o.d. solenoid was cleared of shorts) and became operational in 1970.[3] Weighing 1600 tons, it produced 19 kG in the bore and had a stored energy of 80 MJ. After faithful service in Chicago, it eventually migrated to the Stanford Linear Accelerator Center in California for another lease on life; it was succeeded by even larger bubble chamber magnets—devices that, by their very feasibility, were dramatic testimony to the soundness of the cryostability principle. Fermilab's 15-ft air-core chamber magnet (30 kG or 396 MJ) was first energized in 1973, not to be outdone by CERN's Big European Bubble Chamber (BEBC) of about the same size (Figure 13-1) but generating 35 kG at 800 MJ.[4] A conventional magnet for the BEBC would have required 70 MW, whereas the CERN magnet needed only 360 kW of installed power.

Although of impressive scale, these split solenoids had one feature in common with the humbler prototypes of Yntema, Kunzler, et al.: simplicity in winding geometry. Essentially, a solenoid is simply a circular cylindrical winding of rectangular cross section. (For use in bubble chambers the solenoid must be split into two halves in order to allow traverse access of the particle beam to the field volume enclosed by the coil sections.) Considerably more complex and difficult to wind are the transverse field magnets that bend the primary charged particle beams within the accelerator proper or guide primary or secondary beams extracted out of the accelerator to some external experimental area, say to a bubble chamber.[5] In such accelerator or "beam transport" magnets the field lines are transverse to the long axis of the magnet, not parallel to the axis as in a solenoid. Such a field distribution is produced most simply by a pair of racetrack-shaped coils—an impractical geometry, however, because clear access to the bore is required at either end of the magnet. Furthermore, the field distribution is highly nonuniform, whereas both accelerator and beam transport magnets require a field uniformity better than one part in 1000. In practice, most coil configurations in magnets of this type are designed in the form of a long slender saddle with a coil cross section shaped as an approximation to one of several ideal current density distributions wound on a cylindrical surface. If truly replicated, they produce perfectly uniform fields within the cylindrical volume (in the case of bending magnets or *dipoles*, or uniform field gradients in beam focusing or *quadrupole* magnets).

From the point of view of the superconductor, transverse field magnets differ from their giant bubble chamber counterparts in an important respect: Their small cross section and high operating fields imply a very high current density and therefore rule out the luxury of full copper stabilization.

Among the first operational beam transport systems to utilize superconducting magnets was the "8-degree" magnet system designed under Gordon Danby and his team at Brookhaven.[6] Utilizing a pair of 2-m long 3.7 tesla dipoles in the form of "window frame" iron yokes surrounding dipole coil packages of

Figure 13-1. Erection of the cryostat for the BEBC magnet at CERN, Switzerland. The split-coil magnet stored 800 MJ of energy at the peak operating magnetic field of 3. 5 T. Courtesy European Laboratory for Particle Physics (CERN).

rectangular cross sections (a design philosophy identified with the "Danby school"), the system went on line on the experimental floor of Brookhaven's Alternating Gradient Synchrotron (AGS) in 1973. It may be fairly said to mark the first major commitment of a research laboratory to involve a superconducting magnet system (other than in particle detectors or analyzing systems) in an on-line application where several experimental teams depend on the full time, fail safe reliability of upstream superconducting magnets operating in a radiation environment. Another pioneering system was the beam line of superconducting magnets installed at Argonne's Zero Gradient Synchrotron (ZGS), which became operational in 1976 under John Purcell (who was responsible for the 12-ft bubble chamber magnet as well). That beam line involved 10 1-m dipoles and 2 quadrupoles; they were somewhat unique in that the coils were made by randomly winding the conductor (Purcell's innovation) in the cross-sectional shape of 2 intersecting circles.[7] The same year also saw the commissioning of Brookhaven's beam line for the High Energy Unseparated Beam from the AGS.[8] It featured 4 2.5-m dipoles of 25-cm aperture, each storing one megajoule of energy. These magnets were in fact oversized prototypes of the ISABELLE magnets then under development by William Sampson's group working in parallel with Danby's at Brookhaven.

These and other dc beam transport magnets of the 1970s were the early forerunners of the much more demanding *pulsed* dipole and quadrupole accelerator magnets that could be "ramped" upward over a period of seconds or tens of seconds; these magnets were optimized for operation at the highest possible fields, and they soon became the focus of the state-of-the-art superconducting magnet effort in the United States and abroad. While the multiplicity of conventional beam lines fanning out from the beam ports of the "fixed target" proton synchrotrons of that era were voracious consumers of energy and excellent candidates for superconducting replacements, by far the biggest energy consumers were the accelerator magnets themselves. Moreover, with the cost of real estate soaring and leaner funding levels becoming the new fact of life, upgrading the performance of existing accelerators, particularly their top energy, became an overriding priority in the laboratory directors' long-range planning. Here, however, a fundamental limitation on the effective energy of stationary target accelerators lay in the way; its elegant circumvention, predicated on the concepts of *colliding beams* and storage rings, would eventually usher in a new approach to experimental particle science and obviate the complex system of external beam lines that cluttered the forefront high-energy physics (HEP) facilities of the 1960s and 1970s. Unfortunately, translating these new concepts into reality would require a long and painful international effort to develop high-field, pulsed superconducting accelerator magnets of acceptably low ac energy loss that would meet additional stringent "field quality" (uniformity and reproducibility) requirements of accelerator magnets. Six groups, three in Europe and three in the United States, bore the primary brunt of this initial effort.

The incentive for the European push was originally the possibility of constructing part of CERN's new Super Proton Synchrotron (SPS) with superconducting magnets in a "missing magnet" scenario. The idea, first floated by CERN's Director General John Adams in 1971,[9] was first to fill half of the 2.2-km ring circumference with conventional magnets, enabling 250 GeV (later reevaluated as 300 GeV) to be reached, and then filling the remaining slots with superconducting magnets, making it possible to reach 500 GeV at negligible additional power consumption. In a further phase, 1000 GeV ought to be possible with higher-field superconducting magnets replacing the conventional ones as well. With this scheme in mind, three laboratories with relatively strong ongoing superconducting magnet R&D programs entered into a GESSS Collaboration (Group for European Superconducting Synchrotron Studies). These were Great Britain's Rutherford High Energy Laboratory, the Center for Nuclear Studies at Saclay in France, and the Karlsruhe Institute for Experimental Physics in what was then West Germany.

Working separately toward a common goal, each laboratory designed and constructed several short (1–2 m) model dipoles meeting certain basic accelerator magnet specifications with regard to field level, field quality, and ac losses.[10] The Rutherford group demonstrated a series of pulsed dipoles AC1 through AC5 devised under the leadership of J. H. Coupland, P. F. Smith, and D. B. Thomas; Karlsruhe contributed, among other things, the dipoles D2a and D2b under P. Turowski and colleagues; at Saclay there were the dipoles known as Moby and Alec—the handiwork of G. Bronca, J. Perot, and co-workers. Nevertheless, despite making considerable progress in the art of high-performance superconducting magnets, the hope of deploying superconducting magnets in the SPS was abandoned in 1973. Too many unknown factors loomed large at the time, among them costs, magnet-to-magnet reproducibility of critical parameters, uncertain delivery time scales, and above all the operational reliability of these magnets (mainly the degree of acceptable training, a phenomenon that still haunted magnet builders, and quench protection considerations). Instead, the decision was made to proceed with a second, conventional stage at once by filling the entire SPS circumference with copper–iron magnets, making 400 GeV possible from the outset.

Despite the dissolution of GESSS, the three laboratories managed to keep viable magnet R&D programs going on leaner funding, with the effort shifting to the production of specialized magnets for high-energy physics customers abroad: a British large-bore solenoid dispatched to the TRIUMF Meson Research Facility in Canada, and several more to CERN; dc quadrupoles from Karlsruhe to CERN; French dc dipoles to CERN, and so on. But more and more the groups redirected their effort toward the escalation of national and European initiatives in fusion research. One particular legacy of the early HEP effort would have a lasting impact in experimental particle science, however: the flat, twisted NbTi cable devised by the British, universally known as the Rutherford cable, which is utilized today in virtually all superconducting accelerator magnets. Except for that singular development, European superconducting magnet R&D for high-energy physics would, reasonably, pass from the multidisciplined laboratories at Saclay, Karlsruhe, and Oxfordshire to the HEP laboratories themselves—to CERN and to the subsequent HERA project of the

German Electron Synchrotron Laboratory DESY in Hamburg—and to European industry. But that is anticipating the thread of developments.

13.2 The Painful Path to the Tevatron

As GESSS was being phased out, with essentially limited results to show (mainly on account of marginal enthusiasm and support from CERN), three separate superconducting accelerator initiatives were gaining momentum in the United States, on the East coast, West coast, and in the Midwest. Of the three, LBL's ESCAR project was most modest in scope; its conception at a time when the ascendancy of HEP technology had passed to Brookhaven and Fermilab deserves closer scrutiny than is possible here.

ESCAR, an Experimental Superconducting Accelerator Ring, was conceived by the superconducting magnet and accelerator physics groups at the Lawrence Berkeley Laboratory in the spring of 1974. It was initially intended to be the first serious test bed for a full-blown superconducting accelerator *system*, not simply a beam line with superconducting magnets replacing conventional magnets.[11] The design taking form under the direction of Edward Lofgren (who supervised the design and construction of the Berkeley Bevatron) and other senior colleagues (LBL's new generation of "machine builders" who were carrying on the tradition of Ernest Lawrence) called for a bona fide accelerator/storage ring test bed. Its design specified an average radius of 15.3 m and 4.2-GeV peak energy, capable of accelerating and storing beams of 4×10^{12} protons per pulse. The technical centerpiece of the facility was to be 24 one-meter ("warm iron") pulsed dipoles, producing a maximum operating guide field of 4.6 T, and 32 shorter focusing quadrupoles, all cooled by a two-phase helium refrigeration system.[12] As it turned out, the project "went slower than it should"[13] have to be of timely relevance to the more ambitious superconducting projects under way elsewhere (at Brookhaven and Fermilab) because of limited funding and flagging enthusiasm from the HEP community at large.[14] In mid-stream the scope of the project was drastically scaled down by the Berkeley team to a test of the superconducting magnet system alone, albeit a curtailed one at that. Twelve dipoles were installed in an external beam hall of the Bevatron, with the 1500-W refrigerator and all necessary instrumentation. This truncated system was put through an extensive series of tests ending in the spring of 1978—a test sequence that was not devoid of some electrical problems but it encountered no "show-stoppers"[15]—and the project was officially terminated in June of that year.[16]

From the point of view of accelerator systems, the experience of ESCAR played a minor role in guiding the full-scale superconducting accelerator projects that were then under way at Brookhaven and Fermilab, but it did provide considerable operational experience with superconducting magnet and cryogenic systems. Most important, this experience proved invaluable in broadening the expertise of Berkeley's superconducting accelerator magnet team—expertise that would be put to good use largely by the same team members a decade after ESCAR's ignominious termination.

The contemporary Brookhaven project, begun a decade before ESCAR was launched, was the legacy of a spirited competition between Brookhaven and LBL (then the Lawrence Radiation Laboratory) for the design and construction of a super-energy accelerator, albeit not necessarily a superconducting one. Both laboratories possessed crack teams of accelerator designers whose skills had been honed on Brookhaven's Cosmotron, Berkeley's Bevatron and, most recently, on the Brookhaven Alternating Gradient Synchrotron, and who were the presumed heirs to the next generation of machines. Prospects for both contending teams brightened with the recommendation in 1963 of a scientific advisory panel to the Atomic Energy Commission (AEC) (the influential "Ramsey Panel") that

> construction [be undertaken] by the Lawrence Radiation Laboratory of a high energy proton accelerator of approximately 200 BeV energy, with construction authorized at the earliest possible date. This accelerator should be a major national facility, and in its planning, consideration should be given to the convenience of user groups and visiting scientists from other parts of the country. The site should be adequate for later addition of storage rings to the accelerator.
>
> Intensive and extensive design and experimental studies should be supported at the Brookhaven National Laboratory with a view toward the next major step to higher energy. Request for authorization should be anticipated in about five or six years. The present enthusiasm of the entire community of interested scientists should be exploited in the planning and execution of this truly exciting scientific step. The nature of this accelerator should be determined as a result of the design study, but a proton accelerator of 600–1000 BeV would appear at present to be a reasonable objective.[17]

The matter was clinched by the AEC's subsequent authorization of respective design studies at the two laboratories. (There was a precedent for a friendly rivalry between Brookhaven and LRL. In 1947 discussions between Leland Haworth at Brookhaven, Ernest Lawrence at Berkeley, and the AEC authorities led to the decision that both laboratories, instead of competing for a 10-BeV machine as originally proposed by LRL, would each build a smaller proton synchrotron, one around 3 BeV and one at 6 BeV. Haworth chose the smaller Cosmotron, and LRL got the 6 BeV Bevatron.)

In fact, the Radiation Lab lost the 200-GeV prize (GeV, the abbreviation for giga electron volts, soon would be entrenched in the parlance of high-energy physics and thereby avoided "billion," which represents 10^{12} in British usage), as we shall see, after investing more than 2 years to preliminary design studies. At Brookhaven, early design studies on a conventional 600–1000-GeV accelerator got under way under John Blewett. They were augmented by an ongoing in-house program on pulsed superconducting magnet development initiated by Blewett and George K. Green (prominent Brookhaven machine builders). This R&D program was led by William Sampson with initial enthusiastic participation by Gerald Kruger of the University of Illinois, an old cyclotron builder like

Green, and held the promise of overcoming economical and technical limitations in conventional accelerator designs. Despite its success, these early accelerator studies were scaled back as the Federal budget tightened in the late 1960s. During 1969–1970, they gave way to a more modest scheme for a superconducting 112-GeV conversion of the AGS.[18] In addition to upgrading the maximum energy of the existing accelerator (30 GeV), this facility was visualized as a pilot project (in the spirit of ESCAR) and ultimately as a booster for a very high-energy machine that was still believed to be in the cards sooner or later.

However, other events soon overtook this plan. As a consequence of the imminent completion of a scaled-up version of the 200-GeV machine (albeit not at Berkeley) and the approval of a 300-GeV European counterpart, coupled with the highly successful operation of the Intersecting Storage Rings (ISR) at CERN, a superconducting AGS seemed inadequate for the next major initiative. During 1970–1972 Brookhaven's efforts were redirected once more and now focused on exploiting the new technological challenge represented by colliding beam systems instead of fixed-target accelerators, which represented the end of a technological line. (In fixed-target accelerators with a single beam of particles striking a target at rest, only a small fraction of the projectile energy is useful for producing interesting reactions; the rest is wasted in imparting forward motion to the center of mass of target plus projectile. In colliders, accelerators with two beams of equal energy colliding head on, by symmetry (and intuition) the center of mass of the oppositely moving particles remains at rest and no beam energy is wasted. Colliders are also termed "storage rings" because the counter-rotating beams circulate for long periods (hours) while collisions occur at a chosen point. The first proton–proton collider was the aforesaid Intersecting Storage Ring at CERN; electron–electron and electron–positron colliders came earlier.) Brookhaven's efforts eventually culminated in a preliminary design for a 200-GeV colliding beam accelerator/storage ring devised to take equal advantage of the promising superconducting advances that were being made in Sampson's laboratory and elsewhere. The preliminary design for ISABELLE was completed in 1972[19] and a first formal proposal was submitted to the AEC 2 years later.[20]

In ISABELLE, collisions between two proton beams, each of 200-GeV energy (dictated by what was judged to be the maximum prudent dipole field and by available real estate on the Brookhaven site), would yield 400 GeV in the center-of-mass, which is equivalent to protons of 86,000 GeV (86 TeV) striking protons at rest. ISABELLE would extend the center-of-mass energy range then available almost by a factor of 10 and provide a luminosity of a factor of 50 higher than that available at the ISR.[21] The collider's design was based on separate superconducting bending and focusing magnets distributed around two rings (each ring 2.9 km long, initially arranged in an eightfold symmetric configuration, which was essentially that of a circle expanded by eight long experimental straight sections), where the proton beams intersected. Proton pulses of 30 GeV from the AGS would be stacked sequentially in each ring until a specified circulating current had accumulated, at which time the two beams could be slowly accelerated to full energy.

The main features of the proposed superconducting magnet system for ISABELLE were as follows: Each of the two interlaced accelerator/storage rings would contain 264 dipoles, each unit 4 1/4-m long and capable of 4 tesla, and 216 1 1/2-m quadrupoles producing gradients of somewhat over 5 T/m. The inner diameter of the single-layer dipole coil was 12 cm—an unprecedented size for accelerator or beam transport magnets except for the HEUB magnets noted earlier. The coils were wound from a flat conductor braided from approximately 100 multifilamentary NbTi wires; it was filled with an InPb alloy for additional stability, and spiral wrapped with fiberglass epoxy tape insulation. The assembled coil was clamped in a (helium-cooled) yoke of laminated low-carbon steel by a clever shrink-fitting technique that was devised to ensure that the coils remained under a predetermined compression during cooldown.

The superconducting magnet development program supported by BNL's ISABELLE Division had its origins in the early 1960s, or shortly after Kunzler's path-breaking niobium-tin solenoid. Sampson, like Charles Laverick at Argonne, had wasted no time in exploiting the state-of-the-art technology. Sampson went out on a limb, placing the initial emphasis of his fledgling group on the application of stabilized commercial Nb_3Sn tape conductors in compact high-field devices, such as small-bore Helmholtz coils and solenoids in the 100-kG range, as well as in more complicated forerunners to deflecting and focusing saddle coils for particle beam transport application.[22] In 1968, when intrinsically stable filamentary NbTi superconductors rather suddenly appeared on the scene,[23] the effort was immediately diverted to the development of state-of-the-art accelerator dipole and quadrupole magnets incorporating this promising conductor in a novel high-current geometry. This phase began with a fresh set of solenoids, and then proceeded through several series of progressively larger saddle coil magnets starting around 1970: initially 0.35-m long 4 tesla pulsed dipoles, followed by 1-m long model ISABELLE dipoles producing 4 tesla over an 80-mm bore, and culminating in the first full-scale ISABELLE prototype dipole (Mk I) in the spring of 1975 (Figure 13-2). BNL's unique and controversial braided conductor, primarily the handiwork of A. D. McInturff, dates from 1970–1971.[24] The famous Mk V reached 5 tesla in 1976, thus exceeding the design field by a full tesla in a performance which, paradoxically, ultimately played a part in ISABELLE's demise.

One year later an advisory panel to ERDA (Energy Research and Development Administration), the short-lived successor to the AEC, pronounced ISABELLE "the next natural step beyond the only other existing device of its kind, the ISR at CERN," and judged "that the technical level of achievement of the ISABELLE research and development program is such that the project is ready for funding."[25] That year, 1977, ISABELLE was officially funded as a construction project.

A technical decision with disastrous consequences was made at Brookhaven one year later (1978) when, partly on basis of promising short dipoles culminat-

Figure 13-2. Mk I, first full-length prototype dipole for ISABELLE. The magnet was 4.25 m long, designed for an operating field of 4 T over an 80-mm bore. Courtesy Brookhaven National Laboratory.

ing swiftly in the performance of Mark V, but under concomitant pressure from other developments in particle science, the beam energy of the collider was doubled from 200 to 400 GeV corresponding to a center-of-mass energy of 800 GeV. Indeed, arguments for higher collision energy had first circulated in Brookhaven circles in 1974, chiefly, that Fermilab's fixed target center of mass energy, approximately 25 GeV, and even that of CERN's ISR proton–proton storage ring, 50 GeV, was too low to ensure "new physical phenomena"—for example, detection of the W boson.[26] Support for the idea of a higher energy ISABELLE by the high-energy physics community at large was hinted in the report of the 1974 HEPAP Subpanel on New Facilities.[27] Recommending that funds be provided in the fiscal year 1976 allowing Brookhaven to complete its superconducting magnet prototype program with the goal of making a request for early construction of a proton–proton collider with "a beam energy of *at least* 200 GeV" [italics added], the subcommittee added lamely that

> … the choice of 200 GeV is based on the desire to reach a new important region of energy at a cost within the scope of a realistic program. A higher energy would be more desirable but the 200 GeV level promises to open up many new horizons.[28]

The need for higher energy was readdressed off and on in terms of purely machine-technical grounds over the next several years.[29] Moreover, by 1978 it was abundantly clear that Fermilab's superconducting magnet program was making excellent progress, and there was mounting pressure to convert the Doubler to a proton–antiproton collider with the potential for yielding 1000 GeV in the center of mass.

The Brookhaven revision required raising the dipole operating field from 4 to 5 tesla, which was achieved at the expense of reducing the operating temperature, lengthening the magnets somewhat, and "permitting a limited amount of training."[30] Alas, there then followed a painful period of several years dominated by disappointing magnet performance. The singular performance of Mk V could not be replicated, with the higher design field barely attainable after excessive training to a quench plateau that left scant operating "margin" with the NbTi superconductor at hand. Not only that, the metal-filled, braided conductor was prone to excessive eddy current effects that could not be controlled despite ingenious metallurgical efforts.

The way out of this protracted technical quagmire was shown in 1981 by the Anglo-American physicist Robert Palmer of Brookhaven. Palmer, a particle physicist and somewhat of an outsider to the magnet community (but not lacking in technical ingenuity and drive), was struck by the markedly improved performance shown by the Fermilab (Doubler) magnets at that time—magnets incorporating certain design features that were seemingly at cross-purposes with BNL's venerable design philosophy—and he proposed to capitalize on this fact. To sort out the design differences, let us backtrack and recall the origins and progress of a Midwestern magnet program which, in time, became a serious challenger to Brookhaven's long-unrivaled program.[31]

The superconducting magnet program at the new National Accelerator Laboratory (as it was initially known) at Batavia, Illinois, had its start nearly a decade after Brookhaven's and from its inception had a much more focused goal. The idea of adding a superconducting ring to the recently authorized 200-BeV machine at NAL, possibly for a "beam stretcher" (to lengthen the time of "spill" of extracted beams) or as a storage ring, was first suggested in 1967 in discussions among the accelerator design staff that was temporarily housed at Oak Brook, a suburb of Chicago. Several years earlier, mounting political pressure for a truly national accelerator (but one in a Midwestern setting) had undermined LBL's hopes to build the 200-BeV accelerator in California.[32] Subsequent events included the formation of the Universities Research Association for building, maintaining, and operating the 200-BeV machine under contract with the AEC, appointment of an Academy Site Evaluation Committee that screened more than 200 proposed sites in 46 states, and ultimately the naming of Weston, Illinois from 6 finalists. Ground was broken at the site on 1 December 1968 by Robert R. Wilson (who got his start under Ernest Lawrence), renowned machine builder and newly appointed laboratory director, and AEC Chairman Glenn T. Seaborg (formerly front rank Berkeley chemist). Meanwhile as the design work, which started from scratch, gathered momentum at Oak Brook, Wilson felt obliged to prohibit active work on superconducting magnets until the design of the conventional accelerator was in hand.[33] Wilson was, however, sufficiently prescient to insist that adequate space be left free in the machine tunnel for the eventual addition of a second ring of superconducting magnets of an unspecified purpose.

The concept of a superconducting energy "Doubler" was fairly widespread by 1971 when NAL's 200-GeV main ring was approaching completion. That summer the AEC requested that Wilson and his staff "perform the necessary work in the coming fiscal year to clearly define the scope of this undertaking and to ascertain whether the inclusion of energy doublers can be achieved within the $250 million authorized for this project."[34] The same summer the first model Doubler magnet was constructed.[35] A first estimate of the level of effort required for implementation of a Doubler project, although it was woefully off the mark, was prepared by William Fowler and Paul Reardon in early 1972. With the attainment of 200 GeV that spring, Wilson could revoke his standing edict. In September, he established an informal working group that met on a weekly basis to review various technical issues. Soon two more model dipole magnets were in the making, the 3-ft "Mk I Pancake" (denoting the style of coil winding) and the 1-ft "Mk I Shell"; both were tested in 1973. At the same time, refrigeration tests with a 400-ft helium loop were under way under Peter VanderArend in the "Protomain" area of NAL. The Doubler group was formally incorporated into the Accelerator Division under Reardon's direction the same year. However, Wilson continued to dominate the discussions, and the Doubler's priority remained rather low. And although the AEC did approve NAL's request for support of an expanded superconducting magnet R&D program, it declined a request to spend the remaining balance of the $250 million that was authorized for the original 200-GeV accelerator on the Doubler.

Despite the various bureaucratic hurdles, two particularly important technical decisions were made at this time. The first was the adoption of a dipole design based on "warm" iron and a "cold" bore—that is, one in which the ferromagnetic yoke is mounted outside the magnet cryostat and consequently is at room temperature, while the beam vacuum tube is at helium temperature. The LBL group also opted for warm iron and a cold bore for ESCAR the very same year. In stark contrast, ISABELLE's more conservative magnets utilized a helium-cooled yoke in close proximity to the coil winding, and a warm bore. It was facetiously said that the warm iron option for the Doubler made possible rapid warm up and replacement of failing magnets—a bitter lesson from experience with the conventional magnets of the 200-GeV main ring. In point of fact, warm iron magnets produce less field for the same excitation current, because of the intervening cryostat structure, but by the same token, they are practically free of field distortions from iron saturation effects. An advantage of cold iron designs, in addition to their higher "transfer function" (gauss per ampere), is the inherent simplification and ease of fabrication made possible by relying partially (or entirely) on the massive yoke for structural coil support. A cold bore (which minimized coil size and cost) was somewhat controversial at the time, and was seen in some quarters as a potential source of vacuum problems.

The second design decision reached in 1974 was to adopt (as had ESCAR) the so-called Rutherford cable for the conductor in a two-layer coil design. The Rutherford cable (which nowadays is the universally favored conductor configuration for superconducting accelerator magnets) derives its name from the laboratory where it was invented. It consists of a flat cable produced from an initially hollow cable of twisted multifilamentary superconducting strands, which is shaped by compaction into a rectangular cross section. The primary reason for Brookhaven's eschewal of the Rutherford cable at the time was the importance attached by Sampson's group to a *single-layered* coil winding, well clamped (dictating cold iron, as noted), something only possible with a wide ribbon-like conductor of very high aspect (width-to-thickness) ratio—a ratio that was considered too high to be feasible with the Rutherford cable.

A milestone in the Doubler magnet saga was reached when Alvin Tollestrup took charge of the program in the spring of 1975. Tollestrup initially came on board while on sabbatical from the physics department of the California Institute of Technology at a time when several key physicists were preparing to leave the laboratory (which was renamed Fermilab in 1974). In short order, he introduced several more technical innovations that proved to be pivotal. One was the novel scheme of interlocked, stainless steel "collars" for providing well-controlled mechanical coil prestress, replacing the ubiquitous metal clamping rings or bands in universal use until then. The original collars were, in fact, solid; laminated collars were mainly the handiwork of Will Hansen, Henry Hinterberger, and Karl Koepke.[36] Crucial to the success of this scheme was the co-discovery of the feasibility of keystoning the Rutherford cable slightly without degrading its superconducting properties. Another modification seemed rather more mundane when it was proposed, but in hindsight it proved to be perhaps the single most important feature of the reworked Fermilab magnets; it also provided the key to the poor performance of their Brookhaven counterparts. Up to this time the bare Rutherford cable (and the ISABELLE braid) had been spirally wrapped with fiberglass tape insulation impregnated with epoxy resin. (The role of the epoxy, cured after coil winding, was as an agent to bond the turns of the coil prior to final magnet assembly.) Still concerned about electrical shorts in the Doubler coils, however, Tollestrup introduced a layer of Kapton film insulation, spirally wrapped around the base cable *before* application of the fiberglass-epoxy tape. Behold, not only did the Kapton improve the electrical performance, *but the training performance improved drastically as well.* Apparently, the reasoning went, direct contact of the superconductor with epoxy resin tends to promote microcracks and heat generation by friction associated with conductor (wire) movement even on the micrometer scale, whereas an intervening layer of Kapton or Mylar shields the superconductor from such heat and may have other mechanical benefits.

A critical factor in the implementation of these technical features was the adaptation for Doubler R&D of the on-site facility originally established for the assembly of the conventional 200-GeV magnets; this facility, suitably revamped, allowed construction and testing of short model and full-length prototype superconducting magnets on a turn-around time scale sufficiently short to make possible the testing of only one feature or parameter at a time. Once the design was frozen, this facility became the in-house Doubler magnet production facility. (In contrast, Brookhaven's approach was again more conservative, namely, to craft all or many of the mutually interacting engineering features in a smaller number of prototype magnets before addressing mass production issues.) The prodigious output of R&D magnets fed gradually lengthening strings of magnets in the B-12 area of the laboratory.

> The tense, isolated, and continuous effort burned out physicists unused to "working where the wolves are howling and the blizzards are blowing, and where there is this long string of magnets which at any moment might do terrible things."[37]

Still, the project was by no means out of the woods. With funding support merely dribbling in, a frustrated Wilson went out on a limb in 1977 by threatening the DOE with his resignation unless "the laboratory's present dreary expectations for the future" improved.[38] Alas, by 1978 the patience of the Fermilab Trustees wore thin as well and his resignation was accepted. Wilson stepped down and after a hastily organized search Leon Lederman of Columbia University was named Director Designate, with Philip Livdahl staying on as Acting Director in the interim. Lederman brought a new style of leadership to the project; whereas Wilson had a tendency to make design decisions on his own and rely on a handful of intimate colleagues for consultation, Lederman was much more open to the technical community at large.[39] Lederman immediately faced a hard choice: whether to scrap the Doubler and embark instead on a project aimed at obtaining proton–antiproton collisions in the main Fermilab ring, as urged earlier and voicefully by Carlo Rubbia and coworkers.[40] Fermilab's Program Advisory Committee had rejected Rubbia's proposal 2 years earlier, but the idea lingered. Lederman quickly convened a marathon session of the Fermilab staff to thrash over the question. The upshot after 18 hours of debate was a decision to go for a $p\bar{p}$ collider, but only after completion of the Doubler, which now became Fermilab's highest priority despite the DOE's demural. The trauma of shortening the 22-ft long Doubler magnets by 1 foot, with 130 22-footers already in hand[41] helped in fact to unite the whole laboratory behind the project.[42] At last the DOE acquiesced, and raised the project status from that of an R&D project to a fully authorized ("Energy Saver") construction project in 1979.

One last technical hurdle remained in the way: a lack of stability of the vertical magnetic field component, which was traced to inadequacies in the cryostat and "cold mass" support system. The problem was solved by Richard Lundy (who took charge of magnet production in 1979) with beefed-up support and his "smart bolt" coil-centering fix. In September of 1980, the dipole design was finally frozen. A first sector of magnets was successfully tested in the spring of 1982, and magnet installation began in earnest that summer with

Figure 13-3. The main tunnel at Fermilab, housing both the older (top) ring of conventional magnets and the newer (bottom) ring of superconducting magnets. In the collider mode of operation, the superconducting ring receives injections of protons and antiprotons from the conventional ring. The two beams are accelerated in opposite directions around the ring to their final energy of 1 TeV, and then guided into collisions at certain interaction points. Courtesy Fermi National Accelerator Laboratory.

operation of the main ring suspended. Installation was completed in March of 1983 (Figure 13-3) under the supervision of Peter Limon, Thornton Murphy, and Laurence Sauer. Cooldown began in May and commissioning followed in June with injected beam and the magnets ramped, the process going smoother by far than had commissioning the conventional ring (which was now relegated to the role of injector or source of protons). On 3 July the beam was accelerated to 512 GeV, and a month later a 700-GeV beam was extracted out of the ring to the external target switchyard. That month, August 1983, ISABELLE was terminated.

As noted earlier, the problems plaguing the ISABELLE magnets were overcome in a dramatic development in 1981 (or 1 year after the Doubler magnet design was frozen), by an inspired initiative of Palmer and a small team around him in Brookhaven's physics department. Palmer's inspiration lay in adopting the Rutherford/Fermilab cable, unaltered in every respect *including its Kapton wrap*, for the ISABELLE dipoles while retaining as many of the latter's features as practical. By chance, Palmer had observed that the radial width of the cable was very nearly half of the width of the braid, enabling a two-layer coil wound from the cable to be neatly substituted for the original single-layer coil without otherwise perturbing the magnet cross section significantly. The properties of the multifilamentary NbTi wire and the cable were well understood, so virtually no further R&D was necessary. Various lesser modifications were introduced, some intended to minimize heat associated with conductor motion, others to maximize the field homogeneity, but features such as a cold iron yoke and a warm bore tube were retained. One additional major change was introduced, however: the abandonment of the awkward shrink-fitting coil-in-yoke insertion procedure in favor of a simpler assembly system in which, essentially, the laminated iron yoke, now split on the midplane, captured the coil by simply bolting together the two yoke halves.

Two short model "Palmer magnets" were hastily wound from surplus Fermilab cable and assembled during a nerve-wracking 6-month period during the spring of 1981. They were tested during July–August of 1981; both exceeded 5 tesla on the first quench and showed no signs of training.[43] So did a full-length (4.6-m) prototype dipole ("LM1") that October. By March of 1982, a total of 6 prototype dipoles had demonstrated equally fine performance: all reached the design field, 5 T, without training (the design field was subsequently raised slightly), exhibited a quench "plateau" of 5.5 T at the normal liquid helium bath temperature, and reached 6 T by pumping on the bath. Thus rejuvenated, the Magnet Division of the renamed Colliding Beam Accelerator project forged ahead. By the summer of 1983, 30 prototype dipoles (and 11 quadrupoles) were in hand (including several preproduction units). Their performance, including that of 10 "field quality" dipoles built in as nearly identical a manner as possible using the construction techniques envisioned for production magnets, demonstrated that the machine could be constructed, and had been so certified by a Department of Energy review panel that April. A string of magnets had been subjected to a major test sequence with satisfactory results in the largely completed 3.8-km CBA tunnel on the Brookhaven site; and the magnet production and testing facilities at Brookhaven were ready for use.[44] But unfortunately, it was too late.

13.3 The Superconducting Super Collider

The demise of the CBA, née ISABELLE, had its roots in two related workshops on "possibilities and limitations of accelerators and detectors" sponsored by the newly organized International Committee for Future Accelerators (ICFA), one at Fermilab in 1978 and another the following year at Les Diablerets, Switzerland in 1979. ICFA had grown out of earlier discussions, primarily between European and Soviet physicists, on the future of high-energy physics and international collaboration that tended to focus on the on again-off again concept of an international Very Big Accelerator (VBA for short). In any event, the VBA concept was rendered passé by the championing of new regional facilities, but the seed had been planted for consideration of some kind of new facility of an unprecedented scope.

Exotic possibilities for behemoth accelerators and storage rings dwarfing projects then in the making[45] were aired at the 1978–1979 ICFA workshops, including proton–proton colliders in the 20-TeV beam energy range, without inquiring too closely into economics or engineering details. The U.S. high-energy physics community rallied behind a major new initiative at the 1982 Summer Study of the American Physical Society's Division of Particles and Fields on "elementary particle physics and future facilities" organized at Snowmass, Colorado; the Superconducting Super Collider (SSC), which emerged from the building ground swell of support and would far exceed current efforts had its origins in the heated discussions that took place amidst the craggy Snowmass peaks.

The formidable technical issues involved in such an undertaking were addressed in greater depth by U.S. and European experts in accelerator science and technology at a 20-TeV Hadron Collider Technical Workshop at Cornell University in the spring of 1983—a banner year for the future of "big" science. The reassuring conclusion reached at Ithaca was that the key to success, superconducting magnet technology, had indeed reached the requisite stage of maturity to be viable—thanks largely to the recent advances at Fermilab and Brookhaven. The time was ripe for in-depth development of several promising candidate superconducting magnet systems; only then would designs and cost estimates allow narrowing cost uncertainties and more sharply define construction methodologies. Additional DPF-sponsored workshops on collider detectors and other accelerator issues, respectively, were held the same year at Lawrence Berkeley Laboratory and at the University of Michigan at Ann Arbor.

The mounting enthusiasm for a national multi-TeV collider in the high-energy community led to the formation in February of 1983 of a Subpanel on New Facilities to DOE's influential High Energy Physics Advisory Panel, with the specific charge to "consider and make recommendations relative to the scientific requirements and opportunities for a forefront United States High Energy Physics Program in the next five to ten years." The subpanel, chaired by Stanley Wojcicki of Stanford University, held a series of briefing meetings at the three major U.S. accelerator laboratories that were vying as well for future facilities of their own: Fermilab, Brookhaven, and the Stanford Linear Accelerator Center; these meetings were accompanied by "town meetings" arranged by the local accelerator "user's organizations." Wojcicki and his group completed their report in the course of two lengthy deliberative meetings at Woods Hole, Massachusetts, and Columbia University's Nevis Laboratory during June–July, 1983.[46] Among other things, the subpanel unanimously recommended "the immediate initiation of a multi-TeV high-luminosity proton–

proton collider project with the goal of physics experiments at this facility at the earliest possible date." For good measure, the report emphasized that

> the Subpanel recommends at the highest priority that a major new project be initiated to design and build a proton–proton colliding beam facility exploiting our superconducting magnet technology with an energy goal of 10 to 20 TeV per beam and completion in the first half of the 1990's. To meet this ambitious time scale, the project must begin in FY 1984 with an intensive research and development phase to establish the precise specifications for the facility.
>
> This project, referred to here as the Superconducting Super Collider (SSC), has captured the minds of high energy physicists everywhere. The conjunction of compelling physics arguments for exploration of the mass region up to a few TeV with the hard won success of large-scale superconducting accelerator technology creates a unique opportunity for the U.S. The SSC is ambitious, but practical. It has immense physics potential, with the very real possibility of truly major advances in our growing understanding of the patterns of particles and the unification of the forces of nature. The SSC can assure a forefront U.S. high energy physics program into the 21st century.[47]

A second recommendation was definitely more controversial, namely "... that the Colliding Beam Accelerator (CBA) project at Brookhaven not be approved." Reading between the lines of the report reveals that there was obvious tension among subcommittee members on reaching this negative recommendation, with a "narrow" majority voting against CBA approval. Among the arguments cited *for* approval, in addition to the facility's intrinsic scientific promise, were the interim experimental opportunities afforded and valuable technical experience promised in advance of the SSC itself. Arguments *against* approval included diminished CBA physics potential resulting from the costly 2-year hiatus associated with the late technical hurdles, coupled with dramatic developments taking place elsewhere in the field of elementary particle science. (For example, other developments included recent discoveries of the charged and neutral intermediate bosons in CERN's $p\bar{p}$ collider, which became operational in 1981; Fermilab's own collider was expected to be operational one to two years before CBA and at over twice its energy.)

The momentous recommendation to proceed with the SSC was unanimously endorsed by HEPAP. In his letter transmitting the subpanel report to the director of DOE's Office of Energy Research (Alvin W. Trivelpiece), HEPAP Chairman Jack Sandweiss added that the SSC "... has fired the imagination of high energy physicists everywhere. The SSC would be the forefront high energy facility of the world and is essential for a strong and highly creative United States high energy physics program into the next century."[48] Following DOE's review and concurrence, HEPAP, on DOE's request, set up a new subpanel to advise DOE on planning the initial round of R&D. With Congressional approval, funds were reallocated to support such R&D. In December of 1983 DOE and the directors of the high energy laboratories (Leon Lederman of Fermilab, Boyce McDaniel of Cornell, and Nicholas Samios of Brookhaven) chartered a Reference Designs Study to review in detail the technical and economic feasibility of a range of design options for creating the SSC facility.

As a result, some 150 scientists and engineers from the national laboratories and various universities converged on Lawrence Berkeley Laboratory, which had offered to host the study, starting in February of 1984.

In chartering the study, the laboratory directors suggested a set of primary design objectives in terms of maximum beam energy (20 TeV), luminosity (10^{33} cm^{-2} sec^{-1}, implying 100 million collisions per second), and other parameters that would henceforth remain the sine qua non of the collider's technical specifications. They also suggested that a range of superconducting magnet options be considered because the system of choice seemed sure to hinge on some economic optimum between magnetic field strength and machine tunnel circumference.

Thus challenged, the design study team went to work under the leadership of Maury Tigner of Cornell University, a prominent accelerator builder and expert on radiofrequency superconductivity. (Tigner had coordinated the Cornell collider workshop and was both a member of the Wojcicki subpanel and a HEPAP member.) Three dipole designs were chosen to embody a variety of design concepts and span a range of field strengths from 3.0 to 6.5 T. The high field end (Design A) was represented by a magnet concept resurrected from one pursued in CBA's waning days: a two-in-one dipole of 6.5 T central field with two closely spaced, horizontally aligned beam tubes and their coils mounted in a common circular iron yoke and suspended in a single cryostat.[49] The medium-field design (B) was a 5 T "warm iron" (room temperature) magnet with each beam tube and coil in its own cryostat. The low field, iron-dominated 3 T magnet (C), a so-called superferric design, consisted of separate beam tubes, coils, and yokes in a common cryostat. The required main ring circumference for the three designs was no less than 90, 113 and 164 km, respectively, or not surprisingly, in approximately inverse proportion to their field strengths. With regard to the magnet option of choice, the conclusion drawn from the study, which was completed in May of 1984, was that each of the three reference magnet styles could no doubt serve as the foundation for a "technically feasible" collider of 20-TeV beam energy.[50] To pin down the matter more closely, however, a vigorous R&D program of approximately three years duration was deemed necessary—to refine the design, cost estimate, and manufacturability of one of the magnet types, assuming that the magnet options could be narrowed to a single one during an early phase of the R&D program.

While the reference designs study was still going on, the DOE had contracted with the Universities Research Association, then a consortium of 56 leading research universities that had managed Fermilab from its inception under contract with the AEC (later DOE) to oversee the R&D phase of the SSC program. URA, in turn, established a Central Design Group (CDG) to lead and coordinate the far-flung effort, the burden of which would obviously fall mainly on the National Laboratories. LBL extended its hospitality to the CDG, and the group settled in "on the hill" with the indefatigable Tigner continuing at the helm. By 1 October 1984, not only was the CDG open for business, but work on model magnets for the SSC had already started at Brookhaven, Fermilab, LBL, and the newly established Texas Accelerator Center at the Woodlands,

Texas. In the remainder of this section we do not intend to trace the general chronology of the SSC in all its variegated aspects, but we will merely recall the principal milestones in the early stages of the R&D program for the SSC superconducting magnet system.

Five different magnet styles were under active development at the various participating laboratories by early 1985: the three original reference design contenders A, B, and C, now joined by C* and D.

Design A, the original high-field 2-in-1 contender slightly modified, was the result of a collaborative effort between Brookhaven and LBL. LBL contributed both superconductor expertise and R&D input from its well-established model magnet program. Brookhaven was primarily responsible for the overall magnet design (in its mature form incorporating concepts from both the Tevatron and CBA magnets), and short (1 m) and longer (4.5 m) demonstration models were fabricated at LBL and Brookhaven, respectively. The essence of Design A, as noted, was a pair of stainless steel-collared coils locked in a single circular iron yoke and cryostat forming a 2-in-1 configuration with horizontally aligned bores.[51] (The stainless steel collars were added in a revision of the original Design A.) Each coil member was a two-layer "cosine theta" coil of 40 mm aperture wound from the highly successful NbTi Rutherford cable. The four models actually constructed at Brookhaven featured a coil aperture of 32 mm—a record low for accelerator magnet apertures—and a reusable iron yoke. (The higher the beam energy in an accelerator, the less the circulating particles tend to oscillate transversely about their "equilibrium orbit" within the confining vacuum chamber, allowing a smaller chamber cross section and hence smaller overall magnet cross section.) Their coil ends were flared out conically ("dog boned"), in anticipation of eventual use of the more challenging Nb_3Sn conductor and the concomitant need to limit strain-induced reduction in critical current by increasing the minimum bending radius in the coil ends; flared ends also helped to minimize the peak field in this coil region. (These flared ends had their legacy in much earlier saddle-coils dating from the first decade of practical high-field superconductivity when the ubiquitous flat niobium-tin ribbon conductors predominated.) A model dipole of type A under assembly is depicted in Figure 13-4.

The prominent feature of Fermilab's medium-field contribution, Design B, was the avoidance of iron in close proximity to the coils. The essential element of the Tevatron dipole, the *collared coil*, formed the basis for Design B, with the cosine theta coils constrained by aluminum collars.[52] The warm iron at a large radius was merely sufficient to shield one coil from the field in the other. In addition, it served as the vacuum vessel for the cryostat. The design operating field, 5 T, was chosen because it led to coil dimensions similar to those of the proven Tevatron magnets with cable available in production quantities. The combination of warm iron and collared coil resulted in a design offering minimum cold mass, which ensured rapid cooldown and warmup, ease of assembly, low heat leak, and little iron saturation. Finally, Reference Design B, in contrast to A and C, incorporated fully decoupled magnet rings; the conservative 1-in-1 feature allowed great flexibility in machine operation and in the

Figure 13-4. First stage in two-in-one dipole assembly. Two side-by-side "bottom" coil subassemblies have been positioned in the lower yoke assembly. Note flared coil ends. Courtesy Brookhaven National Laboratory.

spacing of the magnet rings. A drawback of the nearly iron-less design was the high electromagnetic force between the coil package and iron-walled vacuum vessel, requiring a robust coil support structure which negated to some extent the low heat leak.

Most of the research on the iron-dominated, low-field 2-in-1 "superferric" dipole magnet—the antithesis of Design B—was carried out at the Texas Accelerator Center near Houston, with short models constructed in-house and full-length (28 m!) prototype dipoles fabricated at General Dynamics in San Diego. (Three of these gargantuan units were completed and transported to Texas by tractor-trailer before TAC's program in this area was phased out.) The term "superferric" was coined relatively recently,[53] but the concept exploited one long-championed by Gordon Danby in his "window frame" approach to superconducting magnet design at Brookhaven.[54] The term superferric implies that the ferromagnetic iron yoke contributes a dominant fraction (about two-thirds in TAC's design) of the magnetic field, and that the field "shape" (homogeneity) is dictated mainly by the profile of the iron pole face by (1) limiting the central field and (2) exploiting a particular conductor geometry in the form of simple current sheets delineating the vertical boundaries of an approximately rectangular good-field region. The inspiration for adopting the superferric principle for SSC magnets was first provided by Robert Wilson and Russel Huson at the DPF Snowmass meeting in 1982.[55] TAC was established 1 year later, and Design C was basically frozen in early 1985.[56] TAC's superferric version consisted of separate beam tubes, coils, and (more or less) separate yokes in an over-under configuration in a common cryostat. The conductor proposed for the magnet was the cable developed for the Tevatron's so-called low-beta quadrupoles. Below 2 T the field of such a magnet is fully determined by the iron. Above 2 T, iron saturation gradually sets in, and nonlinear field contributions would require substantial corrections. An anticipated method for alleviating the burden of the correction system foresaw the use of gradually saturating pole face "crenelations" for effectively tailoring the pole face with increasing field (an old practice with conventional magnets). This design is more sparing in the use of superconductor and iron than its high-field, conductor-dominated counterparts, and because of the over-under arrangement of the two coil-in-yoke assemblies, there would be little flux linkage between the two magnet rings. Its proponents envisaged major cost savings by installing such magnets in very long units (up to 35 m each), thus minimizing the number of magnet ends and interconnecting cryostat sections, which tend to be particularly costly. (Similar arguments dictated the maximum credible lengths of dipoles A and B.) However, because 2-in-1 magnets (including Design A) have the drawback of enhanced complexity in magnet assembly and loss of flexibility in accelerator operation, a 1-in-1 variant (design C^*) was subsequently and prudently pursued at TAC as a fallback option. Although most design effort continued to be expended on Design C, which after all embodied most of C^*'s features anyway, a majority of the model magnets that were actually constructed and tested were in fact 1-in-1 magnets.

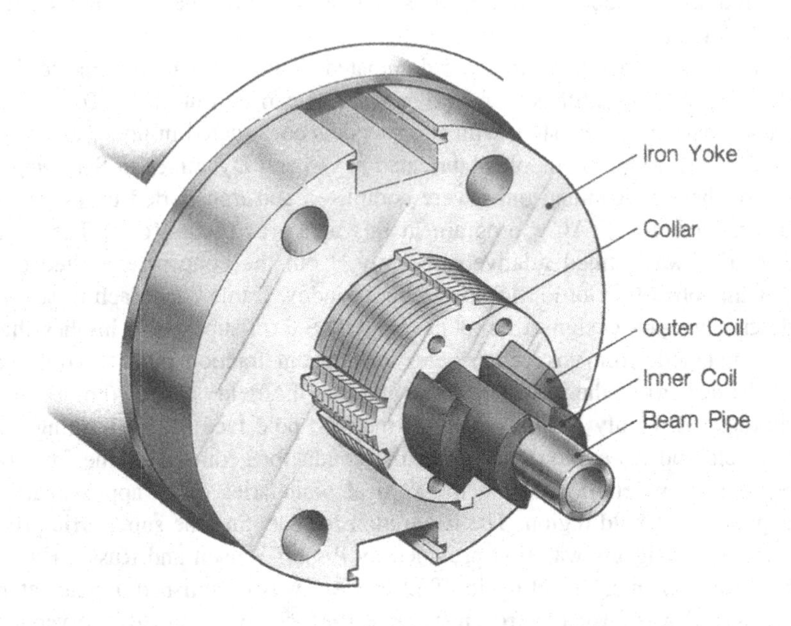

Figure 13-5. Schematic depiction of SSC dipole cold mass configuraiion. Courtesy SSC Laboratory and Lawrence Berkeley Laboratory.

For much the same reason, LBL, Fermilab, and Brookhaven joined forces in the waning weeks of 1984, "in order to minimize technical options and to more efficiently use scarce R&D funds,"[57] by adopting a common high-field design incorporating the most promising features of Designs A and B. The result, Design D, was a 1-in-1 dipole featuring a two-layer cosine theta coil clamped with stainless steel collars (the Tevatron's legacy, as noted), and the collared subassembly captured in a split iron yoke also at helium temperatures, as depicted in Figure 13-5.[58] (Cryogenic studies had satisfied the Fermilab contingency that cooldown and warmup times for cold-iron magnets would be reasonable.) The conductor remained the venerable Rutherford cable, and the inner coil diameter was now 40 mm—a compromise between Design A's skimpy 32 mm and Design B's conservative 50 mm. The magnetic length, approximately 17 m, was the maximum length deemed practical in view of numerous contraints, many of them nonquantifiable, including accelerator physics issues, magnet design aspects, and handling and transportation considerations.

Thus the choice was essentially narrowed down to two magnet options, representing highly diverse applications of superconducting magnet technology with concomitant economic and technical implications. Very high-field magnets would imply a relatively short, and hence cheap, machine tunnel, but inherently costly and technically challenging magnets. Low-field magnets are presumably relatively inexpensive requiring modest R&D (or so the argument went); in this case the accelerator cost is dominated by the civil construction of a correspondingly longer tunnel.

To pave the way for the selection of a particular design for "further development to a stage where a responsible SSC conceptual design could be prepared," the CDG now convened a Technical Magnet Review Panel under the chairmanship of Alvin Tollestrup. The panel would review the status of all designs in the offing and the R&D program generally and recommend modifications in the program to ensure that enough technical information would be available to choose fairly a magnet style by the end of the summer of 1985; the CDG also set up several task forces to collect and evaluate specialized information pertinent to the selection. The actual selection, to be made by Maury Tigner himself, would rely heavily on the recommendation of an SSC Magnet Selection Advisory Panel appointed by Tigner and chaired by Frank Sciulli of Columbia University.

The Sciulli panel, charged with the task of submitting " ... a report to the Director [of the Central Design Group] containing the Panel's recommendation in the form of an ordered list of these five basic magnet styles ... ," met for 4 days in late August of 1985 at the CDG in Berkeley, and heard testimony from proponents of the various styles as well as presentations on technical, operational, and cost issues by CDG staff. Though requested to consider all five basic magnet types, most of the discussion centered on Designs C and D, as these were "most strongly advocated by the proponents," thereby obviating the need for an "ordered list or ranking." The panel's final report to the director, Maury Tigner, is dated 9 September 1985. Of "the two principal magnets placed in competition," the panel was "unanimous in recommending D as the design basis for an SSC dipole element."

> The hope of two years ago that the superferric magnet would provide a less complicated and less costly SSC has not become true. The Panel believes that there is a real possibility that an SSC based on superferric magnets would be more costly than currently estimated. On the other hand, the cos θ design has proceeded smoothly and met all expectations. For these and other reasons ... the Panel believes that the cost of an SSC based on cos θ magnets is predictable.[59]

The panel justified its recommendation on two different grounds. The *cosine theta style magnet* was favored for being "a well understood magnet with reliably predictable costs and production schedules. The reliability of the predictions rest on a large base of data developed in building and operating one large accelerator and developing models and prototypes for several others." *High field* was recommended in light of machine operational as well as capital cost penalties associated with the larger circumference ring based on magnets of style C.

> The work with the superferric style [adds the panel] has shown it to be more complex than foreseen two years ago. As a result only models of one-meter length have been extensively tested so far. In the panel's opinion, this style has not

displayed the simplicity and ease of construction and operation it originally promised. Of course, if the superferric style held out the promise of substantial construction cost saving to outweigh the additional R&D costs required to develop it for production, it would be a strong candidate despite its less mature status of development. Although its proponents argued that this was the case, the panel is convinced that, as cost studies done by the CDG suggest, the contrary is true. Use of the superferric style, in the panel's opinion, is likely to result in higher total project costs, exclusive of R&D.

In view of these findings, the panel recommends that the choices between the two styles be made now and that the cos θ Style D be chosen.[60]

On 13 September Maury Tigner concurred with the Sciulli Panel's recommendation, cautioning that what had been selected was "a style, not a finished design. Production of an optimized and industrially producible design is the next order of business."[61] The stage was set for the various R&D teams to proceed in unison henceforth. By mutual agreement (in the official CDG language), LBL would concentrate on superconducting cable development and short model magnets, BNL on longer models and the fabrication of full-length prototypes, FNAL on cryostat development and testing of full-length magnets, and TAC on quench protection and magnet correction systems (as well as phasing out the superferric magnet program).[62] Concurrently, the Central Design Group and the four laboratories would initiate a multiphased magnet technology transfer program involving industry, leading to industrial engineering, production tooling, demonstration of preproduction fait accompli, and ultimately culminating in mass production of the 10,000-odd magnets that would be required.[63] For that matter, industrial partners (in consultation with the University of Wisconsin) had been heavily involved since the SSC's inception in the technological heart of the magnets, the basic NbTi composite wire. Steady improvements were being realized in the current-carrying capacity and other properties over the conductor performance characteristic of the older Tevatron or CBA magnets (e.g., 50% improvement in J_c by 1986).

The first visible fruit of the focused R&D program set in motion by the magnet selection process was a series of six 4.5-m long Type D model magnets constructed at Brookhaven and tested there with good results between June and November of 1985. All reached a stable quench "plateau" at the SSC design operating field (6.6 T) in pool boiling helium—albeit the conductor limit achievable at that time—with modest training, and displayed field uniformity by and large meeting SSC accelerator specifications.[64] Cooldown at Fermilab of the first 17-m prototype "cold mass," trucked in from Brookhaven (Figure 13-6) and encased in a Fermilab cryostat, occurred in July of 1986. To be sure, once cooled the performance of D0001 proved to be flawed compared to that of the shorter forerunners, requiring excessive training to reach the design field. A full 2 years would elapse before a sixth full-length prototype (DD0012 in the subsequent nomenclature) exhibited performance more nearly approaching the excitation requirements demanded of true production magnets[65]—the payoff of a strenuous mutual interlaboratory development effort that was gradually gaining momentum.

Figure 13-6. SSC dipole DOOO1 prior to shipment to Fermilab in refrigerated container. At Fermilab, the 17-m cold mass will be mounted in a cryostat and will undergo acceptance tests in liquid helium. Courtesy Brookhaven National Laboratory.

The last word had not been heard from the proponents of superferric magnets when, in April of 1986, as D0001 was being readied for its journey to Illinois, TAC submitted "new information" to DOE. Based on their by now largely completed magnet program, the TAC protagonists maintained "that the superferric design is the best and least expensive choice for the SSC," and that "we are prepared to demonstrate that the superferric magnet is industrialized and ready for construction, and that the project cost can be reduced by $1 billion compared to the high field design."[66] Accordingly, the DOE was compelled to set in motion a reexamination of TAC's program and consider the wisdom of reopening the magnet selection process with the aid of yet another HEPAP subpanel, this time chaired by Burton Richter of the Stanford Linear Accelerator Center. However, it required the subpanel but a month of investigation to convince itself "that [TAC's] claim to be able to do the job better, cheaper, and faster with superferric magnets is not borne out by the information supplied to the Subpanel." More specifically, the subpanel concluded:

> The TAC magnet R&D program is still about two years behind the high-field magnet program. The TAC magnet is not ready for industrialization.
>
> Cost uncertainties in TAC technical components are large, and superferric magnet costs cannot be estimated with confidence at the present state of R&D.

> We do not believe that the new information presented warrants reexamination of the CDG magnet-selection decision.[67]

With the superferric option finally laid to rest, all available resources, as they were, could be brought to bear on "the SSC dipole" (the designation Design D was no longer pertinent). Even so, the SSC dipole was not yet out of the woods, but the path out was well defined. The year 1987 is remembered mainly for the unremarkable performance of two new long magnets, neither of which did substantially better than D0001 or its successor D0002, for a new series of healthier short magnets for quantifying performance questions, for a blur of numerous Magnet Systems Integration Meetings, and for the tireless goading of the Central Design Group's "gang of four" (Victor Karpenko, Peter Limon, Edwin Goldwasser, and Tigner himself). The years 1988 and 1989 saw a succession of 1-meter LBL models, 1.8-meter BNL models (the odd length dictated by available tooling), and 17-meter protoypes, albeit fewer in number, incorporating BNL-fabricated cold masses and Fermilab cryostats. Of the four prototypes tested in 1988, two passed muster; 1989 did better with four out of five qualifying. The development of a finite element mechanical analysis capability, coupled with refined diagnostic instrumentation in the later magnets, was finally paying off. The shorter models served as vehicles for expeditiously proofing design features and modifications suggested by the mixed performance of the full-length prototypes, the latter naturally requiring much longer lead time to test new ideas. Design concepts were honed by engineering analysis at Brookhaven, Fermilab, and Berkeley, and thrashed out in wearisome Magnet Systems Integration Meetings that rotated on a regular basis between the laboratories.

While the dipole program received the lion's share of the technical attention, the next major SSC project milestone following the Reference Designs Study of 1984 and the subsequent magnet selection in 1985 was the issuance of the SSC Conceptual Design Report (CDR) in the spring of 1986[68]—the massive, multi-volumed tome under J. D. Jackson's editorship that was the handiwork of scores of individuals and institutional efforts, augmented by task forces, review panels, and workshops. The effort paid off in the DOE's subsequent favorable review of the conceptual design, which concluded that:

> ... the design set forth in the CDR is technically feasible and properly scoped to meet the requirements of the U.S. high energy program in the period from the mid-1990s to well into the next century.[69]

In January of 1987, President Ronald Reagan personally endorsed construction of the collider. Three months later the Department of Energy initiated the SSC site selection process by soliciting site proposals nationwide. Forty-three responses narrowly met the deadline, each consisting of the requisite numbered thirty sets of eight volumes each, ranging from executive summary to socioeconomic considerations. Thirty-five proposals that met the DOE's stringent criteria were then evaluated and whittled down to a Best Qualified List of select sites by a prestigious screening committee assembled by the National

Academy of Sciences and the National Academy of Engineering, in a scenario not unlike the hunt for a site for the 200-BeV accelerator 22 years earlier—which, like the SSC, wound up with six finalists! At long last, in November of 1988 DOE Secretary John Herrington announced the Dallas-Fort Worth site in rural Ellis County centered on historic Waxahachie as the winner.

Concurrently, the DOE issued a call for proposals for management of the construction and subsequent operation of the SSC Laboratory. In January of 1989 the successful bidder, a triad consisting of Universities Research Association teaming with two industrial partners, was officially designated "M&O Contractor" for the SSC with Roy F. Schwitters of Harvard appointed as director of the laboratory. With these politically charged milestones behind, the SSC entered a period of transition—from Berkeley to the Dallas suburbs and from the preconstruction mission of the CDG to the completion of a site-specific design and the actual construction of the 54-mile behemoth collider ring and its world-class laboratory. But that, including the eventual disposition of the superconducting magnet system for the SSC, remains unfinished history as this volume goes to press, and it is a history better left to others to complete.

CHAPTER 14

Aftermath

14.1 High-Temperature Superconductivity

Superconductors, it will be recalled, were first brought out of the liquid helium temperature range in wartime Germany in 1941, when Ascherman, Friedrich, Justi, and Kramer found that niobium nitride is superconducting at about 15 K. It took 13 years to raise their critical temperature record by a paltry 3 additional degrees (Nb_3Sn in 1954), and 19 more years to top it 4 additional degrees to 23.2 K for Nb_3Ge in 1973.[1] In fact, niobium-germanium had also been investigated in 1954 as a potential superconductor (by Hardy and Hulm) and had been known to be one for about a decade when John Gavaler revisited the compound—considered promising at the time for its beta-tungsten crystal phase, like niobium-tin.

"The intriguing question," ends Gavaler's 1973 paper, "remains whether the present ... technique can succeed in raising T_c even further in the Nb-Ge or in some other metallurgical system. This question, of course, can only be resolved by further experiment"[2] [the BCS theory was providing scant guidance on this question]. But for 13 years experiments got nowhere, and the next most promising intermetallic compound, Nb_3Si, exhibited a disappointing T_c value of 20 K, not 30 K as had been confidently projected[3]—not until early 1986, that is. The spectacular experimental breakthrough by Johannes Georg Bednorz and Karl Alex Müller at the IBM Zürich Research Laboratory in Ruschlikon, Switzerland that suddenly opened up an entirely new era in superconductivity was the result of a dogged 4-year search in a quite different direction in "the conviction that the efforts in intermetallic compounds should not be pursued further."[4] Once again history repeated itself when Bednorz and Müller returned to a venerable class of compounds, albeit nonmetallic ones. We recall that Hulm and Matthias had been intrigued by the perovskite-type compounds as far back as the late 1940s.[5] For that matter, in 1937 Fritz London had speculated on the existence of supercurrents in nonmetallic systems (organic molecules).[6] At any rate, the first superconducting metallic oxide, $SrTiO_3$, was discovered in 1964 by J. F. Scholey and colleagues following discussions on the possible existence of superconductivity in semiconductors.[7] To be sure, T_c was only 0.3 K, eventually tweaked at Ruschlikon by Bednorz and Gerd Binning to a mere 0.7 K by doping the oxide with niobium. However, rapid progress was made in 1973 when David Johnston of the University of California, San Diego, detected superconductivity as high as 13.7 K in the oxide $LiTiO_3$; two years later followed Arthur Sleight's Ba–Pb–Bi–O, a perovskite with slightly higher T_c (14 K). Immediately following these promising developments came a rash of exaggerated claims and irreproducible results, topped by the Russian report in 1977 of superconductivity in CuCl at the preposterous temperature of 140 K![8] None proved credible.

In 1983, with the flurry of exorbitant claims not yet subsided, Bednorz, joining Müller, instituted a new search for superconductivity in the perovskites, first in nickel-containing perovskites but without success. In 1985 their search shifted to copper-containing compounds, in particular oxides with mixed copper valence.[9] Again the hunt got off to a poor start, but before long they were alerted to the fact anticipated in an old paper by the French team of Er-Rakho, Michel, Provost, and Raveau on the conductivity of the mixed perovskite $La_4BaCu_5O_{13.4}$, that emphasized the very property of mixed valency. "The presence of copper in these compounds [wrote the French authors in 1981], simultaneously with two oxidation states, Cu(II) and Cu(III), suggests interesting electrical properties, which will be investigated as a function of oxygen amounts."[10] Losing no time, Bednorz and Müller set to work on reproducing the French mixed oxide with continuous variations in valency, all the while keeping a careful watch for superconductivity. Sure enough, dc resistivity measurements in January of 1986 revealed the onset of a sharp drop in resistivity at a temperature somewhat dependent on current density but as high as 35 K when corrected for this dependency, interpreted "as possible high-T_c superconductivity of a percolative nature." The presumed superconductivity was subsequently isolated to a layer-like oxide phase in the sample—a phase which, as the researchers modestly put it, "would, even without the lucky preparation conditions, have been found sooner or later."[11] Bednorz and Müller were weary of premature publication in view of the hubris of earlier unfounded claims to high-T_c superconductivity, and their paper did not appear in print until the following September.[12] And even then, they hedged on their results by adding the cautious title "Possible high T_c Superconductivity in the Ba–La–Cu–O System," because they lacked a suitable magnetometer for performing the definitive test for the Meissner effect.

Once the news broke, however, Bednorz and Müller's tantalizing finding was quickly confirmed by several other groups that had, in fact, been in close pursuit of oxide superconductivity for some time. By chance, this subject had been a hot topic from the time of Arthur Sleight's unveiling of superconductivity in Ba–Pb–Bi–O at 14 K in 1975. Leading runners-up in the building race were Shoji Tanaka of the University of Tokyo, Paul Chu of the University of Houston, and Bertram Batlogg of AT&T Bell Laboratories at Murray Hill, New Jersey. First into print was Tanaka's Tokyo group. Tanaka had been patiently pursuing the superconductivity of Ba–Pb–Bi–O ever since Sleight's discovery at Du Pont.[13] But by mid-1986 his group had essentially exhausted this material, and was casting about for other possible superconducting oxides. On reading Bednorz and Müller's paper in November, they lost no time in confirming the results on Ba–La–Cu–O, throwing in a positive test for the Meissner effect for good measure. The Tokyo results were rushed into publication within the month (22 November) and Koichi Kitazawa reported them at the 4 December meeting of the Material Research Society in Boston.

Paul Chu and colleagues were not far behind. Chu, former student of Matthias when he was at the University of California in San Diego, was an authority on the effect of high pressure on the superconducting transition. His team, too, quickly confirmed the Ruschlikon results, and then ascertained that the temperature marking the onset of superconductivity rises under hydrostatic pressure by about one degree Kelvin per kilobar, reaching 40 K at 13 kbar.[14] Not only that, one sample exhibited $T_c > 56$ K but "was later destroyed and could not be reproduced." In a note added in proof, however, Chu et al., state that detailed examination of the results from this particular sample indicated an "onset temperature" of 70 K and a sharp drop in resistance near 60 K. To enhance the pressure effect, they contrived to substitute strontium for barium. In the event, they found, now in collaboration with M. K. Wu at the University of Alabama at Huntsville, that substituting strontium produced an onset of resistivity near 42 K *at ambient pressure*. Close on Chu's heels, Batlogg and colleagues skipped barium in favor of the strontium substitution without further ado. They, too, found the onset for ambient-pressure La–Sr–Cu–O to occur near 40 K—actually 36.5 K.[15]

The role of Maw-Kuen Wu, a former student of Chu, vis-à-vis Chu himself in the ongoing developments that would swiftly propel superconductivity out of the liquid hydrogen domain in early 1987, was obscured in the hoopla accompanying Y–Ba–Cu–O that spring. According to Pool, Chu ran into Wu at the meeting of the Materials Research Society in Boston where he told his protégé about their recent superconductivity results in Houston and the two agreed to collaborate in a search for other high-temperature superconductors.[16] In particular, Chu suggested that Wu look into the effect of substituting some smaller element for barium, say strontium. By mid-December Wu's group in Alabama had prepared a La–Sr–Cu–O sample that they quickly confirmed to be superconducting in the low 40 K-range. Their appetite wetted, they turned to different compositions of the strontium system and various methods of preparation, and this occupied them through December and into January. At the same time, the Houston group was exploring the possibility of superconductivity in various other materials related to the original La–Ba–Cu–O oxide, and in some they even found tantalizing hints of superconductivity as high as 100 K. Although these results could not be reproduced, Chu went ahead and filed, on 12 January

1987, a patent for a wide range of materials including, significantly, a particular composition of the oxide Y–Ba–Cu–O.

Meanwhile, Wu and his students were also looking into several other substitutions for barium, but with mixed results. Thus, calcium produced a superconductor, but with a lower T_c than with strontium. If nothing else, these results appeared to underscore the special role of La–Sr–Cu–O in promoting T_c. Then, on 17 January Wu's graduate student Jim Ashburn, on a hunch, did some sketchy calculations on substituting for *both* lanthanum and strontium; according to his scribbles "on the back of some homework,"[17] substituting yttrium for lanthanum and barium for strontium should yield a particularly good candidate superconductor. Wu agreed. Unfortunately, there was no yttrium at hand in the laboratory. A hurried search located some at the nearby Marshall Space Flight Center, and on 28 January a sample of Y–Ba–Cu–O was prepared by Ashburn and fellow student Chuan-Jue Torng according to Ashburn's recommended starting composition and baked overnight. The rest is history, as they say. On 29 January resistance measurements in liquid helium revealed an onset temperature of 90 K! To be doubly sure of the bath temperature calibration, the run was quickly repeated in liquid nitrogen with identical results. There was no doubt. The world's first superconductor with a transition temperature well above the liquid nitrogen boiling point, 77 K, was a fact, as testified in Figure 14-1. Wu called Chu the same afternoon with the breathtaking news: "We've hit the jackpot."[18] He and Ashburn, sample in hand, flew to Houston the next day where susceptibility measurements confirmed the Meissner effect as well. The results were announced at a joint Houston–Alabama press conference on 16 February and published jointly in two back-to-back letters in *Physical Review Letters* on 2 March (submitted to PRL on 5 February).

> For the first time [reads their first letter], a zero-resistance state ($\rho < 3 \times 10^{-8}$ Ω-cm, an upper limit only determined by the sensitivity of the apparatus) is achieved and maintained at ambient pressure in a simple liquid-nitrogen Dewar.[19]

Critically important to exploiting the landmark discovery, but missing, was the precise composition of the superconducting phase of the Huntsville sample. Knowledge of Ashburn's starting composition before baking was of little help, because the processed sample appeared to be a mixture of two distinct chemical compounds: transparent "green stuff" and a black, opaque phase—suspected to be the superconductor—and so intergrown as to defy analysis or separation by routine techniques available to Chu or Wu. Time was of the essence, so Chu dispatched their material to Robert Hazen, an experimental mineralogist, and his group at the Geophysical Laboratory at the Carnegie Institution of Washington. There, the elemental ratios of the two phases were soon deduced by electron microprobe analysis; the green phase turned out to be Y_2BaCuO_5, and the all-important black phase an "oxygen-deficient" perovskite $YBa_2Cu_3O_7$—the legendary "1-2-3" superconductor.[20] Initial x-ray diffraction and subsequent neutron diffraction revealed the 1-2-3 structure to be a cubic perovskite arrangement of atoms with strong two-dimensional layering because of the oxy-

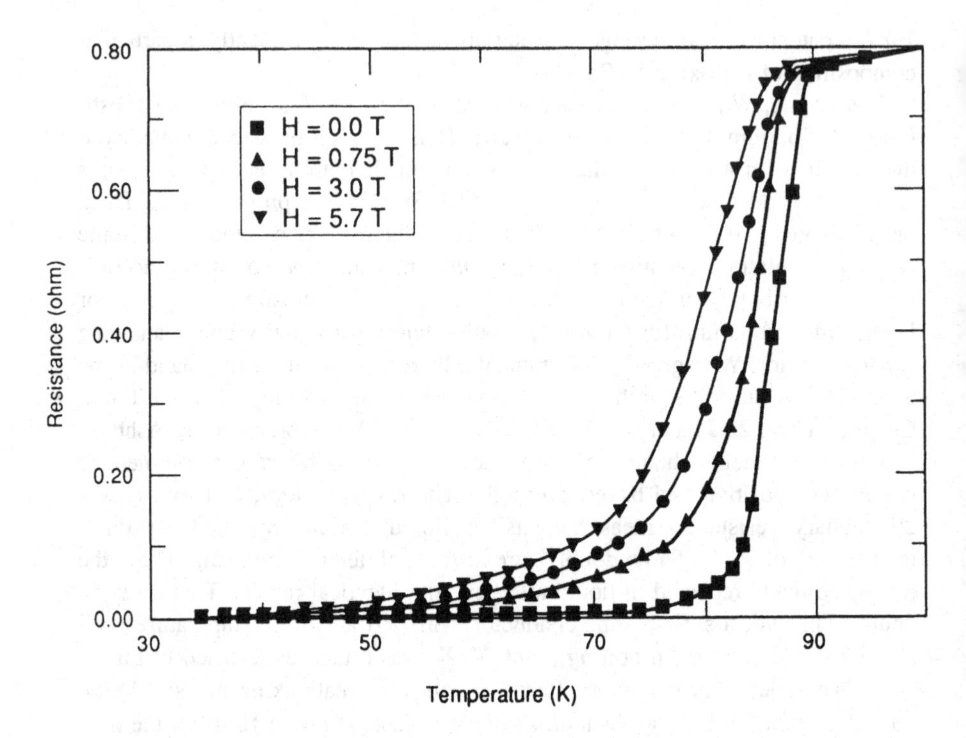

Figure 14-1. Temperature dependence of resistance of Y–Ba–Cu–O in liquid nitrogen, as a function of ambient magnetic field. Wu et al., PRL, **58** *(1978), 909.*

gen defects; the basic repeating unit consisted of three cubes stacked one above another with layers of copper and oxygen atoms in planes separated by yttrium and barium atoms, as depicted in Figure 14-2. On advice of their patent attorney, however, Chu withheld announcement of the structure for the time being.

The discovery of Y–Ba–Cu–O was made independently by Zhongxian Zhao and co-workers at the Institute of Physics of Academia Sinica in Beijing and soon confirmed in laboratories worldwide. The climax to the eventful opening weeks of 1987 was the March meeting of the American Physical Society at the New York Hilton. One of its featured sessions was an evening symposium on 17 March on the history of superconductivity,[21] sponsored by the society's History of Physics Division. In fact, a historical milestone in superconductivity proved to be the next evening's "Woodstock of Physics" when 2000 scientists squeezed into a Hilton ballroom and just as many crowded around television monitors in adjoining rooms, even the main lobby, struggling for space, in a hastily organized session on the new ceramics that lasted well past three a.m. (Of the 3080 abstracts contributed in advance to the March meeting, only *one* covered high-T_c superconductivity, namely an IBM paper on Ba–La–Cu–O!) A hushed audience strained to hear as group leaders from Ames and Brookhaven, Los Alamos, AT&T Bell Labs, Tokyo, Huntsville–Houston, and the Carnegie Institution reported superconductivity above 90 K in a broad class of com-

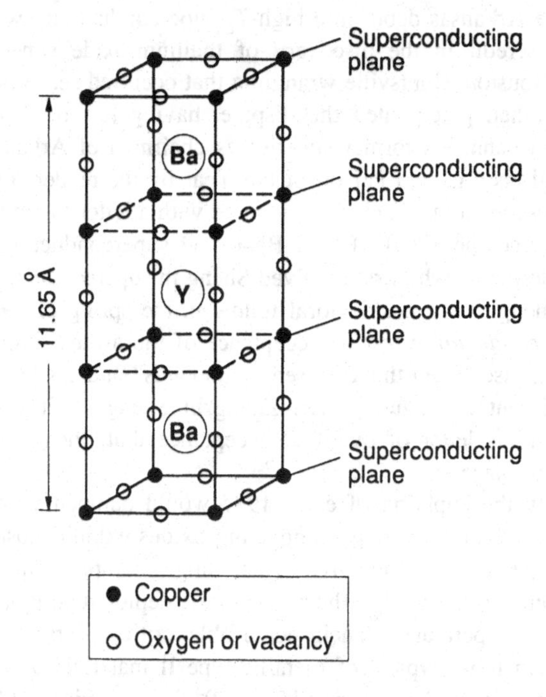

Figure 14-2. Crystalline structure of yttrium-barium-copper-oxide.

pounds with chemical composition $RBa_2Cu_3O_{9-y}$, where R respresents a transition metal or rare earth ion.[22]

A year later, just when the superconducting onset temperature for the ceramics seemed to have settled down to around 95 K, it took another—albeit more modest—leap with the announcement in January of 1988 by Japanese researchers headed by Hiroshi Maeda at the National Institute for Metals in Tsukuba, Japan of superconductivity near 110 K in bismuth-strontium-calcium-copper-oxygen. Within days, Chu announced that his group had been working on this material as well. Not only that, in less than a month a second "triple-digit" superconductor followed, this time a thallium-barium-calcium-copper-oxygen compound that showed onset near 120 K, which was announced at the World Congress on Superconductivity in Houston in February by Allen Hermann and Zhengzhi Sheng of the University of Arkansas. The Arkansas team had actually discovered the thallium compound in October and forwarded samples to several other laboratories for corroboration, among them Timir Datta's group at the University of South Carolina. At "Woodstock II" during the March 1988 meetings of the American Physical Society in New Orleans, Datta commended Hermann's group for its courtesy and for not rushing their results into print as was so often done.[23] Shortly after the Arkansas announcement, their claim was confirmed by Uma Chowdry of DuPont, Zhao at Beijing, and by Paul Grant at IBM-Almaden Research Center; this last group reported a record onset of 125 K.

Actually, the Arkansas debut into high-T_c superconductivity was marred by a dispute over credit in the discovery of thallium-oxide superconductivity, much like the Houston–Huntsville wrangling that occurred scarcely 2 years earlier. Zhengzhi Shen precipitated the dispute, having become upset when he learned that Hermann, his former superior as chairman of Arkansas's physics department, had been named Person of the Year by the readers of the high-T_c newsletter *Superconductor Week*; he countered with a letter to the newsletter on "How I discovered the 120 K Tl–Ca–Ba–Cu–O superconducting system." In point of fact, Hermann, who first involved Sheng in superconductivity measurements while Sheng was a postdoctoral fellow in the spring of 1987, had conveyed to *Superconductor Week* his acceptance of the award simply "as leader of the team that discovered the Tl-based compounds" and urged "special mention to my brilliant colleague ... Dr. Zhengzhi Sheng."[24] It was from Hermann's copy of his letter of qualified acceptance that Sheng learned of the award in the first place.

Unfortunately the euphoria of early 1987 would dampen with the gradual realization that, although the superconducting oxides exhibit most of the characteristics of type II superconductivity (including incomplete Meissner effect), they differ in one respect vital to their large-scale deployment in applications at liquid nitrogen temperatures: Their achievable critical current densities are vastly lower than those typical of metallic type II materials. A representative low-field critical current density for NbTi is 10^4 A/mm^2 (Figure 14-3), or about an order of magnitude lower than the BCS electron depairing critical current density.[25] Even the very first Nb_3Sn conductor of 1961 carried 10^3 A/mm^2 at 8.8 tesla! By contrast, J_c-values for bulk polycrystalline oxide materials remain typically 10 A/mm^2 after 2 years of effort, or not much better than copper. The oxides also exhibit greater fall in J_c with increasing field than do J_c's for conventional superconductors. In addition, they are, by their very nature, highly anisotropic. Not only is J_c in unoriented bulk samples of $YBa_2Cu_3O_7$ an order of magnitude lower than for oriented films and single crystals,[26] but the ratio $J_c(\|)/J_c(\perp)$, where the comparison is with respect to the Cu–O plane of the crystal structure (a structural feature of the Bi or Tl-based materials as well), ranges between one and two orders of magnitude. Heuristically, the anisotropy reflects the fact, anticipated by the French in 1981, that the superconductivity is confined to the Cu–O plane, the intervening rare earth atoms serving simply as the bonding agent for the structure. A physical clue to the anisotropy was provided by the finding that the (upper) critical field H_{c2} exhibits similar behavior. This anisotropy and the low absolute magnitude of J_c are believed to stem from the anisotropy and smallness of the coherence length, which is related to H_{c2}.[27] If not bad enough, large-scale technological application is equally frustrated by the notorious brittleness of the oxides which are, after all, simply brittle ceramic shards—a problem even more vexing than that posed by the A-15 compounds such as Nb_3Sn.

High-T_c superconductivity is a young and developing discipline, however. New developments during 1988–1989 appear to have thrown new light on the critical current question that is so important from the viewpoint of applications.

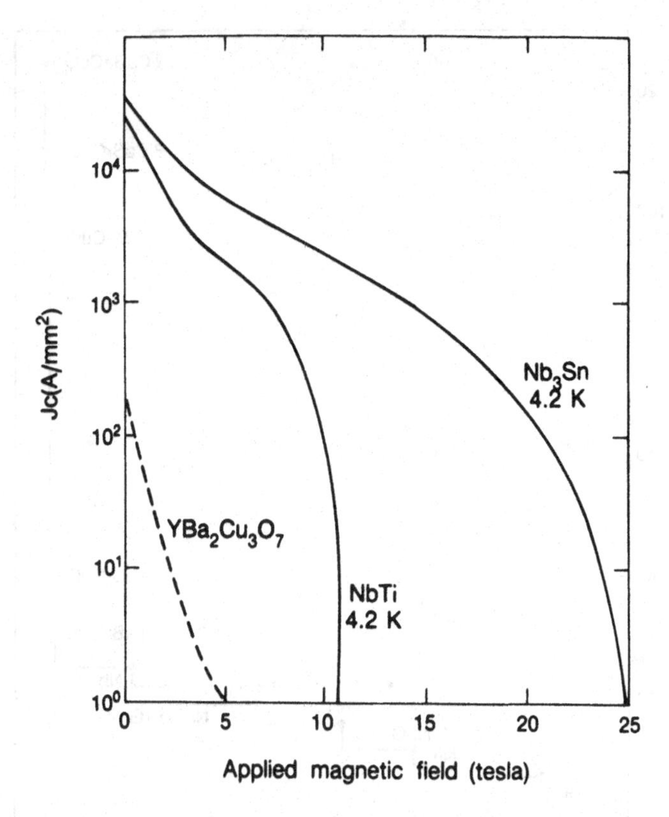

Figure 14-3. Variation of critical current density with applied field for NbTi, Nb_3Sn, and $YBa_2Cu_3O_7$.

Experiments on single crystals, particularly at IBM, Yorktown, and AT&T Bell Laboratories, have ranged from magnetization and resistivity measurements to direct observations of the Abrikosov flux lattice by "decorating" surfaces of thin superconducting layers with ferromagnetic particles—a rather old technique rendering fluxoids (as well as dislocations and other imperfections) visible. Collectively they suggest that the behavior of the oxide superconductors in a magnetic field may very well be fundamentally different from that of conventional type II superconductors.[28] In conventional type II materials, resistivity due to flux creep (flux motion from thermal activation under conditions when the pinning force exceeds the Lorentz force exerted by the current on the flux lines) is very small due to the strong pinning. However, in the superconducting oxides, the flux creep appears larger by an order of magnitude; low pinning energy and appreciable flux creep may be an intrinsic property of these materials, and a consequence both of their high-critical temperatures (i.e., more thermal energy for flux motion) and small coherence length (smaller barriers).[29] Not only that, but because the high flux creep arises from an unusual degree of thermal activation, it will most likely increase with temperature.

Indeed, it has become customary to speak of the "melting" of the flux lattice

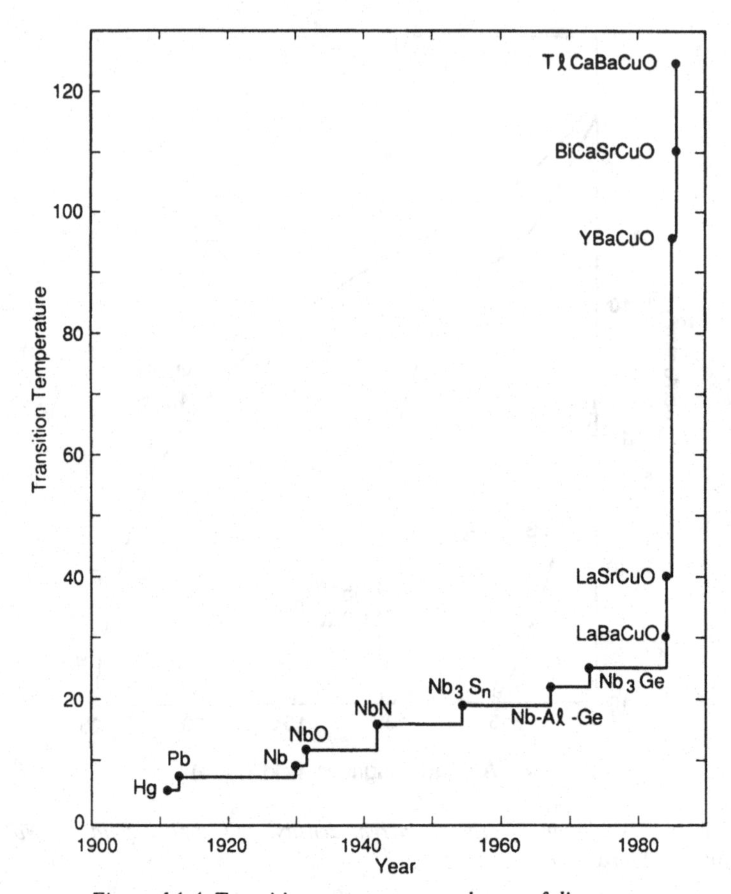

Figure 14-4. Transition temperature and year of discovery.

above a certain temperature and at a field strength smaller than H_{c2}, or before the superconductivity is destroyed. For conventional type II superconductors, the lattice melting temperature is comfortably above the critical temperature. However, according to David Bishop and colleagues at Bell Labs, the lattice of Y–Ba–Cu–O melts slightly below the temperature of LN_2, and the bismuth and thallium-based materials have lattice melting temperatures around 30 K—far below their T_c.[30] This alarming feature suggests "severe implications for potential applications."[31]

Despite the sobering outlook for high-field/high-current density applications, and the concomitant realization that the new oxides are not a panacea to boundless commercial opportunities, exciting technological opportunities seem virtually assured in view of the formidable industrial, governmental, and academic resources and concerted materials effort that is being brought to bear on the problem worldwide—most evidently in thin film applications where the superconducting and materials properties attain optimum values or are most amenable to manipulation.

Nor should preoccupation with technological payoff obscure the fundamental importance and startling nature of the phenomenon of high-temperature superconductivity itself. With recurring hints of superconductivity at even higher temperatures than those being reproducibly realized today, or simply the trend shown in Figure 14-4, the dream of room-temperature superconductivity can no longer be summarily dismissed. And with a theoretical mechanism to account for high-T_c superconductivity remaining as vexing as the material properties involved, the challenge is again manifold. If events of the past several years are a guide, we can be confident that wondrous developments in experimental, applied, and theoretical superconductivity will take place in the years to come, just as was true for the elemental and metallic type II superconductivity of past years. Whether they might be developments engendered in the research tradition stretching from Kamerlingh Onnes to Meissner, from Lindemann through the Londons to Bardeen and colleagues, from Mendelssohn to Kunzler and those in his footsteps, and certainly the singular school of Matthias, aspects of an exciting chapter in the history of solid state physics are once more being written.

Abbreviations

The following abbreviations are used extensively in the Notes:

AHQP	Archive for History of Quantum Physics
AP	*Annalen der Physik*
APC	*Annalen der Physik und Chemie*
BHS	Becker, Heller and Sauter
BSM	Bohr Scientific Manuscripts
CR	*Cryogenics*
DSB	Dictionary of Scientific Biography
ETH	Library, Eidgenössische Technische Hochschule, Zürich, History of Science Collection
HSPS	*Historical Studies in the Physical Sciences*
HSPBS	*Historical Studies in the Physical and Biological Sciences*
JAP	*Journal of Applied Physics*
KO	Kamerlingh Onnes Collection, Museum Boerhaave, Leiden
NA	*Nature*
PH	*Physica*
PLC	Leiden, University, Physical Laboratory, *Communications*
PLS	Leiden, University, Physical Laboratory, *Supplements to the Communications*
PM	*Philosophical Magazine*
PR	*Physical Review*
PRL	*Physical Review Letters*
PRS	Royal Society of London, *Proceedings*
PT	*Physics Today*
PTR	Physikalisch-Technische Reichsanstalt
PZ	*Physikalische Zeitschrift*

PZS	*Physikalische Zeitschrift der Sowjetunion*
RMP	*Reviews of Modern Physics*
RN	Akademie der Wetenschappen, Amsterdam, *Proceedings*
SB	Königlichen Akademie der Wissenschaften zu Berlin, *Sitzungsberichte*
SC	*Science*
SON	Deutsches Museum, Munich, Sondersammlungen
TM	*IEEE Transactions on Magnetics*
TNS	*IEEE Transactions on Nuclear Science*
WM	Walther Meissner Papers, Deutsches Museum, Munich
ZK	*Zeitschrift für die gesamte Kälte-Industrie*
ZP	*Zeitschrift der Physik*

Notes

Notes to Chapter 1

1. Dewar, James, "Liquid gases," *The Encyclopædia Britannica*, 11th ed., vol. XVI (Cambridge, 1911), 751.
2. *Het Natuurkundig Laboratorium der Rijks-Universiteit te Leiden in de jaren 1904–1922. Gedenkboek aangeboden aan H. Kamerlingh Onnes, Directeur van het laboratorium, bij gelegenheid van zijn veertiwarig professoraat op 11 November 1922.* (Leiden, 1922).
3. Material concerning the 1922 celebration is found in Archive 113, Kamerlingh Onnes Collection, Museum Boerhaave, Leiden.
4. Cohen, E., "Kamerlingh Onnes Memorial Lecture," Chemical Society, *Journal*, **1** (1927), 1208.
5. Armstrong, H.E., "Sir James Dewar, 1842–1923," Obituary notices, Chemical Society, *Journal*, **131** (1928), 1066–1076 on 1070–1071; Koopman, J.F.H., "The late professor Dr. H. Kamerlingh Onnes," *Cold Storage*, XXIX, no. 33 (1927), 129–130.
6. Cohen, E., "Kamerlingh Onnes memorial lecture," Chemical Society, *Journal*, **1** (1927), 1195.
7. Ibid., 1195.
8. Ibid., 1197.
9. Mendelssohn, K., *The Quest for Absolute Zero* (New York, 1966), 51; E.G.D. Cohen, "Toward absolute zero," *American Scientist*, **65** (1977), 752–758 on 754; E.M. Codlin, *Cryogenics and Refrigeration: A Bibliographical Guide* (New York, 1968).
10. Dahl, P.F., "Kamerlingh Onnes and the discovery of superconductivity: The Leyden years, 1911–1914," *HSPS*, **15** (1984), 1.
11. Dewar, J., "Recent researches on meteorites," Notices of the Royal Institution of Great Britain Weekly Evening Meetings, 20 Jan. 1893; *Collected Papers of Sir James Dewar*, ed. Lady Dewar (Cambridge, 1927).

12. Lord Rayleigh, "Some reminiscences of scientific workers of the past generation, and their surroundings," *PRS*, **48** (1936) 228.
13. Dewar to Onnes, KO, Archive 8. Onnes's problem was a chief subject of correspondence between himself and Olszewski in these years (Onnes to Olszewski, KO, Archive 8); Z. Wojtaszek, "The first years of cryogenics in the light of Olszewski's correspondence," *Actes*, **13**, no. 7 (1971), 137–138. Onnes, interestingly, avoided all reference to this incident in later years. "Perhaps," notes Gorter and Taconis, "he did not wish the meddlesome town council to receive credit for the observation of the magnetic splitting of the sodium spectral lines by his collaborator Zeeman at the time of the incident." C.J. Gorter and K.W. Taconis, "The Kamerlingh Onnes laboratory," *CR*, **4** (1964), 346.
14. Keesom, W.H., "Prof. Dr. H. Kamerlingh Onnes. His life-work, the founding of the cryogenic laboratory," *PLS*, **57** (1926), 3.
15. Cohen (note 6).
16. Keesom (note 14), 4.
17. Keesom, W.H., "Prof. Dr. H. Kamerlingh Onnes. His life-work, the founding of the cryogenic laboratory," *PLS*, **57** (1926), 5–21.
18. Gorter and Taconis (note 13) put it more bluntly: "[Onnes] gradually pushed his colleagues out of the building and the central administration also left." Casimir claims that Lorentz advised Onnes against concentrating exclusively on his laboratory's efforts on research at low temperatures—a domain not very promising in the eyes of a classical physicist. H.B.G. Casimir, "Superconductivity" in *History of Twentieth Century Physics*, Proceedings of the International School of Physics "Enrico Fermi," Course LVII, ed. C. Weiner (New York, 1977), 172.
19. Kamerlingh Onnes, H., "The liquefaction of helium," *RN*, **11** (1909), 170 [*PLC*, **108** (1908)]. The helium liquefier is exhibited at the Museum Boerhaave.
20. Onnes (note 19), 170.
21. By Onnes's own account, the gas "was exploded with oxygen, cooled with liquid air, and, compressed, led over charcoal at the temperature of liquid air. Then it was burned over CuO. Then it was compressed over charcoal at the temperature of liquid air, after which it was under pressure led over the charcoal at the temperature of liquid hydrogen several times til the gas which had been absorbed in the charcoal and then separately collected no longer contained any appreciable admixtures." Onnes (note 19), 179.
22. Ibid., 180–181.
23. Casimir, H.B.G., "Low temperatures," in *Haphazard Reality: Half a Century of Science* (New York and London, 1938), 162–163, 177; Rayleigh (note 12) 232–235. Tributes to Flim, dated February 1933, are expressed in W.H. Keesom, W.J. de Haas, C.A. Cromelin, E.C. Wiersma, and W. Tuyn in *Toesspraken bij jubilea van technici*, KO, Archive 94.
24. Maxwell, in editing Cavendish's manuscripts on electrical researches, is said to have been especially attracted to an experiment in which Cavendish had been his own galvanometer and had estimated the strength of the current by the shock it gave him. Visitors to the laboratory had currents passed through them to see whether or not they were good galvanometers. Quoted in J.J. Thomson, *Recollections and Reflections* (New York, 1975), 104–105.
25. Davy, Humphrey, "Farther researches on the magnetic phenomena produced by electricity; with some new experiments on the properties of electrified bodies in their relations to conducting power and temperature," Royal Society of London, *Philosophical Transactions*, **111** (1821), 424–439.
26. Lenz, E., "Ueber die Leitungsfähigkeit der Metalle fur die Elektricitat bei

verschiedenen Temperaturen," *APC*, **34** (1835), 418–437; "Ueber die Leitungsfähigkeit des Goldes, Bleis und Zinns fur die Elektricitat bei verschiedenen Temperaturen," *APC*, **45** (1838), 105–121.

27. Arndtsen, A., "Ueber den Galvanischen Leitungswiderstand der Metalle bei verschiedenen Temperaturen," *APC*, **104** (1858), 1–57. Arndtsen's work on metallic conduction, begun at Christiania, was completed during his tenure with Wilhelm Weber at Göttingen from 1857 to 1858.
28. Becquerel, E., "Untersuchung uber das elektrische Leitungsvermogen starrer und flussiger Körper, *APC*, **70** (1847), 243–254. Edmund Becquerel was the son of Antoine-Cesar Becquerel. In the same month that Ohm's first paper was published, an extract appeared in Ferussac's ***Bulletin*** of the elder Becquerel's and Peter Barlow's work on the electrical conductivity of metals.
29. "Der Leitungswiderstand der einfachen Metalle im festen Zustande name proportaional der *absoluten Temperatur*." R. Clausius, "Ueber die Zunahme des elektrischen Leitungswiderstandes der einfachen Metalle mit der Temperatur," *APC*, **104** (1858), 650.
30. Matthiessen began his work on the electrical conductivity of metals while he worked in Bunsen's Laboratory at Heidelberg during the period of 1853–1857. The chief defect of the earlier method of measuring resistance, comparing its current measured by a galvanometer, with that in a standard resistance substituted in the circuit, was its dependence on the constancy of the galvanic battery. This defect was obviated by two null methods, the first based on the differential galvanometer (introduced by E. Becquerel), which was superseded by methods based on the Wheatstone bridge (actually invented by S.H. Christie).
31. Benoit, R., "Sur la Résistance électrique des Métaux," *Compte rendus* (1873), 342–346.
32. The concept of specific resistance was not new, having been introduced by Davy and developed in the form we essentially know today by Becquerel. It is defined by expressing the electrical resistance of a cylindrical conductor of length l and cross section A by $R = \rho(l/A)$, where the specific resistance ρ is by definition independent of the dimensions of the conductor but dependent on the nature of the material, as well as on temperature.
33. Cailletet, L., and F. Bouty, "Sur la conductibilité électrique du mercure et des métaux purs aux basses températures," *Journal de physique*, **6** (1885), 297–304.
34. The temperature coefficient of resistance is usually defined by

$$\alpha = \frac{1}{\rho_0} \frac{\rho_{100} - \rho_0}{100}$$

where ρ_0 and ρ_{100} are the specific resistances at 0 °C and 100 °C, respectively. More precisely, since the temperature function is usually far from linear, the instantaneous temperature coefficient,

$$\frac{1}{\rho} \frac{d\rho}{dT},$$

is more appropriate. If specific resistances were exactly proportional to the absolute temperature, α would have a value of ~1/273 or approximately 0.0037—the rule of Clausius.

35. Wroblewski, S.F., "Sur la résistance électrique du cuivre à la température de 200 ° au-dessous de zéro, et sur le pouvoir isolant de l'oxygène et de l'azote liquides," *Compte rendus*, **101** (1885), 161.
36. Thomson (note 24), 131.

37. Callendar, H.L., "On the construction of platinum thermometers," *PM*, **32** (1891), 109.
38. Heilbron, J.L., "Physics at McGill in Rutherford's time," in *Rutherford and Physics at the Turn of the Century*, ed. M. Bunge and W.R. Shea (New York, 1979).
39. Ibid., 45.
40. Cited in Mendelssohn (note 9), 26.
41. Dewar, J., and J.A. Fleming, "On the electrical resistance of pure metals, alloys, and non-metals at the boiling point of oxygen," *PM*, **34** (1892), 330–332. This and other papers coauthored with Fleming were actually written by Fleming. See also Chilton, D., and N.G. Coley, "The laboratories of the Royal Institution in the nineteenth century," *Ambix*, **XXVII** (1980), 191.
42. Dewar and Fleming (note 41), 334.
43. Matthiessen, A., "Ueber die elektrische Leitungsfähigkeit der Legirungen," *APC*, **110** (1860), 190.
44. Dewar and Fleming (note 41), 336.
45. Ibid.
46. Dewar, J., and J.A. Fleming, "Electrical resistance of metals and alloys," *PM*, **36** (1893), 271–299.
47. Olszewski, K., "On the liquefaction of gases," *PM*, **39** (1895), 188–189.
48. Olszewski, K., "Determination of the critical and the boiling temperature of hydrogen," *PM*, **40** (1895), 202–210, on 203.
49. Callendar, H.L., "Notes on platinum thermometry," *PM*, **47** (1899), 195.
50. Callendar (note 49), 204–208. Dickson, J.D.H., "On platinum temperatures," *PM*, **44** (1897), 445–459. Dickson, then a fellow at Peterhouse, Cambridge, was, among other things, to be coadjutor to Lady Dewar in the publication of the collected papers of James Dewar. The same year that saw Callendar's definitive paper in print, 1899, the Committee on Electrical Standards, headed by Lord Rayleigh, accepted Callendar's proposal for a standard scale of temperature based on the platinum thermometer. At the General Conference of Weights and Measures held in Paris in 1927, it was agreed to adopt Callendar's old value for the boiling point of sulfur as one of the fixed points. Three years later, in an obituary to Callendar, Sir Richard Glazebrook referred to Callendar's lasting paper (actually, to an earlier and still more detailed version of 1897) as follows: "Nowadays his paper may seem long and unnecessarily detailed, but he had a difficult case to prove, and so the data are given in full, and the reader who has the patience can verify his results himself." Obituary notice, *PRS*, **134A** (1932), xviii–xix.
51. Callendar (note 49), 221–222.
52. Dewar, J., "The nadir of temperature, and allied problems," Abstract of Bakerian Lecture, *PRS*, **68** (1901) 360–366, on 362–364.
53. Dewar, J., "Electrical resistance thermometry," *PRS*, **73** (1904), 244.
54. Ibid.
55. The first constant volume hydrogen thermometer was employed at Leiden in 1896. The first practical thermometer for continuous low-temperature measurements was a thermoelement after Wroblewski.
56. Het Natuurkundig Laboratorium der Rijks-Universiteit te Leiden in de jaren 1882–1904. Gedenkboek aangeboden aan den hoogleeraar H. Kamerlingh Onnes, Directeur van het Laboratorium, bij Gelegenheid van zijn 25-jarig doctoraat op 10 Juli 1904 (Leyden, 1904).
57. Holborn, L., and W. Wien, "Ueber die messung tiefer Temperaturen," *APC*, **59** (1896), 213–228. Holborn, "Untersuchungen über Platinwiderstände und Petroläthermometer," *AP*, **6** (1901), 242-258.

58. Meilink, B., "On the measurement of very low temperatures. IV. Comparison of the platinum thermometer with the hydrogen thermometer," *RN*, **7** (1902), 495–500 [*Communication* **77**].
59. Meilink, B., "On the measurement of very low temperatures. VII. Comparison of the platinum thermometer with the hydrogen thermometer," *RN*, **7** (1905), 290–299 [*Communication* **93**].
60. Ibid., 299.
61. Meilink, B., "On the measurement of very low temperatures. VIII. Comparison of the resistance of gold wire with that of platinum wire," *RN*, **7** (1905), 302 [*Communication* **93** cont'd].
62. Onnes and J. Clay, "On the measurement of very low temperatures. XI. A comparison of the platinum resistance thermometer with the hydrogen thermometer," *RN*, **9** (1907), 207–213 [*Communication* **95c**]; "On the measurement of very low temperatures. XIII. Comparison of the platinum resistance thermometer with the gold resistance thermometer," *RN*, **9** (1907), 213–216 [*Communication* **95d**].
63. Travers, M.W., and A.G.C. Gwyer, "On the platinum and normal scales of temperature," *PRS*, **74** (1905), 528–538.
64. Weber, W., "Ueber die Bewegungen der Elektricicät in Körpen von molecularer Constitution, *APC*, **156** (1875), 1–16.
65. The last of his *Elektrodynamische Maassbestimmungen*; see *Wilhelm Webers Werke*, 6 vols. (Berlin, 1892–1894).
66. This was expounded particularly in his Faraday lecture of 1881. Superficially, Maxwell speaks of a "molecule of electricity" in his *Treatise*, though he was publicly indifferent on the physical nature of electricity, to the consternation of many of his Continental readers. For a thorough discussion of the codevelopment and final merging of the English (Maxwellian) field theory approach and the largely Continental action-at-a-distance and particle school, see J.Z. Buchwald, *From Maxwell to Microphysics: Aspects of Electromagnetic Theory in the Last Quarter of the Nineteenth Century* (Chicago, 1985).
67. Giese, W., "Grundzüge einer einheitlichen Theorie der Electricitätsleitung," *APC*, **37** (1889), 576–577.
68. Schuster, A., "Experiments on the discharge of electricity through gases. Sketch of a theory," *PRS*, **37** (1884), 318.
69. Feffer, S.M., "Arthur Schuster, J.J. Thomson, and the discovery of the electron," *HSPBS*, **20**, no. 1 (1989), 38.
70. "Conduction of electricity through gases," Ch. 10 in Chalmers, T.W., *Historic Researches: Chapters in the History of Physical and Chemical Discovery* (New York, 1952).
71. Rieke, E., "Zur Theorie des Galvanismus und der Wärme," *APC*, **66** (1898), 353–389, 545–581.
72. Drude, P., "Zur Elektronentheorie der Metalle," *AP*, **1** (1900), 566–613.
73. The term electron was introduced by G.J. Stoney in 1900. The existence of a single mobile particle in metals, the negative electron, was first postulated by J.J. Thomson in 1900. He based his assumption on the manifold evidence for the existence of a negative electron, of negligible mass compared with that of chemical atoms, in contrast to positive electricity that seemed invariably limited to combinations with masses of the order of atomic masses. J.J. Thomson, "Indications relatives à la constitution de la matière fournies par les recherches récentes sur le passage de l'électricité à travers les gaz," Rapports on Congres de Physique, Paris, vol. 3 (1900), 138.
74. Drude (note 72), 566.

75. Riecke, E., "Ueber das Verhältnis der Leitfähigkeiten der Metalle für Wärme und für Elektricität," *AP*, **2** (1900), 837–838.
76. Bohr, N., *Studier over metallernes elektrontheori* (Copenhagen, 1911); reprinted with English translation in *Niels Bohr Collected Works*, vol. 1, ed. L. Rosenfeld and J.R. Nielsen (Amsterdam, 1971).
77. Wiedemann, G., and R. Franz, "Ueber die Wärme-Leitungsfähigkeit der Metalle," *APC*, **89** (1853), 531. The well-known law appears on p. 531. Note that while Wiedemann and Franz showed that the ratio between the conductivities is the same for all metals at the same temperature, the Danish physicist Ludvig V. Lorenz proved that this ratio is proportional to the absolute temperature—the Law of Lorenz.
78. Reinganum, M., "Theoretische Bestimmung des Verhältnisses von Wärme-und Elektricitätsleitung der Metalle aus der Drude'schen Elektronentheorie," *AP*, **2** (1900), 398–399. Reinganum obtained his doctorate at Göttingen in 1899. By 1900 he had joined Onnes's laboratory; subsequently he became professor of physics at the University of Freiburg. He died from a shell splinter in 1914 near Le Menil in the Vosges. Marx, E., "Macimilian Reinganum," *PZ*, **16** (1915), 1–3.
79. Jäger and Diesselhorst, "Wärmeleitung, Electricitätsleitung, Wärmecapacität und Thermokraft einiger Metalle," *SB* (1889), 719–726.
80. Hoddeson, L., and G. Baym, "The development of the quantum mechanical electron theory of metals: 1900–28," *PRS*, **A371** (1980), 9; Bohr (note 76), 338.
81. Lorentz, H.A., "The motion of electrons in metallic bodies, I," *RN*, **7** (1905), 449.
82. Lorentz, "The motion of electrons in metallic bodies, II and III," *RN*, **7** (1905), 590–593, 684–691. The Hall effect was sometimes positive for certain metals, implying that the carriers of charge are positive, not negative. Lorentz's emphasis on the positive Hall effect in his papers in 1905 perhaps induced Walther Nernst to discard entirely the idea of free electrons. Ehrenfest always pointed out in his lectures that the occurrence of a positive Hall effect in metals was "one of the major mysteries." H.B.G. Casimir, "Development of solid-state physics," in *History of Twentieth Century Physics* (note 18). The changing sign of the Hall effect is also discussed in Walter Kaiser, "Early theories of the electron gas," *HSPBS*, **17**, Part 2 (1987), 290–292.
83. Lorentz (note 81), 450.
84. Bohr's doctoral dissertation on the electron theory of metals played a very minor role in these developments, because he failed in his attempts to have it translated into English from the Danish. His analytical contribution to the problem of electrical conductivity, in particular, was a force between the metal atoms and electrons varying inversely with the nth power of the distance (letting n approach infinity in his formula for σ corresponds to elastic, "hard sphere" collisions and yields Lorentz's formula).
85. Riecke, E., "Die jetzigen Anschaugen über das Wesen des metallischen Zustandes," *PZ*, **10** (1909), 508, 511–512.
86. Lorentz (note 82), 592.
87. Drude, P., "Zur Ionentheorie der Metalle," *PZ*, **1** (1900), 161.
88. Schuster, A., "On the number of electrons conveying the conduction currents in metals, *PM*, **7** (1904), 151. The subject was further treated by Drude, in light of Schuster's analysis, with which he disagreed on several points, in his "Optische Eigenschaften und Elecktronentheorie," *AP*, **14** (1904), 936–961. Schuster might have anticipated J.J. Thomson in the discovery of the electron, had he not made an incorrect determination of the velocity of cathode rays.
89. Thomson, J.J., *The Corpuscular Theory of Matter* (London, 1907), 79. Thomson

himself, in fact, obtains a $T^{1/2}$ dependence, due to a calculational error; Bohr (note 76), 352.

90. Koeningsberger, J., and O. Reichenheim, "Ueber ein Temperaturgesetz der elektrischen Leitfähigkeit fester einheitlicher Substanzen und einige Folgerungen daraus," *PZ*, **7** (1906), 507–578; J. Koeningsberger, and K. Schilling, "Ueber Elektrisitätsleitung in festen Elementen und Verbindungen," *AP*, **32** (1910), 179–230.
91. Reprinted in Lord Kelvin, "Aepinus Atomized," *PM*, **3** (1902), 272–274.
92. Ibid., 272.
93. Ibid., 274.
94. Reprinted in translation in Onnes, *PLS*, **9** (1904), 28.
95. Becquerel, J., and Onnes, "The absorption spectra of the compounds of the rare earths and the temperature obtainable with liquid hydrogen, and their change by magnetic field," *RN*, **10** (1908), 597 [*PLC*, **103**, 3–16].
96. Ibid., 597–598.
97. Ibid., 598.
98. Onnes and Clay, "On the measurement of very low temperatures, XII. Comparisons of the platinum resistance thermometer with the gold resistance thermometer," *RN*, **9** (1907), 213–216 [*PLC*, **95d**].
99. Ibid., 214.
100. Onnes and Clay, "On the measurement of very low temperatures, XV. Calibration of some platinum-resistance thermometers," *RN*, **10** (1908), 200 [*PLC*, **99b**].
101. Onnes and Clay, "On the change of resistance of the metals at very low temperatures and the influence exerted on it by small amounts of admixtures," *RN*, **10** (1908), 207 [*PLC*, **99c**].
102. The resistance of mercury had previously been measured, first by Cailletet and Bouty down to −100 °C; (note 33), and subsequently by Dewar and Fleming to −204 °C; Dewar and Fleming, "On the electrical resistivity of mercury at the temperature of liquid air," *PRS*, **60** (1896), 76–81.
103. Onnes and Clay (note 101), 208.
104. Onnes, "Further experiments with liquid helium. A. Isotherms of monatomic gases, etc. VIII. Thermal properties of helium. B. On the change in the resistance of pure metals at very low temperatures, etc. III. The resistance of platinum at helium temperatures," *RN*, **13** (1911), 22.

Notes to Chapter 2

1. Onnes, "On the lowest temperature yet obtained," Faraday Society, *Transactions*, **18** (1922), 149.
2. Ibid., 149. Not until 1922 was a further significant reduction in temperature achieved, to "some hundredths of a degree below 0.9 °K" (Ibid., p. 173). The long delay can be partly attributed (Ibid., p. 155) to diversions of more pressing matters that soon cropped up—e.g., superconductivity after 1911 and the problem of the threshold field that vexed Onnes until World War I cut off liquid helium work altogether. Solidifying helium took even longer, accomplished in 1926 (the year Onnes died) by Willem Keesom; it required a pressure of 25 atmospheres at 1 °K and 130 atmospheres at 4 °K. Keesom, W.H., "Solid helium," *RN*, **29** (1926), 1136–1145; *Helium* (Amsterdam, London, New York, 1942), 180–183.
3. Onnes, "Further experiments with liquid helium. A. Isotherms of monatomic gases, etc. VIII. Thermal properties of helium. B. On the change in the resistance of pure

metals at very low temperatures, etc. III. The resistance of platinum at helium temperatures," *RN*, **13** (1911), 1093–1112 [*PLC*, **119**].
4. Ibid., 1095.
5. This portion of the apparatus had no role in the resistance determinations. However, it is worth noting that measurements on the density of liquid helium as a function of temperature, in this series of experiments, seemed to suggest a maximum density near 2.2 °K; whether the density decreases at still lower temperatures was not established. This behavior, nevertheless, greatly puzzled Onnes. "The occurrence of a maximum density in a substance of such simple constitution as helium gives rise to questions of great import from the point of view of molecular theory. With a substance like water it is easy to imagine a particular molecular combination by which some of the parts are more closely united, while others are separated, the whole leading to an increase of volume as the temperature is lowered ... but helium atoms, we are forced to consider as spherical and smooth, and, as appears from the Zeeman-effect for helium, of the simplest possible internal construction; and for their case we seek in the meantime in vain for a basis for a similar explanation." (Onnes, note 3, 1104.) The effect observed was real and would prove to be an aspect of the superfluid properties of helium elucidated many years later at Leiden and elsewhere.
6. Onnes and Clay (chap. 1, note 98).
7. Onnes (note 3), 1108.
8. Onnes and Clay (chap. 1, note 98).
9. The resistance ratios are taken from Table 1 of *Communication* **99c** (chap. 1, note 101).
10. Onnes (note 3), 1109.
11. Table V of *Communication* **99b** (chap. 1, note 100). The importance of this table, comparing Pt_x with various other samples for the present investigation is evident in numerous handwritten corrections in the Boerhaave copy of *Communication* **99b**.
12. Onnes and Clay, "On the change of the resistance of pure metals at very low temperatures and the influence exerted on it by small amounts of admixtures," *RN*, **11** (1909), 345 [*PLC*, **99c**].
13. Onnes (note 3), 1109.
14. Callendar (chap. 1, note 49), 221.
15. Nernst, W., "Ueber die Berechung chemischer Gleichgewichte aus thermischen Messungen," Akademie der Wissenschaften, Göttingen, *Nachrichten*, **1** (1906), 1–40.
16. Dulong, P.L., and A.T. Petit, "Sur quelques points importants de la théorie de la chaleur," *Annales de chimie et de physique*, **10** (1819), 395–413.
17. Boltzmann, L., "Analytischer Beweis des zweiten Hauptsatzes der mechanischen Wärmetheorie aus den Satzen über das Gleichgewicht der lebendigen Kraft," Akademie der Wissenschaften, Wien, *Sitzungsberichte*, **63** (1871), 712–732.
18. Weber, H.F., "Die spezifische Wärme der Kohlenstoffes," *AP*, **147** (1872), 311–319.
19. Dewar, "Studies with liquid hydrogen and air calorimetry. I. Specific heats," *PRS*, **A76** (1905), 325.
20. Ibid., 330–331.
21. Einstein, A., "Die Plancksche Theorie der Strahlung und die Theorie der spezifischen Wärme," *AP*, **22** (1907), 180; Cf. M.J. Klein, "Einstein, specific heats, and the early quantum theory," *SC*, **148** (1965), 173–180. The difficulty of reconciling equipartition with specific heats had been pointed out by Reinganum, while at

Leiden with Onnes, even before Drude published the second part of his basic paper of 1900. Reinganum (chap. 1, note 78). Kaiser (chap. 1, note 82), 285.
22. Einstein (note 21), 184.
23. Mendelssohn, "Walther Nernst: An Appreciation," *CR*, **4** (1964), 131.
24. Ibid., 131. A rather complete bibliography of Nernst's books and papers is given in Viscount Cherwell (F.A. Lindemann) and F. Simon's obituary, Royal Society of London, *Obituary notices*, 4 Nov. 1942, 100–111.
25. Ruhemann, M. and B., *Low Temperature Physics* (Cambridge, 1937), 136.
26. Eucken, A., "Ueber die bestimmung spezifischer Wärme bei tiefen Temperaturen," *PZ*, **10** (1909), 586–589.
27. Nernst, W., "Untersuchungen zur spezifischen Wärme bei tiefen Temperaturen. I and II," *SB* (1910), 247–261 and 262–282 (Part I with F. Koeref and F.A. Lindemann).
28. Ibid.
29. Nernst, W., "Zur Theorie der spezifischen Wärme und über die Anwendung der Lehre von den Energiequanten auf physikalisch-chemische Fragen überhaupt," *Zeitschrift für Elektrochemie*, **17** (1911), 266–275, on 274; "Unterschungen ueber die spezifische Wärme bei tiefen Temperaturen. III," *SB* (1911), 306–315.
30. Cherwell and Simon (note 24), 104.
31. Ibid.
32. Nernst (note 29), 312.
33. Nernst, "Unterschungen über die spezifische Wärme" (note 29).
34. When a beam of radiation with a continuous spectrum is reflected from a substance exhibiting selective reflection (e.g., quartz) at a certain wavelength, the spectrum of the reflected beam will be enhanced at just that wavelength. Heinrich Rubens and associates devised a method for obtaining nearly monochromatic infrared rays by several reflections off polished surfaces of quartz or rock salt; they termed the "residual rays" Reststrahlen. H. Kangro, "Ultrarotstrahlung bis zur grenze elektrisch erzeugter Wellen, das Lebenswerk von Heinrich Rubens," *Annals of Science*, **26** (1970), 235–259 and **27** (1971), 165–200.
35. Lindemann, F.A., "Ueber die Berechnung molekularer Eigenfrequenzen," *PZ*, **11** (1910), 609.
36. Rubens, H., and H. Hollnagel, "Measurements in the extreme infra-red spectrum," *PM*, **19** (1910), 761–782.
37. Magnus, A., and F.A. Lindemann, "Ueber die Abhängigkeit der spezifischen Wärme fester Körper von der Temperatur," *Zeitschrift für Elektrochemie*, **16** (1910), 269–279.
38. Lindemann, F.A., "Untersuchungen über die spezifische Wärme bei tiefen Temperaturen. IV," *SB* (1911), 316–321.
39. Riecke (chap. 1, note 85).
40. Onnes and Clay (chap. 1, note 98); Tables V (lead) and III (silver).
41. Lindemann (note 38), 321.
42. Onnes (note 3).
43. Ibid., 1110.
44. Kramers quoting Uhlenbeck in H.A. Kramers, *Between Tradition and Revolution* (New York, 1987), 475.
45. In a letter to Maria Roosebohm, 27 Feb. 1953. KO, 113; translated in P.F. Dahl, "Kamerlingh Onnes and the discovery of superconductivity: The Leyden years, 1911–1914," *HSPS*, **15** (1984), 36.
46. Onnes (note 3), 1111.

47. Melting points appear in *Communication* **99c** and would crop up again in the years ahead.
48. Einstein in late 1910, apparently unaware of Lindemann's calculations of the eigenfrequency but following up on somewhat similar calculations by William Sutherland, derived a relationship between ν and the compressibility of metals. He used cubical compressibilities from E. Grüneisen. A. Einstein, "Eine Beziehung zwischen dem elastischen Verhalten und der spezifischen Wärme bei festen Körpern mit einatomigen Molekül," *AP*, **34** (1911); W. Sutherland, "The mechanical vibration of atoms," *PM*, **20** (1910), 657–670. The anomalous specific heat behavior of diamond follows from either Einstein's or Lindemann's frequency analysis. Diamond has an abnormally high melting point T_s and low compressibility; hence, from Eq. (2-6) and Einstein's version, which postulates $\nu \sim 1/K^{1/2}$, diamond should have a high eigenfrequency and reach the asymptotic classical value for C_v at corresponding high temperatures. Einstein's results became available after the December meeting of the Amsterdam Academy where Onnes first discussed his new results, but in time for the formal publication in the Dutch Proceedings in Febrary 1911.
49. Onnes (note 3), 1111.
50. Onnes, "Further experiments with liquid helium. E. A helium cryostat," *RN*, **14** (1912), 209–210 [*PLC*, **123a**]. Cf. P. Dahl (note 45), 14.
51. Ibid., 210. Curiously, Lindemann's βν-value not only gives worse agreement with the measured resistance ratio Wt/W_0 for mercury at 13.88 °K, but predicts a lower resistance by an order of magnitude in the temperature range of liquid helium. At 13.88 °K, Lindemann's value inserted in Eq. (2-8) gives $W_t/W_o = 0.019$, whereas Onnes's calculated value is 0.027, as noted. At 3 °K, Lindemann's ratio is 2.08×10^{-5} compared with Onnes's ratio of 2.44×10^{-4}.

Notes to Chapter 3

1. Onnes, "Further experiments with liquid helium. C. On the change of electric resistance of pure metals at very low temperatures, etc. IV. The resistance of pure mercury at helium temperatures," *RN*, **13** (1911), 1274–1276 [*PLC*, **120b** (1911), 17–19].
2. Ibid., 1274.
3. Ibid.
4. Ibid., 1275.
5. In connection with this deduction, Onnes drew attention in a footnote to the fact that a gold-silver thermoelement behaved in liquid helium about as predicted from earlier experiments in liquid hydrogen. Ibid., 1275; Onnes and Clay, "On the measurement of very low temperatures. XXII. The thermo-element gold-silver at liquid hydrogen temperatures," *RN* **11** (1909), 344–345 [*PLC*, **107b** (1908)]. That is, he observed a diminished drop in the electromotive force per degree Kelvin of the thermoelement in the liquid helium range, somewhat analogous to the variation of normalized specific resistance of most pure metals, if plotted as ρ/T against T. Such a plot is constant at high temperatures, but as the temperature is lowered the curve rises. This would seem to agree with the Kelvin notion that the substance becomes an insulator. Of course, it merely reflects the approach to a constant ρ; the curve rises simply because T is falling. G.J. Gorter quotes Onnes to the effect that the expected increase in ρ/T at lower temperatures was only half the observed

ρ/T at liquid hydrogen temperatures in the case of mercury. Onnes, he states, remarked that "this mercury perhaps must be an exception to the rule of Lord Kelvin." I have been unable to locate the source for these remarks. G.J. Gorter, Lectures on magnetism and superconductivity, Tata Institute of Fundamental Research, Bombay (1963), 14–15.

6. Onnes (note 1), 1275.
7. Ibid., 1276.
8. Onnes, "Further experiments with liquid helium. D. On the change of the electrical resistance of pure metals at very low temperatures, etc. V. The disappearance of the resistance of mercury," *RN*, **14** (1912), 113–115 [*PLC*, **122b** (1911), 13–15].
9. Ibid., 114.
10. Ibid., 114–115.
11. Onnes (chap. 2, note 50).
12. Planck to Nernst, 11 June 1910. J. Pelsener, "Historique des Instituts Internationaux de Physique et de Chimie Solvay," 6–7. A copy of this unpublished manuscript is found in AHQP.
13. The participants at the first Solvay Congress were as follows. From Germany: Walther Nernst (Berlin), Max Planck (Berlin), Heinrich Rubens (Berlin), Arnold Sommerfeld (Munich), Emil Warburg (Charlottenburg), and Wilhelm Wien (Wurzbourg); from England: James Hopwood Jeans (Cambridge) and Ernest Rutherford (Manchester); from France: Marcel Brillouin, Marie Curie, Paul Langevin, Jean Perrin, and Henri Poincaré (all from Paris); from Austria: Albert Einstein (Prague) and Friedrich Hasenöhrl (Vienna); from Holland: Heike Kamerlingh Onnes and Hendrik Lorentz (Leiden); from Denmark: Martin Knudsen (Copenhagen). Lorentz was chairman of the congress, and Robert Goldschmidt (Brussels), Maurice de Broglie (Paris) and Frederick A. Lindemann (Berlin) performed the functions of scientific secretaries. Ernest Solvay was assisted by E. Herzen and G. Hostelet (Brussels). Lord Rayleigh (London) and J.D. van der Waals (Amsterdam) did not attend, but were listed as offical participants. The proceedings of the congress were published as *La théorie du rayonnement et les quanta, rapports et discussions de la réunion tenue à Bruxelles, du 30 octobre au 3 novembre 1911 sous les auspices de M.E. Solvay*, ed. P. Langevin and M. de Broglie (Paris, 1912), hereafter referred to as *Solvay 1911*.
14. Sharing the secretaryship with de Broglie, in Lindemann's opinion, was the apex of his career as a research chemist. Kurt Mendelssohn, *The World of Walther Nernst: The Rise and Fall of German Science* (London, 1973), 168. The proceedings, while still in the proof stage and being edited by Maurice de Broglie, so excited the younger Louis that he too resolved to become a physicist. Louis de Broglie, *Louis de Broglie: Un itinéraire scientifique*, ed. G. Lochak (Paris, 1987).
15. Quoted in J. Mehra and H. Rechenberg, *The Historical Development of Quantum Theory* (4 vols., New York, 1982), **1**:27.
16. Nernst, W. "Application de la théorie des quanta à divers problèmes physicochimiques," *Solvay 1911*, 259–276.
17. Nernst, W., and F.A. Lindemann, "Spezifische Wärme und Quantentheorie," *Zeitschrift für Elektrochemie*, **17** (1911), 817–827.
18. The notion of two quanta did not affect the black-body radiation formula, because only the kinetic energy of bound atoms would be in equilibrium with the radiation field. This interpretation met with difficulty, however, in explaining thermal conductivity at low temperatures. In this domain the atomic vibrations should cease, implying very small heat conduction, in sharp disagreement with Arnold Eucken's ongoing measurements.

19. Einstein to Nernst, 11 June 1911, Pelsener, "Historique des Instituts," 12. Quoted in M.J. Klein (chap. 2, note 21), 178.
20. A. Einstein in "Discussion du rapport de M. Nernst," *Solvay 1911*, 299. In looking back many years later, Peter Debye expresssed his disdain for the new formula as follows: "In order to represent it [the measured deviations from Einstein's formula] they introduced half quanta. So they had a formula with 50 per cent *h* and 50 per cent *h*/2. I did not stomach that." Interview with P. Debye on 3 May 1962, by T.S. Kuhn and G. Uhlenbeck, AHQP.
21. Einstein, A., "Rapport sur l'état actuel du problèmes des chaleurs spécifiques," *Solvay 1911*, 407–435, on 416–417. The success of the Nernst–Lindemann two-term approximation is probably connected with the fact that it happens to be close to the best fit obtainable with two terms of the Einstein type to the continuous spectrum of frequencies; P.M. Keesom and N. Pearlman, "Low temperature heat capacity of solids," *Encyclopedia of Physics*, vol. XIV, ed. S. Flugge (Berlin, 1956), 282–337.
22. Einstein, A. (chap. 2, note 48).
23. Nernst, W., "Ueber ein allgemeines Gesetz, das Verhalten fester Stoffe bei sehr tiefen Temperaturen betreffend," *PZ*, **12** (1911), 976–979.
24. Nernst (note 16).
25. On 12 September 1911, the prominent physicist H. Nagaoka extended his congratulations by letter to Onnes on the discovery of superconductivity. KO, Archive 113g, item 131.
26. Nernst (note 23), 978.
27. Onnes, "Sur les résistances électriques," *Solvay 1911*, 304–310.
28. The two mercury measurements in liquid hydrogen (at 14 °K and 20 °K) were obtained by allowing the pressure under which the liquid helium evaporated to increase (by closing the tap connecting the cryostat and liquefier). Their purpose was to answer the still nagging question of whether a point of inflection occurs in the resistance curve somewhere between the melting point of hydrogen and the boiling point of helium, although by now its theoretical importance should have faded. About all one can say from Fig. 3-4 is that the resistance ratio exhibits a smooth diminution over this interval, essentially as predicted by Onnes's formula, with a slight leveling trend below 10 °K.
29. No change in volume, in the absence of a magnetic field, has been observed at the superconducting transition, nor a change in the elastic properties and the thermal expansion. D. Shoenberg, *Superconductivity* (Cambridge, 1965), 7.
30. Planck, M., "La loi du rayonnement noir," *Solvay 1911*, 110–111. Planck first discussed the new hypothesis at meetings of the French Physical Society and the Prussian Academy that year; see T.S. Kuhn, *Black-body Theory and the Quantum Discontinuity 1894–1912* (Oxford, 1978), 236–246.
31. Onnes, in "Discussion du rapport de M. Planck," *Solvay 1911*, 129.
32. Onnes, "Further experiments with liquid helium. G. On the electrical resistance of pure metals, etc. VI. On the sudden change in the rate at which the resistance of mercury disappears," *RN*, **14** (1912), 818–821 [*PLC* **124c**].
33. The figure is erroneously printed in a companion *Communication*: Onnes, "Further experiments with liquid helium. F. Isotherms of monatomic gases, etc. IX. Thermal properties of helium," *RN*, **14** (1912), following p. 682 [PLC **124b**]. This paper, incidentally, reports the boiling point of helium "at 4.26 °K, or better in round numbers 4.25 °K, with an accuracy greater than 0.1 °." Ibid., 682.
34. Onnes (note 32), 820.
35. Ibid., 821.

36. Bei den Widerstand abnahm der Quecksilber zeigt sich noch besonders Rätselhaft. Onnes to Vöigt, 22 December 1911, SON, item 5703.

Notes to Chapter 4

1. Onnes, "Further experiments with liquid helium. H. On the electrical resistance of pure metals etc. VII. The potential difference necessary for the electric current through mercury below 4.19 °K," *RN*, **15** (1913), 1406–1427 [*PLC*, **133a**, 3–26].
2. Ibid., 1406.
3. Onnes (note 1). *PLC*, 4–5.
4. Ibid., *PLC*, 5.
5. Ibid.
6. *Communication* **133a** erroneously places the experiments of *Communication* **122b** in June of 1911.
7. Onnes (note 1). *PLC*, 10.
8. Ibid. *PLC*, 12–13.
9. On another occasion, that of the Congress of Refrigeration at Chicago in 1913, Onnes described the stimulation as analogous to the generation of waves on water when the velocity of the wind exceeds a critical value. In 1912 Philipp Lenard had concluded on the basis of the Leiden measurements on pure metals at helium temperatures that Ohm's law is only valid within narrow limits for metals at very low temperatures. Onnes (note 1), 1415; P. Lenard, "Ueber Elektrizitätsleitung durch freie Elektronen und Träger, I," *AP*, **40** (1913), 413–414.
10. Onnes and B. Beckman, "On the Hall-effect, and on the change in resistance in a magnetic field at low temperatures. VI. The Hall-effect for nickel, and the magnetic change in the resistance of nickel, mercury and iron at low temperatures down to the melting point of hydrogen," *RN*, **15** (1913), 981–982 [*PLC*, **132a**].
11. Onnes (note 1). *PLC*, 16.
12. Ibid. *PLC*, 17.
13. Onnes, "Further experiments with liquid helium. H. On the electrical resistance of pure metals etc. VII. The potential difference necessary for the electric current through mercury below 4.19 °K (continuation)," *RN*, **15** (1913), 1427–1430 [*PLC*, **133b,** 29].
14. Ibid., 1427–1428 [*PLC*, **133b**, 29–30].
15. Ibid. *PLC*, 30.
16. Ibid. *PLC*, 31.
17. Ibid.
18. Onnes, "Further experiments with liquid helium. H. On the galvanic resistance of pure metals etc. VII. The potential difference necessary for an electric current through mercury below 4.19 °K (continuation)," *RN*, **16** (1914), 113–124 [*PLC*, **133c**, 35–48].
19. Ibid. *PLC*, 38.
20. Onnes, "Further experiments with liquid helium. H. On the electrical resistance etc. (continued). VIII. The sudden disappearance of the ordinary resistance of tin, and the super-conductive state of lead," *RN*, **16** (1914), 673 [*PLC*, **133d** (1913), 51].
21. Ibid., 674. The first reference to the superconducting state of tin (and lead) is found in *Communication* **133a** (note 1).
22. The hardening tends to be accompanied by an increase in the resistance and decrease in the temperature coefficient. For gold or platinum, this is corrected by annealing, but heating tin causes further increase in resistance.

23. Onnes (note 20). *PLC*, 53.
24. Ibid., 683–684.
25. They are cataloged as items C12 and C13 in the Boerhaave collection. In the catalogue for the Lorentz–Kamerlingh Onnes Exhibition mounted at the Boerhaave during 20 June–30 August 1953, they are briefly described as item 130. KO, Archive 113g.
26. The coils were 1 cm long. The tin coil contained 300 turns of 1/70 mm wire in a layer 7 mm thick, and the lead coil contained 1000 turns of wire nominally the same size in a layer of 10 mm thickness. The Boerhaave catalog (Ibid.) implies 1000 turns in both coils; the Boerhaave staff is of the impression that one coil actually carried approximately 600 turns (private communication, S. Engelsman, June 1984).
27. Onnes (note 20), 685.
28. Archive 287g, KO, contains extensive correspondence between Onnes and the Congress organizers.
29. Onnes to H.S. Miner, 31 March 1913; quoted in R. deBruyn Ouboter, "Superconductivity: Discoveries during the early years of low temperature research at Leiden," *TM*, **MAG-23** (1987), 370. Anton Welsbach was an inorganic chemist and specialist on the rare earths. Among his many lucrative inventions was the Welsbach mantle based on thorium oxide, responsible for the illumination of the gaslight era.
30. Onnes, "Report on the researches made in the Leiden cryogenics laboratory between the second and third international congress of refrigeration: Superconductivity," *PLS*, **34b** (1913), 55–70; reprinted from *Notes on the work of the section for physics, chemistry and thermometry of the first international commission of the Association Internationale du Froid.* The first congress was held in Paris in October 1908, a few months after the liquefaction of helium. The association was formally proposed by Onnes at the Paris congress. The second congress took place in Vienna in 1910.
31. Ibid., 62.
32. Onnes, "On the lowest temperature" (chap.2, note 1), 173.
33. J. Perrin had discussed his scheme at a meeting of the French Physical Society on 19 April 1907. C. Fabry, "Production de champs magnétiques intenses au moyen de bobines sans fer," *Journal de physique*, **9** (1910), 129–134.
34. Onnes (note 30), 64.
35. A less ambitious application of superconducting windings is pointed to in chap. 4, note 20, 684; namely, as auxiliary coils for augmenting the field in Weiss-type electromagnets. (Weiss himself had considered cryogenic coils for the same purpose.) The field contributed by the coil would have to be greater than the field sacrificed by enlarging the "interferrum" to make room for the cryostat.
36. Kuenen, J.P., "Het cryogeen laboratorium als Internationale instelling voor Het natuurkundig onderzoek bij lage temperaturen," *Het natuurkundig laboratorium* (chap. 1, note 1), 33. A printer's error dates the Chicago congress in 1918 instead of 1913.
37. See note 20, 681.
38. Onnes (note 30), 68.
39. Onnes (note 20), 686–687. P. Lenard had exploited H. Hertz's discovery that thin metal foils transmit cathode rays, enabling the rays to be passed out of the discharge tube into air or into a second evacuated experimental chamber.
40. Experiments on the absorption of β-rays and cathode rays in lead and tin foils in the superconducting state carried out in Toronto and Berlin revealed no measurable

discontinuity in the absorption at the transition temperature. Similarly, no measurable discontinuity was observed in the results of the measurements on the photoelectric effect, or in the results of those on the coefficient of absorption of light waves when lead films deposited on glass plates were passed through the transition temperature. J.C. McLennan, "Electric supra-conduction in metals," *Supplement to Nature*, **130** (1932), 882.

41. Gavroglu, K. and Y. Goudaroulis, *Methodological Aspects of the Development of Low Temperature Physics 1881–1956: Concepts Out of Context(s)* (Dordrecht, 1989), 93.
42. *Nobel Lectures, including Presentation Speeches and Laureates' Biographies; Physics 1901–1921* (New York, 1967), 306–336.
43. *Dagens nyheter*, Wednesday, 10 December 1913. Extensive material pertaining to Onnes's award is contained in the Boerhaave archives: communications from the Swedish Academy, formal programs, congratulatory letters and telegrams from friends and colleagues including Planck and Wien (14 November) and Nernst (20 November). KO, Archive 8, 113c and 113d.
44. *Nobel Lectures* (note 42), 305.
45. Ibid., 306–336.
46. Cohen, E. (chap. 1, note 6), 1201.
47. Van der Waals himself received the prize in 1910. Elisabeth Crawford points to the importance the Nobel committee attached to the relationship between van der Waals and Onnes. This is particularly evident, she observes, in the general report of the committee in 1909. Van der Waals's award "served as the wedge for that of Kamerlingh Onnes." E. Crawford, *The Beginnings of the Nobel Institution: The Science Prizes, 1901–1915* (Cambridge, 1984). Onnes was first nominated for the Nobel Prize in 1909, and in 1912 he had the highest number of nominations except for Planck, yet the prize went to Dalén that year.
48. *Nobel Lectures* (note 42), 336.
49. Ibid., 334–335.

Notes to Chapter 5

1. Onnes and B. Beckman, "On the Hall-effect and the change in the resistance in magnetic field at low temperatures. I. Measurements on the Hall-effect and the change in the resistance of metals and alloys in a magnetic field at the boiling point of hydrogen and at lower temperatures," *RN*, **15** (1913), 307 [*PLC*, **129a**].
2. Onnes and K. Hof, "Further experiments with liquid helium. N. Hall-effect and the change of resistance in a magnetic field. X. Measurements on cadmium, graphite, gold, silver, bismuth, lead, tin and nickel, at hydrogen and helium-temperature," *RN*, **17** (1915), 523 [*PLC*, **142b**].
3. Onnes and B. Beckman (chap. 4, note 10).
4. The Boerhaave archives contain considerable material concerned with the procurement of the new magnet, including correspondence with technical specifications from Oerlikon to Onnes from late 1912 to mid-1914 and sketches and magnetostatic calculations by Onnes. KO, Archive 56. The magnet itself is presently in storage at the Boerhaave, to be exhibited when the museum occupies its new quarters in Leiden.
5. Onnes, "Further experiments with liquid helium. I. The Hall-effect, and the magnetic change in resistance at low temperatures. IX. The appearance of galvanic

resistance in supra-conductors, which are brought into a magnetic field, at a threshold value of the field," *RN*, **16** (1914), 988 [*PLC*, **139f**, 66].

6. Ibid., 988.
7. Cf. tables XII and IX, respectively, of *Communication* **133d** (chap. 4, note 20).
8. Onnes (note 5), 988.
9. Ibid., 989.
10. Ibid.
11. Ibid.
12. Ibid., 990.
13. Ibid., 991.
14. Ibid., 991–992.
15. Ibid., 992.
16. Onnes, "Further experiments with liquid helium. J. The imitation of an Ampere molecular current or of a permanent magnet by means of a supra-conductor," *RN*, **17** (1915), 12 [*PLC*, **140b** (1914), 9–18].
17. Ibid., 13 [*PLC*, 10].
18. Ibid., 17.
19. Ibid., 19–20.
20. Ehrenfest, P. to H.A. Lorentz, 11 April 1914, quoted in M.J. Klein, *Paul Ehrenfest, Volume I: The Making of a Theoretical Physicist* (Amsterdam, 1970), 214. Ehrenfest succeeded Lorentz in the chair of theoretical physics at Leiden in 1912, when Lorentz took up his new professorship at Haarlem.
21. Onnes, "Further experiments with liquid helium. J. The imitation of an Ampere molecular current or of a permanent magnet by means of a supra-conductor. (Cont.)," *RN*, **17** (1915), 278–283 [*PLC*, **140c** (1914), 21–26]. One of the lead coils used in the 1914 experiments is preserved in the Boerhaave collection. It is listed and described briefly as item 132 in the 1953 Lorentz–Kamerlingh Onnes exhibition catalog. KO, 113g (chap. 4, note 25).
22. Onnes's reports on the subject do not support Ehrenfest's claim to have witnessed an experiment in which a current was maintained for over 24 hours, as indicated by M.J. Klein, *Paul Ehrenfest*, (note 20).
23. Onnes (note 21), 280.
24. Onnes, "Further experiments with liquid helium. L. The persistence of currents without electro-motive force in supra-conducting circuits (Continuation of J)," *RN*, **17** (1915), 514–519 [*PLC*, **141b** (1914), 15–21]. In a footnote Onnes takes pleasure in acknowledging a Mr. Taudin Chabot of Württemberg, who wrote him shortly after publication of his paper on the disappearance of resistance in mercury but, he later learned, at the time Chabot was only familiar with Onnes's results indicating that the resistance of perfectly pure gold and platinum would probably disappear altogether at some very low temperature [chap 2, note 3]. Chabot suggested the possibility of inducing a permanent current in a ring, for example of gold, if indeed it could be brought to such a state of zero resistance, but Onnes was too preoccupied with exploratory experiments with superconducting mercury to take him up on this.
25. Ibid., 517.
26. Ibid., 519.
27. Onnes (chap. 4, note 18), 120–121.
28. Onnes and G. Holst, "Further experiments with liquid helium. M. Preliminary determination of the specific heat and of the thermal conductivity of mercury at temperatures obtainable with liquid helium, besides some measurements of ther-

moelectric forces and resistances for the purpose of these investigations," *RN*, **17** (1915), 760 [*PLC*, **142c**].

29. Holst received his doctorate at Zürich in the summer of 1914; the subject of his thesis was the thermodynamic properties of ammonia and methyl chloride.
30. Onnes (chap. 4, note 18), 119.
31. Ibid., 120.
32. Stark, J., "Ueber elektrische und mechanische Schussflächen in Metallen, *PZ*, **13** (1912), 585–589.
33. Stark was one of the earliest defenders of the quantum hypothesis, and remained in the forefront of research until 1913; inexplicably, he then turned vehemently against quantum theory. A. Hermann, "Johannes Stark," in *DSB*, ed. C.C. Gillispie (New York, 1975), vol. XII, 614.
34. Onnes (chap. 4, note 18). *PLC*, **133c**, 43–44.
35. Bohr, Niels to Harald Bohr, 28 May 1912 (chap. 1, note 76), 553.
36. Bohr, N., "Note concerning a paper by J. Stark," dated 1912 (chap. 1, note 76), 438.
37. Wien, W., "Zur Theorie der elektrischen Leitung in Metallen," *SB*, **5** (1913), 184–200.
38. The first time was in his acceptance speech for the Nobel Prize in 1911: "On the laws of thermal radiation," *Nobel Lectures including Presentation Speeches and Laureates' Biographies: Physics 1901–1921* (Amsterdam, 1967), 281–286. Another occasion was in his "Theorie der Wärmestrahlung" in *Kultur der Gengenwart*, pt. 3, sect. 3, vol. 1 (Berlin, 1915), 217–220; H. Kangro, "Wilhelm Carl Werner Otto Fritz Franz Wien," *DSB*, ed. C.C. Gillispie (New York, 1975), vol. XIV, 340.
39. The possibility that the electron energy might differ from 3/2 kT appears to have been first suggested by Koeningsberger in a lecture at the 83rd meeting of the Society of Sciences at Karlsruhe; J. Koeningsberger, Deutsche Physikalische Gesellschaft, *Verhandlungen*, **13** (1911), 934; quoted in K.F. Herzfeld, "Zur Elektronentheorie der Metalle," *AP*, **41** (1913), 29. Herzfeld learned of Wien's work as his own paper on metallic conduction went to press.
40. Wien (note 37), 188.
41. Interview with P. Debye (chap. 3, note 20).
42. Wien (note 37), 197–200.
43. Keesom, W., "Ueber die Zustandsgleichung eines idealen einatomigen Gases nach der Quantentheorie," *PZ*, **14** (1913), 665–670; "On the equation of state of an ideal monatomic gas according to the quantum-theory," *RN*, **16** (1914), 227–236 [*PLC*, **30a** (1913), 9–17]. H. Tetrode was independently occupied with calculating the energy distribution among the vibrations of gases and liquids, adopting Debye's calculational technique for solids. H. Tetrode, "Bemerkungen über den Energiegehalt einatomigen Gase und über die Quantentheorie für Flüssigkeiten," *PZ*, **14** (1913), 212–215.
44. Keesom, W., "Zur Theorie der freien Elektronen in Metallen," *PZ*, **14** (1913), 670–675; "On the theory of free electrons in metals," *RN*, **16** (1914), 236–245 [*PLS*, **30b** (1913), 17–26].
45. Kuhn, T., *Black-body Theory* (chap. 3, note 30).
46. Wien to Keesom, KO, Archive 93. Planck himself, Thomas Kuhn notes, had not expected zero-point energy to be "experimentally consequential." Ehrenfest, according to his first doctoral student Johannes M. Burgers, also felt that Planck's first version of the quantum theory was better than his second. "On the other hand, Ehrenfest did feel that there was some good experimental evidence for a zero-point

energy and was therefore far from ready to dismiss Planck II out of hand." T.S. Kuhn, *Black-body Theory* (chap. 3, note 30), 246–247, and 319; Interview with Burgers by T.S. Kuhn, 9 June 1962, AHQP.

47. Lindemann, F.A., "Note on the metallic state," *PM*, **29** (1915), 128.
48. Ibid., 129.
49. Thomson, J.J., *The Corpuscular Theory of Matter* (London, 1907), Chapter V.
50. Thomson, J.J., "Conduction of electricity through metals," *PM*, **30** (1915), 192–193, 197.
51. The dipole theory is, in fact, formally equivalent to assuming a force law $f(r)$ acting upon the electrons inversely proportional to the third power of the distance; N. Bohr (chap. 1, note 76), 322.
52. Interview with Johannes Burgers on 9 June, 1962 by T.S. Kuhn and M. Klein, AHQP.
53. Ibid.
54. Einstein's application for a position with Onnes in 1901 survives in the form of a postcard from Zürich with return postage affixed (ignored by Onnes). Einstein to Onnes, 12 March 1901, KO, Archive 55; reprinted in B.A. van Proosdij, "Some letters from Albert Einstein to Heike Kamerlingh Onnes," *Janus*, **XLVII** (1959), 133.
55. Onnes and Holst, "On the electrical resistance of pure metals, etc. IX. The resistance of mercury, tin, cadmium, constantin, and manganin down to temperatures, obtainable with liquid hydrogen and with liquid helium at its boiling point," *RN*, **17** (1915), 508–513 [*PLC*, **142a**].
56. Holst's thesis concerned the equation of state and thermodynamic properties of ammonia and methyl chloride. Casimir (chap. 1, note 23), 212.
57. Schopman, Joop, "Industrious science: Semiconductor research at the N.V. Phillips' Gloeilampenfabrieken, 1930–1957," *HSPBS*, **19** (1988), 148–157.
58. Casimir (chap. 1, note 23), 165. Holst was elected to the academy in 1926, the year Onnes died. In 1929 Holst became a special professor of physics at Leiden, although his directorship at Phillips would last until 1946.
59. Ibid., 232.
60. Ibid.
61. Ibid., 233.
62. Ibid.
63. Ibid., 237–238.
64. Burgers (note 52).
65. von Smoluchowski, M., to P. Ehrenfest, 9 October 1912; quoted in Klein (note 20), 194.
66. Klein (note 20), 196.
67. Burgers (note 52).

Notes to Chapter 6

1. Heilbron, J.L., *The Dilemmas of an Upright Man: Max Planck as Spokesman for German Science* (Berkeley and Los Angeles, 1986). J.J. Thomson, as President of the Royal Society, played a similar role to Planck's (Ibid.), 80.
2. Klein (chap. 5, note 20), 302.
3. Onnes to Vöigt, June 1916, 27 July 1918, SON; Onnes to Sommerfeld, 6 June 1916, 18 May 1917, SON.
4. Mendelssohn, K., *The World of Walther Nernst* (chap. 3, note 14), 80.

5. Nernst, W., *Die theoretischen und experimentellen Grundlagen des neuen Wärmesatzes* (Halle-Salle, 1918).
6. Birkenhead, The Earl of, *The Professor and the Prime Minister: The Official Life of Professor F.A. Lindemann, Viscount Cherwell* (Boston, 1962), 79.
7. Mendelssohn, "The Clarendon Laboratory, Oxford," *CR*, **6** (1966), 129.
8. Silsbee, F.B., "A note on electrical conduction in metals at low temperatures," Washington Academy of Sciences, *Journal*, **6** (1916), 597–602.
9. Ibid., 598.
10. Ibid., 599.
11. I.e., the coil experiments of 1913 (chap. 4, note 20).
12. Silsbee (note 8), 602. Silsbee published a slightly reworked "note" on electrical conduction in 1917, including reference to a private communication by Langevin on the problem of current distribution in a superconducting wire. F.B. Silsbee, "Note on electrical conduction in metals at low temperatures," Bureau of Standards, *Bulletin* (1917), 301–306. The reference to Langevin occurs on pages 303–304.
13. Onnes and W. Tuyn, "Further experiments with liquid helium. Q. On the electric resistance of pure metals etc. X. Measurements concerning the electrical resistance of thallium in the temperature field of liquid helium," *RN*, **25** (1923), 443, [*PLC*, **160a**].
14. Onnes, "Demonstration of liquid helium," *PLS*, **43c** (1919), 13–19, trans. from Nederlandisch Natuur-en Geneeskundig Congress, XVII (1919), *Handelingen.*
15. Onnes and Tuyn (note 13), 447.
16. Onnes and Tuyn, "Further experiments with liquid helium R. On the electric resistance of pure metals etc. XI. Measurements concerning the electric resistance of ordinary lead and of uranium lead below 14 °K," *RN*, **25** (1923), 451 [*PLC*, **160b**].
17. Hönigschmidt, O. to Onnes, 24 August 1916, KO, Archive 8.
18. Paneth, F. to O. Stern, 5 October 1916, O. Stern Collection, Bancroft Library, 85/96c, carton 2. The translation is mine. Stern was a close postdoctoral associate of Einstein during Einstein's Prague period. His military service consisted of various technical assignments, and he spent the last year of the war at Nernst's Institute in Berlin.
19. Measurements in 1915 had led to the adoption of 4.22 °K as the boiling point of helium (cf. chap. 3, note 33), not far from the accepted value (about 4.216 °K).
20. Onnes and Tuyn (note 16).
21. The isotope effect was a key element in the development of a microscopic theory of superconductivity in the 1950s, as discussed later. The mass dependence implicates the motion of ions and hints that superconductivity must arise from interaction between electrons and vibrations of the ionic lattice. J. Bardeen, "Developments of concepts in superconductivity," *PT*, **1** (1963), 24.
22. Onnes, "Les supraconducteurs et le modèle de l'atome Rutherford-Bohr," *Atomes et électrons* (Paris, 1923), 165–197 [*PLS*, **44a** (1921), 30–50].
23. Lippmann, G., "Sur les propriétés des circuits électriques dénués de résistance," Académie des Sciences, *Comptes rendus*, **168** (1919), 74–75.
24. Bridgman, P.W., "The electrical resistance of metals under pressure," American Academy of Arts and Sciences, *Proceedings*, **52** (1917), 573–646; "Theoretical considerations on the nature of metallic resistance, with especial regard to the pressure effects," *PR*, **9** (1917), 269–289.
25. Bridgman, P.W., "The discontinuity of resistance preceding supraconductivity," Washington Academy of Sciences, *Journal*, **11** (1921), 455.
26. Casimir, H.B.G., "Superconductivity," in *History of Twentieth Century Physics*,

Proceedings of the International School of Physics "Enrico Fermi," course LVII, ed. C. Weiner (New York, 1977), 173–174.

27. Onnes (note 22), 185–187.
28. Silsbee, F.B., "Note on electrical conduction" (note 12); "Current distribution in supraconductors," National Academy of Sciences, *Proceedings*, **13** (1927), 516–518.
29. Brillouin, L., "L'agitation moléculaire et les lois du rayonnement thermique," *J. de phys.*, Part II (1921), 142–155.
30. Einstein, A., and L. Hopf, "Statistische Untersuchung der Bewegung eines Resonators in einem Strahlungsfeld," *AP*, **33** (1910), 1105–1115.
31. Grüneisen, E., "Ueber den Einflufs von Temperatur und Druck auf den elektrischen Widerstand der Metalle," Deutsche Physikalische Gesellschaft, *Verhandlungen*, **15** (1913), 186–200.
32. Einstein, A., "Theoretische Bemerkungen zur Supraleitung der Metalle" (chap. 1, note 1), 429–435.
33. Ibid., 433.
34. Ibid., 434.
35. Tuyn and Onnes, "Measurements concerning the electric resistance of indium in the temperature field of liquid helium," *RN*, **26** (1923), 505–506 [*PLC*, **167a**].
36. Ibid., 509.
37. Lorentz, H.A., "Application de la théorie des électrons aux propriétés des métaux," Institut International de Physique Solvay, *Conductibilité électrique des métaux et problèmes connexes* (Paris 1927), 1–45, hereafter referred to as *Solvay 1924*.
38. Hoddeson, L., G. Baym, and M. Eckert, "The development of the quantum mechanical electron theory of metals: 1928–33," *RMP*, **59** (1987), 312–313.
39. Ibid., 313.
40. Richardson, O.W., "Encore une théorie de la conductibilité métallique," *Solvay 1924*, 137–153.
41. Onnes, "Nouvelles expériences avec les supraconducteurs," *Solvay 1924*, 251–281. See also a reprint of Onnes's presentation in *PLS*, **50a** (1924), 3–34.
42. Ibid., 256.
43. Ibid.
44. Casimir, H.B.G. (note 26), 174.
45. Lorentz, H.A., "On the motion of electricity in a spherical shell placed in a magnetic field," *PLS*, **50b** (1924), 37–40.
46. E.g., see notes 16 and 35.
47. Donnelly, R.J., "Leo Dana: cryogenic science and technology," *PT*, **40** (April 1987), 41–42. Dana measured the specific and latent heats of liquid helium over a range of temperatures. The latent heat measurements suggested a dip near 2.2 °K, hitherto unknown except as a local maximum in the density curve. Dana's recollections are that "neither Kamerlingh Onnes nor members of the scientific staff showed any interest in the data, except perhaps Claude Crommelin." Ibid., 42.
48. Keesom, W.H. and Onnes, "On the question of the possibility of a polymorphic change at the point of transition into the supraconductive state," *PLC*, **174b** (1924), 45.
49. Einstein, "Theortische Bemerkungen" (note 32).
50. Onnes (note 41), 268–269.
51. Bridgman (note 25).
52. Ibid., 457–458. The derivation is most conveniently performed with the aid of a "Magnetic Gibbs function:" e.g., M.W. Zemansky, *Heat and Thermodynamics*, Third ed. (New York, 1951), chap. XVI.
53. A precondition for superconductivity in one class of superconductors was long

thought to be a very low normal electric conductivity. W.J. Hume-Rothery, *The Metallic State* (Oxford, 1931), 337.

54. *Solvay 1924*, 275–279.
55. Sizoo, G.J. and Onnes, "Further experiments with liquid helium. X. On the electric resistance of pure metals etc. XIV. Influence of elastic deformation on the supra-conductivity of tin and indium," *RN*, **28** (1925), 656–666 [*PLC*, **174**].
56. Ibid., 657.
57. Ibid., 662.
58. Shoenberg, D. (chap. 3, note 29), 73–77.
59. Sizoo and Onnes (note 55), 666.
60. Simon, F.E. to Onnes, KO, Archive 8.
61. Kurti, N., Obituary for F.E. Simon in *Biographical Memoirs of Fellows of the Royal Society* (1958), 224–256.
62. The subject for the Habilitation was chosen by an examining committee from a list of titles submitted by the candidate. Simon's list, in addition to the topic selected, foreshadows those areas where he and his group would predominantly make their mark: "On the equation of state of solids," "The verification of Nernst's theorem," "The electric conductivity of metals," and "The region between ultra-violet and X-rays."
63. Arms, Nancy, *A Prophet in Two Centuries: The Life of F.E. Simon* (Oxford, 1966). In 1950, on receiving the first Kamerlingh Onnes Gold Medal instituted by the Dutch Institute of Refrigeration, he remarked in his address that he regretted turning down the offer. He thereby not only forfeited the opportunity of working with the foremost low-temperature physicist of his time, but of availing himself of Leiden's cryogenic technology. Ibid., 35. Nevertheless, despite the stature of Simon's scientific career, his fame rests primarily on his own technological contributions to low-temperature technology.
64. Ibid., 41. The Physikalisches Institut, as well as the Physikalisch-Chemisches Institut, which shared the same massive building on Am Reichstagsufer in Berlin, were destroyed in the closing days of World War II.
65. Simon, F., "Ueber die atomare elektrische Leitfähigkeit der Metalle," *Zeitschrift der Physikalische Chemie*, **109** (1924), 136–142.
66. Simon to Onnes, 22 February 1924; KO, Archive 8.

Notes to Chapter 7

1. Tuyn and Onnes, "Further experiments with liquid helium. AA. The disturbance of supraconductivity by magnetic fields and currents. The hypothesis of Silsbee," *PLC*, **174a** (1926), 21.
2. Tuyn and Onnes, "The disturbance of supra-conductivity by magnetic fields and currents. The hypothesis of Silsbee," Franklin Institute, *Journal*, **201** (1926), 379.
3. Donnelly (chap. 6, note 47), 42.
4. Tuyn, *Weerstandsmetingen in vloeibaar helium*, diss. Leiden, 1924, 131 pages.
5. Tuyn and Onnes (note 1), 13.
6. Ibid., 14.
7. De Haas, W.J., G.J. Sizoo, and H. Kamerlingh Onnes, "Over den invloed van het magneetveld op den weerstand van suprageleiders," *PH*, **5** (1925), 447–448. The footnote in question is note no. 1, in Tuyn and Onnes (note 1), 21.
8. Tuyn and Onnes (note 1), 22.

9. Field measurements on lead in 1914 (*Communication* **139f**; chap. 5, note 5) extended down to $T = 2$ °K, but the fields were produced with the large Weiss-magnet, whereas "with [the present] field intensities [≤ 600 gauss] working with wire-coils is preferable ... " Arms, *A Prophet in Two Centuries* (chap. 6, note 63), 398.
10. Bridgman (chap. 6, note 25), 458.
11. Tuyn and Onnes (note 1), 26.
12. Tuyn and Onnes (note 2), 399. This ambiguity plagues critical current determinations even today: With measurements on fine superconducting filaments (which are invariably used in modern highly "stabilized" superconducting composite conductors), a meaningful critical current value must be defined in terms of the bath temperature, ambient field (if present), and effective resistivity.
13. Tuyn and Onnes (note 1), 29.
14. Onnes and Tuyn (chap. 6, note 13).
15. Tuyn and Onnes (note 1), 33.
16. Silsbee, F.B., "Current distribution" (chap. 6, note 28).
17. Ibid., 517.
18. Tuyn and Onnes (note 1), 37.
19. Silsbee (note 16), 518.
20. Tuyn and Onnes (note 1), 37.
21. Silsbee (note 16), 517.
22. Sizoo, W.J. de Haas, and Onnes, "Influence of elastic deformation on the magnetic disturbance of the superconductivity with tin. Hysteresis phenomena," *RN*, **29** (1926), 221 [*PLC*, **180c**].
23. Ibid., 221.
24. De Haas, Sizoo, and Onnes, "On the magnetic disturbance of the supraconductivity with mercury. I and II," *RN*, **29** (1926), 233 [*PLC*, **180d**].
25. Ibid. This *Communication* contains the first measurements on the critical magnetic fields for mercury.
26. Ibid., 250.
27. The first measurements on single crystal copper wires, supplied by Phillips, had been performed in 1913. [*PLC*, **153a**: 12, and **142a**: 5 (the latter with Holst)].
28. De Haas, Sizoo, and Onnes (note 24), 263.
29. De Haas and Sizoo, "Further measurements on magnetic disturbance of the supraconductivity with tin and mercury," *RN*, **29** (1926), 947–963 [*PLC*, **180**].
30. Sizoo, *Onderzoekingen over den suprageleidenden toestand van metallen*, diss. Leiden, 1926, 119 pages. Sizoo's dissertation is a convenient summary of *PLC*, **174**, **180**, **180a**, **180c**, and **180d**, albeit in Dutch.
31. Tuyn, "Measurements on the disturbance of the supra-conductivity of thallium by magnetic fields, *RN*, **31** (1928), 687–691 [*PLC*, **191b**]; de Haas and J. Voogd, "On the resistance-hysteresis phenomena of tin, lead, indium and thallium at the temperature of liquid helium," *RN*, **32** (1929), 206–213 [*PLC*, **191d**]. (The cited quotation appears on p. 213 of De Haas and Voogd's paper.) These two papers were read before the Amsterdam Academy in May and June of 1928, respectively.
32. *Solvay 1924*, 297.
33. De Haas, Sizoo, and Voogd, "Research about the question whether grey tin becomes supraconductive or not," *RN*, **31** (1928), 352 [*PLC*, **187d**]. A typo in *PLC*, **187d** (p.352) implies grey tin *is* a superconductor.
34. Ibid., 350.
35. De Haas and Voogd, "On the superconductivity of gallium," *RN*, **32** (1929), 214–217 [*PLC*, **193b**].

36. Keesom, "Methods and apparatus used in the cryogenic laboratory. XXII. A cryostat for temperatures below 1 °K," *RN*, **32** (1929), 710–714 [*PLC*, **195c**].
37. De Haas and Voogd, "On the superconductivity of gallium," *RN*, **32** (1929), 733–734 [*PLC*, **199a**]. The footnote appears on p. 733.
38. McLennan, J.C., "The cryogenic laboratory of the University of Toronto," *NA*, **112** (1923), 135.
39. Interview with W. Meissner on 8 February 1963, by T.S. Kuhn, AHQP.
40. Meissner maintained that Planck's selection of the PTR over Göttingen was "not out of friendship with me, but purely objective." He recalled that Planck would cut off arguments on the subject with a simple "that is that." Ibid., 6–7. For an overview of Meissner's early scientific program at the PTR, see Meissner, "Thermische und elektrische Leitfähigkeit einiger Metalle zwischen 20 und 373 ° abs.," *AP*, **47** (1915), 1001–1058. A primary objective of these investigations was testing the laws of Wiedemann–Franz, Lorenz, and Gruneisen.
41. Stark had sought the position, but Planck was senior member of the Reichsanstalt's Kuratorium (Advisory Committee) and had supported Nernst instead. Heilbron (chap. 6, note 1), 119.
42. Meissner, W., "Das neue Kaltelaboratorium der Physikalisch-Technische Reichsanstalt in Berlin," *PZ*, **29** (1928), 617–622.
43. Meissner to Onnes, 9 March 1925 (KO, Archive 8).
44. Heilbron (chap. 6, note 1).
45. The Notgemeinschaft had chosen "metal research" as a prime example of a new "joint program" because of its practical importance to metallurgy. M. Eckert, "Propaganda in science: Sommerfeld and the spread of the electron theory of metals," *HSPBS*, **17**, part 2 (1987), 225.
46. Meissner, "Ueber die Heliumverflüssigungsanlage der Physikalisch-Technische Reichsanstalt und einige Messungen mit Hilfe von flüssigem Helium," *PZ*, **26** (1925), 691–694.
47. Meissner, "Messungen mit Hilfe von flüssigem Helium. II. Widerstand von Metallen. Supraleitfähigkeit von Tantal. Beiträge zur Erklarung der Supraleitfähigkeit. Spezifische Wärme des gasformigen Heliums," *PZ*, **29** (1928), 901.
48. Ibid. Presented at the Versammlung Deutscher Naturforscher und Arzte (Meeting of German scientists and physicians) in Hamburg during 16–22 September 1928.
49. Meissner, "Supraleitfähigkeit von Thorium," *Die Naturwissenschaften*, **17** (1929), 390–391.
50. Meissner and H. Franz, "Messungen mit Hilfe von flüssigem Helium. VIII. Supraleitfähigkeit von Niobium," Physikalisch-Technische Reichsanstalt, *Mitteilung* (1930), 558–559. Meissner presented preliminary measurements on Nb at the Tagung des Deutschen Kaltevereins (German Cryogenic Association) in Stuttgart on 5 June 1930.
51. McLennan, J.C., L.E. Howlett, and J.O. Wilhelm, "On the electrical conductivity of certain metals at low temperatures," Royal Society of Canada, *Transactions*, **23** (1929), 295–296.
52. Van Aubel, E., de Haas, and Voogd, "New super-conductors," *RN*, **32** (1929), 223–224 [*PLC*, **193c**].
53. De Haas and Voogd, "The resistance of alloys at the temperature of liquid hydrogen and helium," *RN*, **32** (1929), 723 [*PLC*, **197b**].
54. De Haas, van Aubel, and Voogd, "Ein aus zwei Nicht-Supraleitern zusammengesetzter Supraleiter," *RN*, **32** (1929), 226–230 [*PLC*, **197a**]; "A superconductor, consisting of two non-superconductors," *RN*, **32** (1929), 730 [*PLC*, **197c**].

55. Meissner, "Messungen mit Hilfe von flüssigem Helium. V. Supraleitfähigkeit von Kupfersulfid," Physikalisch-Technische Reichsanstalt, *Mitteilung* (1929), 571.
56. Bismuth is now known to be a superconductor under very high pressures.
57. De Haas and Voogd, "Disturbance of the superconductivity of the compound Bi_5Tl_3 and the alloys Sn-Sb and Sn-Cd by magnetic fields," *RN*, **32** (1929), 874–882 [*PLC*, **199c**].
58. Ibid., 882.
59. De Haas and Voogd, "The influence of magnetic fields on supraconductors," *RN*, **33** (1930), 262–270 [*PLC*, **208b**].
60. De Haas and Voogd (note 53), 723. *Communication* **208b** erroneously locates measurements of the magnetic disturbance of various superconducting alloys in a companion report [De Haas; van Aubel, and Voogd, "On the supra-conductivity of alloys," *RN*, **33** (1930), 258–261 [*PLC*, **208a**].
61. De Haas and Voogd (note 59), 267.
62. Ibid., 269.
63. Meissner to de Haas, 30 December 1930 (WM, Archive N45, book II).
64. Mendelssohn, K. (chap. 6, note 7).
65. Ibid., 130.
66. Arms, N., *A Prophet* (chap. 6, note 63).
67. The scheme basically involves four steps. First, the paramagnetic salt (substituting for a gas) is magnetized (the magnetic field substituting for a piston), during which heat is developed in the sample. The salt is then successively cooled through a helium bath of around 0.8 °K, insulated from the bath, and finally demagnetized in the course of which the sample now cools off. The scheme was successfully demonstrated almost simultaneously in 1933 at Berkeley by Giauque and McDougal (reaching 0.25 °K with gadolinium sulphate) and at Leiden by de Haas, Eliza Cornelis Wiersma, and H.R. Kramers (initially reaching 0.13 °K with cerium flouride) and a year later by Simon and Kurti at Oxford. Actually, Onnes and H.R. Woltjer (conservator in the Leiden laboratory) had unwittingly laid the foundation for magnetic cooling in the twenties in their study of paramagnetic saturation in the first of the working substances, gadolinium sulphate. [*Communication* **167c**].
68. This was the first time liquid helium was produced in the U.S. Simon even used his apparatus for a simple demonstration of superconductivity at Berkeley. Arms, N., *A Prophet* (chap. 6 note 63), 62.
69. Van Beelen, H., A.J.P.T. Arnold, H.A. Sypkens, J.P. van Braam Houckgeest, R. de Bruyn Ouboter, J.J.M. Beenakker, and K.W. Taconis, "Flux pumps and superconducting solenoids," *PH*, **31** (1965), 415–416 [*PLC*, **342a**].
70. Mendelssohn, "Production of high magnetic fields at low temperature," *NA*, **132** (1933), 602.
71. Mendelssohn, "Prewar work on superconductivity as seen from Oxford," *RMP*, **36** (1964), 8.
72. Kurti, N., "Cryomagnetic research at the Clarendon Laboratory," *Search and Research*, ed. J.P. Wilson (London, 1971). Note that the size and weight, and hence the cost, of copper-iron electromagnets depends both on the volume required in the pole-tip gap and on the intensity of the field. As the required volume grows, the iron cross section must increase to avoid iron saturation, and the coils must provide correspondingly more ampere-turns. At Bellevue, Simon was able to establish the thermodynamic scale in the new temperature range achieved by magnetic cooling, or down to 0.02 °K. Once that was in hand, it became possible not only to study the thermal and magnetic properties of paramagnetic salts and other sub-

stances as a function of temperature, but to extend the study of specific heats and the remarkable properties of liquid helium itself. Arms, N., *A Prophet* (chap. 6, note 63), 77–78.

73. Keesom, "On the disturbance of supraconductivity of an alloy by an electric current," *PH*, **2** (1935), 36.
74. Ibid.
75. Mendelssohn (note 71), 8.
76. De Haas and Voogd, "Further investigations on the magnetic disturbance of the supraconducting state of alloys," *RN*, **34** (1931), 56–58, [*PLC*, **214b**]. A primer on the metallurgy of lead–bismuth eutectics.
77. Kunzler, J.E., "Superconductivity in high magnetic fields at high current densities," *RMP*, **33** (1961), 501; "Recollections of events associated with the discovery of high field-high current superconductivity," *TM*, **MAG-23** (1987), 396–402.
78. Ibid., 501.
79. De Haas and Voogd, "The magnetic disturbance of the supraconductivity of single-crystal wires of tin," *RN*, **34** (1931), 63–69 [*PLC*, **212d**].
80. E.g., de Haas and Voogd, "Measurements on the electrical resistance of pure indium, thallium, and gallium at low temperatures and of the magnetic disturbance of the supraconductivity of thallium," *RN*, **34** (1931), 52 [*PLC*, **212d**, unfortunately numbered the same as *PLC*, **212d** of note 79]; de Haas and Voogd, (note 76).
81. De Haas and Voogd (note 79), 64–65.
82. Ibid., 68.
83. Ibid., 68–69.
84. Ibid., 68.
85. Meissner and B. Voigt, "Messungen mit Hilfe von flüssigem Helium. XI. Widerstand der reinen Metalle in tiefen Temperaturen," *AP*, **7** (1930), 761–797.
86. De Haas and Voogd, "On the steepness of the transition curve of supraconductors," *RN*, **34** (1931), 192–203 [*PLC*, **214c**].
87. Ibid., 200.

Notes to Chapter 8

1. Gorter, C.J., "Superconductivity until 1940 in Leiden and as seen from there," *RMP*, **36** (1964), 4; Casimir (chap. 1, note 23), on 167–175.
2. Ibid., 4.
3. Onnes and Holst (chap. 5, note 28), 764.
4. Ibid., 766.
5. Casimir (chap. 1, note 23), 333.
6. De Haas and H. Bremmer, "Conduction of heat of lead and tin at low temperatures," *RN*, **34** (1931), 325–338 [*PLC*, **214d**].
7. Grüneisen, E., and E. Goens, "Untersuchungen an Metallkristallen. V. Elektrizitäts- und Wärmeleitung von ein-und vielkristallinen Metallen des regulären Systems," *ZP*, **44** (1927), 615–642.
8. Peierls, R., "Zur kinetischen Theorie der Wärmeleitung in Kristallen," *AP*, **4** (1929), 1055–1101; "Zur Theorie der elektrischen und thermischen Leitfähigkeit von Metallen," *AP*, **4** (1930), 121–148; "Zwei Bemerkungen zur Theorie der Leitfähigkeit," *AP*, **5** (1930), 244–246. Peierls's thermal conductivity work was a quantum-mechanical generalization of Debye's earlier classical treatment of the

subject that was only valid at high temperatures; Peierls completed it during the summer of 1929 as his thesis under Pauli at the ETH in Zürich.

9. De Haas and Bremmer (note 6), 333.
10. Earlier measurements had been made between hydrogen and oxygen temperatures; de Haas, S. Ayoama, and H. Bremmer, "Thermal conductivity of tin at low temperatures," *RN*, **34** (1931), 75–77 [*PLC*, **214a**].
11. Meissner to Kronig, 11 November 1932 (ETH Hs. 1045:316).
12. Hoddeson, L., G. Baym, and M. Eckert (chap. 6, note 38), 313.
13. Bloch, F., "Memories of electrons in crystals," *PRS*, **371A** (1980), 27. A more radical form of "Bloch's second theorem" was "Superconductivity is impossible!" (Ibid., 27).
14. Landau, L.D., "Zur Theories der Supraleitfähigkeit," *PZS*, **4** (1933), 43–49.
15. Hoddeson, Baym, and Eckert (chap. 6, note 38), 314.
16. Kronig, R. de L., "Zur Theorie der Supraleitfähigkeit," *ZP*, **78** (1932), 744–750.
17. McLennan to Kronig, 8 July 1933 (ETH, Hs. 1045:231).
18. McLennan to Kronig, 2 October 1932 (ETH, Hs. 1045:227).
19. Kronig, R. de L., "Zur Theorie der Supraleitfähigkeit II," *ZP*, **80** (1933), 203–216.
20. Hoddeson, Baym, and Eckert (chap. 6, note 38), 317, quoting Kronig to Bohr, 25 November 1932, in Bohr scientific manuscripts 1932 (hereafter referred to as BSM) in AHQP.
21. Hoddeson, Baym, and Eckert (chap. 6, note 38), 315, quoting Bohr to Bloch, 15 June 1932 (BSM).
22. Ibid., 42. N. Bohr, "Zur Frage der Supraleitung," manuscript dated June 1932, submitted to *Die Naturwissenschaften* and accepted for 11 July issue of 1932 but withdrawn by Bohr in proof stage. The proofs, with the heading "Naturwissenschaften, 11.7.32 (Art. 492. Bohr)" are preserved in BSM; quoted in Ibid., 42. McLennan, "Electric supra-conduction" (chap. 4, note 40), 884. Note by Bloch, BSM, quoted by Hoddeson, Baym, and Eckert (chap. 6 note 38), 315.
23. Bohr to Kronig, 27 December 1932, BSM.
24. Hoddeson, Baym, and Eckert (chap. 6, note 38), 317, quoting Bloch to Bohr, 30 December 1932, BSM.
25. De Haas and Bremmer, "Thermal conductivity of indium at low temperatures," *RN*, **35** (1932), 135 [*PLC*, **220b**], communicated at the Amsterdam meeting on 27 February 1932.
26. Hulm, J.K., "The thermal conductivity of tin, mercury, indium and tantalum at liquid helium temperatures," *PRS*, **204** (1950), 109–110.
27. Mendelssohn, K., "Heat conduction in superconductors," *Progress in Low Temperature Physics*, I, ed. C.J. Gorter (New York and Amsterdam, 1955), 184–201; Shoenberg (chap. 3, note 29), 78–82.
28. Keesom and J.N. van den Ende, "The specific heat of solid substances at the temperatures obtainable with the aid of liquid helium. II. Measurements of the atomic heats of lead and of bismuth," *RN*, **33** (1930), 243 [*PLC*, **203d**]; "The specific heat of solid substances at the temperatures obtainable with the aid of liquid helium. III. Measurements of the atomic heats of lead and bismuth. A correction," *RN*, **34** (1931), 210–211 [*PLC*, **213c**].
29. Keesom and van den Ende, "The specific heat of solids at temperatures obtainable with liquid helium. IV. Measurements of the atomic heats of tin and zinc," *RN*, **35** (1932), 149–152 [*PLC*, **219b**].
30. Keesom and J.A. Kok, "On the change of the specific heat of tin when becoming supraconductive," *RN*, **35** (1932), 744–748 [*PLC*, **221e**].
31. Mendelssohn, K., and F. Simon, "Ueber den Energieinhalt des Bleies in der Nähe

des Sprungpunktes der Supraleitfähigkeit," *Zeitschrift für Physikalische Chemie (B)*, **16** (1932), 76.

32. Keesom and J.A. Kok, "Measurements of the specific heat of thallium at liquid helium temperatures," *PH*, **1** (1934), 178 [*PLC*, **230c**].
33. Rutgers, A.J., "Note on supraconductivity," *PH*, **1** (1934), 1055–1058. A more general derivation of this formula was subsequently published by C.J. Gorter and H.B.G. Casimir the same year: "On supraconductivity I," *PH*, **1** (1934), 308–313. Rutgers's derivation was first drawn attention to in a note added in proof to P. Ehrenfest's "Phasenumwandlungen in üblichen und erweiterten Sinn, classifiziert nach den entsprechenden Singularitäten des thermodynamischen Potentiales," *RN*, **36** (1933), 157 [*PLS*, **75b**].
34. Keesom and K. Clusius, "Ueber die spezifische Wärme des Flüssigen Helium," *RN*, **35** (1932), 307–320; "Die Umwandlung Flüssiges Helium-I—Flüssiges Helium II unter Druck," *RN*, **34** (1931), 605–609 [*PLC*, **216b**].
35. Keesom to McLennan, 9 January 1933 (Boerhaave Archive, 128).
36. Keesom and Kok, "Measurements of the latent heat of thallium connected with the transition, in a constant external magnetic field, from the supraconductive to the non-supraconductive state," *PH*, **1** (1934), 503–512 [*PLC*, **230e**].
37. De Haas and Voogd (chap. 7, note 79), 53–55.
38. Keesom and Kok (note 36), 506.
39. Keesom and Kok, "Further calorimetric experiments on thallium" *PH*, **1** (1934), 596 [*PLC*, **232a**, 596].
40. Ibid. The dilemma did not prevent Ehrenfest from introducing, in his last year, the notion of a novel type of second-order phase transition for the transition in the absence of a field. This is a transition in which there is no discontinuity in the internal energy, but a discontinuity in the specific heat, in contrast to the transition *in* a field, which is an ordinary phase change of the first kind (including a latent heat). Ehrenfest's treatment was the starting point for Rutgers's derivation of his formula for the specific heat discontinuity. Ehrenfest and Rutgers, respectively (note 33).
41. Boerhaave Archive, 128.
42. Martin, T., *The Royal Institution* (London, 1961), 43–44.
43. Ibid., 32.
44. Keesom to McLennan (note 41).
45. Commander William Francis-Sempill (1893–1965), the 19th Baron of Sempill and well-known aviator. Wing Commander and Colonel in World War I; advisor on aeronautics and organizer to several governments, among them Imperial Japanese Naval Air Service; frequent King's Cup Air Racer; Chairman and President of Royal Aeronautical Society; at the time of the helium flight he was Deputy Chairman of the Council of the London Chamber of Commerce; he was active in numerous engineering and aviation committees, institutions, and societies; etc.
46. McLennan to Keesom (note 41).
47. McLennan, "Electrical conductivity of metals at the lowest temperatures," *The Royal Institution Library of Science: Physical Sciences*, vol. 9, ed. W.L. Bragg and G. Porter (Amsterdam, 1970), 419–420.
48. Armstrong, Professor H.E. in a letter to *The Times* after McLennan's death, quoted in H.H. Langton, *Sir John Cunningham McLennan: A Memoir* (Toronto, 1939), 67.
49. Master of Sempill to Keesom (note 41).
50. As quoted in the Toronto *Globe*; ibid., 68.
51. Keesom to McLennan (note 41).

52. Ibid.
53. Boerhaave Archive; ibid.
54. Simon, F., "The approach to the absolute zero of temperature," *The Royal Institution Library of Science*, vol. 10, 71–86.
55. Arms, N., *A Prophet* (chap. 6, note 63), 75.

Notes to Chapter 9

1. De Haas and Voogd, "On the steepness" (chap. 7, note 86), 192.
2. Gavroglu, K., and Y. Goudaroulis, *Methodological Aspects* (chap. 4, note 41), 60.
3. Ibid., 119.
4. Hermann, Armin, "Max von Laue," *DSB*, vol. VIII, 50.
5. Meissner, W., "Max von Laue als Wissenschaftler und Mensch," Bayerische Akademie der Wissenschaften, *Sitzungsberichte*, **9** (1960), 109. After World War II Laue was partly instrumental in the reestablishment of the Physikalisch-Technische Bundeanstalt in Burnswick, the successor to the PTR in Charlottenburg.
6. Von Laue, M., "Zur Deutung einiger Versuche über Supraleitung," *PZ*, **33** (1932), 793–795.
7. De Haas and Voogd (chap. 7, note 79).
8. E.g., Joos, *Theoretische Physik*, 2nd edition, (1934), 175.
9. Historical references and outlines of the derivation are found in E.C. Stoner, "The demagnetizing factors for ellipsoids," *PM*, **36** (1945), 804.
10. De Haas and Voogd (chap. 7, note 86).
11. De Haas and Voogd (chap. 7, note 79).
12. De Haas, "Supraleiter im Magnetfeld," *Leipziger Vorträge 1933, Magnetismus*, ed. P. Debye (Leipzig, 1933), 65–69.
13. De Haas, Voogd, and J.M. Jonker, "Quantitative Unterschung über einen möglichen Einfluss der Achsenorientierung auf die magnetische Uebergangsfigur," *PH*, **1** (1934), 285 [*PLC*, **229c**].
14. Shoenberg (chap. 3, note 29), 113. The dashed curve in Figure 9-3 is due to Shoenberg; ibid., 111.
15. Serin, B., "Superconductivity. Experimental part," ed. S. Flügge, *Encyclopedia of Physics* (Berlin, 1956), 228–229; E.R. Andrew, "The intermediate state of superconductors II. The resistance of cylindrical superconductors in transverse magnetic fields," *PRS*, Ser. A, **194** (1948), 80.
16. De Haas (note 12), 65.
17. Ibid., 68.
18. Ibid.
19. Ibid.
20. Casimir, "Superconductivity and superfluidity," *The Physicist's Conception of Nature*, ed. Jagdish Mehra (Amsterdam and Boston, 1973), 487.
21. De Haas, Voogd, and Jonker (note 13), 290.
22. Ibid.
23. Meissner to Kronig, 10 December 1932 (ETH, Hs. 1045:317).
24. Hartmann, Hans, *Schöpfer des neuen Weltbildes* (Bonn, 1952).
25. Breit, G., "Transients of magnetic field in supraconductors," *RN*, **26** (1923), 540–541.
26. De Haas-Lorentz, G.L., "Iets over het mechanisme van inductiever schijnselen," *PH*, **5** (1925), 384. Geertruida Luberta de Haas wrote her dissertation on Brownian motion—that is, on the microscopic zig-zag motion of colloidally suspended parti-

cles due to collisions with molecules of the surrounding fluid. Gorter describes her as "very clever—much more clever at mathematics than [her husband]." Interview with Gorter on 13 November 1962, by J.L. Heilbron, AHQP. We have already met two other women at Leiden, both experimental physicists: Anna Petronella Keesom, W.H. Keesom's daughter, and Joshina Jonker, who married H.G.B. Casimir. In addition, women scientists came and went at Leiden, among them Mme Curie, Anna Beckman (wife of Bengt Beckman, who worked on magnetoresistivity), and Olga Trapeznikova (who assisted her husband, L.V. Schubnikow, with his magnetoresistance studies at Leiden). The Leiden laboratory by no means had a monopoly on women scientists, although it probably attracted more than most. The Clarendon laboratory had Judith Moore (later Judith Hull), who collaborated with K. Mendelssohn. Nernst's laboratory had Clara von Simson, who collaborated with F. Simon on several papers, in spite of Nernst himself, "who refused to take women physicists seriously." Arms, N., *A Prophet* (chap. 6, note 63), 42.

27. Becker, R., G. Heller, and F. Sauter, "Ueber die Stromverteilung in einer supraleitenden Kugel," *ZP*, **85** (1933), 779–784.
28. E.g., D. Shoenberg (chap. 3, note 29), 180–181; see also Casimir, "Superconductivity and superfluidity" (note 20).
29. Casimir (note 20), 488.
30. Interview with H.B.G. Casimir on 6 July 1963 by T.S. Kuhn, L. Rosenfeld, A. Bohr, and E. Rudinger, AHQP.
31. Meissner, "Ueber die Heliumverflüssigungsanlage" (chap. 7, note 46), 693. Meissner's early preparation for the forthcoming experiments that were to culminate in those with Ochsenfeld in 1933 includes his laboratory notebook of 1920 entitled *Permeabilität*, which is simply a handwritten compilation of literature citations on this subject. (WM, Archive N45, Notebook *PTRIIa)* (July 1920).
32. Meissner (chap. 7, note 47), 901–902.
33. Ibid., 902.
34. Meissner to Kronig (chap. 8, note 11).
35. Meissner to Kronig, 10 December 1932 (ETH Hs. 1045:317).
36. Referred to in Hoddeson, Baym, and Eckert (chap. 6, note 38), 317.
37. Meissner to Kronig, 17 January 1933 (ETH Hs. 1045:318).
38. Meissner, "Arbeiten des Charlottenburger Kältelaboratoriums über Supraleitfähigkeit und über das Wasserstoff-Isotop," *Helvetica Physica Acta*, **6** (1933), 414–418. The correspondence between Meissner and Scherrer concerning Meissner's forthcoming ETH lecture is found in WM, Archive N45, Scientific Papers, book III.
39. Meissner to P. Scherrer, 33 June 1933 (WM, Archive N45, book I).
40. Meissner (note 38), 416.
41. Von Laue and F. Möglich, "Ueber das magnetische Feld in der Umgebung von Supraleitern," Preussischen Akademie der Wissenschaften, *Sitzungsberichte*, **16** (1933), 544–565. Submitted for publication 25 July 1933. Meissner (note 38), 416.
42. Notebooks VIII and IX (WM, Archive N45, Notebook carton).
43. Notebook VIII (note 42).
44. *Vorträge und Diskussionen des IX. Deutschen Physikertage in Würzburg, vom 18-22 September, 1933*, *PZ*, **34** (1933), 777–790, 809–846, 849–880, 889–897.
45. Debye, P. to Meissner, 2 November 1933 (WM, Archive N45, book III).
46. Meissner to Debye, 6 November 1933 (WM, Archive N45, book III).
47. Meissner and R. Ochsenfeld, "Ein neuer Effect bei Eintritt der Supraleitfähigkeit," *Die Naturwissenschaften*, **21** (1933), 787–788.
48. Meissner to A. Berliner, 16 October 1933 (WM, Archive N45, book III).

49. Meissner and Ochsenfeld (note 47), 787.
50. Ibid., 788.
51. Ibid.
52. As already discussed, the modified current distribution cannot be strictly confined to the surface; otherwise the current density would be infinite.

Notes to Chapter 10

1. McLennan to Meissner, 12 November 1933 (WM, Archive N45, book III).
2. Meissner to McLennan, 16 November 1933 (WM, Archive N45, book III).
3. McLennan to Kronig, 20 November 1933 (ETH, Hs. 1045:232).
4. By this time Joshina M. Jonker, a physicist, had married Hendrik B. G. Casimir.
5. De Haas and Casimir-Jonker, "Unterschungen über den Verlauf des Eindringens transversalen Magnetfeldes in einem Supraleiter," *PH*, **1** (1934), 291–296 [*PLC*, **229d**].
6. Kohlrausch, F.W.G., *Lehrbuch der Praktischen Physik.*, 9th ed., (Leipzig, 1901), 466.
7. De Haas and Casimir-Jonker (note 5), 296.
8. Gorter, C.J., "Superconductivity until 1940" (chap. 8, note 1), 4.
9. Ibid.
10. Ibid.
11. Gorter, C.J., "Some remarks on the thermodynamics of supraconductivity," Musée Teyler, *Archives*, **7** (1933), 378–386.
12. Gorter, C.J., "Theory of supraconductivity," *NA*, **132** (1933), 931.
13. Ibid.
14. Gorter to Meissner, 27 November 1933 (WM, Archive N45, book III).
15. Meissner to Gorter, 4 December 1933 (WM, Archive N45, book III).
16. Braunbek, W., "Die Ausbreitung elektromagnetischer Wellen in einem Supraleiter," *ZP*, **87** (1934), 470–483. Braunbek developed, much earlier, an atomic model for ferromagnetic substances that took into account the motion of the atomic nucleus. Braunbek, "Eine Erklärung des Einstein-de Haas Effektes durch Annahme Rotierender Atomkerne," *PZ*, **23** (1922), 307–309.
17. Meissner to W. Braunbek, 11 December 1933 (WM, Archive N45, book III). See also Meissner to Braunbek, 22 December 1933 and 4 January 1934 (ibid.).
18. Casimir, *Haphazard Reality* (chap. 1, note 21), 147.
19. Ibid., 178.
20. Gorter and Casimir, "On supraconductivity I," *PH*, **1** (1934), 306–320.
21. Ibid., 319.
22. Mendelssohn, K. and J.D. Babbitt, "Persistent currents in supraconductors," *NA*, **133** (1934), 459–460.
23. Mendelssohn, K., "The Clarendon Laboratory" (chap. 6, note 7), 133.
24. Lorentz, H.A., "On the motion of electricity" (chap. 6, note 45).
25. Mendelssohn and Babbitt (note 22), 460.
26. Burton, E.F., J.O. Wilhelm, and F.G.A. Tarr, "Magnetic properties of supraconductors," *NA*, **133** (1934), 684.
27. Mendelssohn and Babbitt (note 22), 459.
28. Keesom and Kok (chap. 8, note 36).
29. Ibid., 503.

30. Ibid., 504–505.
31. Ibid., 511.
32. Joffé, A.F., "Twenty years of soviet physics," *PZS*, **12** (1937), 494.
33. Ibid., 496.
34. Obreimow, I., and L. Schubnikow, "Eine Methode zur herstellung einkristalliger Metalle," *ZP*, **25** (1924), 31–36; Bridgman, P.W., "Certain physical properties of single crystals of tungsten, antimony, bismuth, tellerium, cadmium, zinc, and tin," American Academy of Arts and Sciences, *Proceedings*, **60** (1925), 305–383.
35. Kapitza, P., "The study of the specific resistance of bismuth crystals and its change in strong magnetic fields and some allied problems," *PRS*, **119A** (1928), 360–364.
36. Schubnikow, L., "Ueber die herstellung von Wismutheinkristallen," *RN*, **33** (1930), 327–331.
37. Schubnikow, L., and W.J. de Haas, "Magnetische widerstandsvergrösserung in Einkristallen von Wismut bei tiefen Temperaturen," *RN*, **33** (1930), 130–133; Schubnikow, L., and W.J. de Haas, "Die Abhängigkeit des elektrischen Widerstandes von Wismutheinkristallen von der Reinheit des Metalles," *RN*, **33** (1930), 350–362; Schubnikow, L., and W. J. de Haas, "Neue Erscheinungen bei der Widerstandsanderung von Wismuthkristallen im Magnetfeld bei der Temperatur von flüssigem Wasserstoff," *RN*, **33** (1930), 363–378, 418–432; Schubnikow, L., and W.J. de Haas, "A new phenomenon in the change of resistance in a magnetic field of single crystals of bismuth," *NA*, **126** (1930), 500.
38. Rjabinin, J.N., and L.W. Schubnikow, "Dependence of magnetic induction on the magnetic field in supraconducting lead," *NA*, **134** (1934), 286–287.
39. Rjabinin, J.N., and L.W. Schubnikow, "Verhalten eines Supraleiters in magnetischen Feld," *PZS*, **5** (1934), 641–643. The short-lived *Physikalische Zeitschrift der Sowjetunion* was modeled on Onnes's *Communications*, even to the extent of publishing most articles in English. Volume 1 appeared in 1932, Volume 13, the last, in 1938.
40. Rjabinin and Schubnikow, "Dependence of magnetic induction" (note 38), 287.
41. Meissner to Erhard Ahrens, 9 December 1933 (WM, Archive N45, book III).
42. Simon to Meisner, 28 January 1934 (WM, Archive N45, book III).
43. W. Gerlach to Meissner, 7 March 1934 (WM, Archive N45, book III).
44. Meissner with R. Ochsenfeld and F. Heidenreich, "Magnetische Effekte bei Eintritt der Supraleitfähigkeit," *Zeitschrift für die gesamte Kälte-Industrie*, **41** (1934), 125–130.
45. Meissner, "Bericht über neuere Arbeiten zur Supraleitfähigkeit," *PZ*, **35** (1934), 931–944.
46. Meissner and F. Heidenreich, "Ueber die Änderung der Stromverteilung und der magnetischen Induktion beim Eintritt der Supraleitfähigkeit," *PZ*, **37** (1936), 449–470.
47. Ibid., 451.
48. The Meissner–Ochsenfeld letter states 140 mm long and 1.5 mm spacing.
49. Von Laue to Meissner, 23 April 1933 (WM, Archive N45, carton, von Laue correspondence).
50. Von Laue and F. Möglich (chap. 9, note 41).
51. Ibid., 544.
52. Meissner, "Magnetische Effekte" (note 44), 127.
53. Meissner, notebook VIII (WM, Archive N45, notebook carton).
54. Meissner and Heidenreich, "Änderung der Strömverteilung" (note 46), 453.
55. Ibid.
56. Meissner with Ochsenfeld and Heidenreich (note 44), 125.

57. Meissner and Heidenreich, "Ueber die Änderung der Stromverteilung" (note 46), 455.
58. Ibid., 458.
59. Meissner, "Bericht" (note 45), 934.
60. Meissner, "Magnetische Effekte" (note 44), 126.
61. Meissner and Heidenreich (note 46), 459.
62. Ibid., 458.
63. Meissner, "Magnetische Effekte" (note 44), 127.
64. Ibid.
65. Ibid., 126.
66. Ibid., 127.
67. Meissner, "Bericht" (note 45), 935.
68. Rose-Innes, A.C. and E.H. Rhoderick, *Introduction to Superconductivity* (New York, 1969), 25–27.
69. Meissner to Grüneisen, 5 June 1934 (WM, Archive N45, book III).
70. Keesom to Grüneisen, 12 June 1934 (WM, Archive N45, book III); Grüneisen to Meissner, 14 June 1934 (WM, Archive N45, book III).
71. Grüneisen to Meissner, 7 June 1934 (WM, Archive N45, book III).
72. Grüneisen to Meissner, 24 June 1934 (WM, Archive N45, book III).
73. De Haas to Grüneisen, 25 June 1934 (WM, Archive N45, book III).
74. Grüneisen to Meissner, 19 July 1934 (WM, Archive N45, book III); Gorter, C.J., "Superconductivity until 1940" (chap. 8, note 1), 5.
75. Grüneisen to Meissner, 2 July 1934 (WM, Archive N45, book III).
76. Laue, M. von to O. Stern, 6 July 1934 (O. Stern collection, Bancroft Library, 85/96C, carton 2).
77. Laue to Meissner, 21 May 1937; Meissner to Laue, 25 May 1937 (WM, Archive N45).
78. Stark, J., "Bemerkung über den Zustand der Elektronen in der Supraleitung," *PZ*, **36** (1935), 515–516; Steiner, K., and P. Grassmann, *Supraleitung* (Braunschweig, 1937), 119–122.
79. Heilbron, H.J., *The Dilemmas* (chap. 6, note 1), 91.
80. Debye, P., "Die magnetische Methode zur Erzeugung tiefster Temperaturen," *PZ*, **35** (1934), 923–928.
81. Clusius, K., "Zwei Vorlesungsversuche mit flüssigem Wasserstoff," *PZ*, **35** (1934), 929–930.
82. Meissner, "Bericht" (note 45).
83. Keesom, W., "Das kalorische Verhalten von Metallen bei den tiefsten Temperaturen," *PZ*, **35** (1934), 939–944.
84. Meissner to Laue, 18 July 1934 (WM, Archive N45, book III). Justi, E., and M. von Laue, "Phasengleichgewichte dritter Art," *PZ*, **35** (1934), 945–963.
85. Gorter, C.J., and H.B.J. Casimir, "Zur Thermodynamik des supraleitenden Zustandes," *PZ*, **35** (1934), 963–966.
86. Gorter's words, in Gorter, (chap. 8, note 1), 5.
87. Meissner to Heidenreich, 5 November 1934 (WM, Archive N45, book IV).
88. Meissner to Heidenreich, 17 December 1934 (WM, Archive N45, book IV).
89. Gorter and Casimir (note 85), 964.
90. Sommerfeld, A., "Zur Elektronentheorie der Metalle," *Die Naturwissenschaften*, **15** (1927), 825–832.
91. Gorter and Casimir (note 85), 965.
92. Gorter, C.J., "The two fluid model for superconductors and helium II," *Progress in Low Temperature Physics*, vol. I (New York and Amsterdam, 1955), 1–16.

93. Gorter, C.J. (chap. 8, note 1), 5.
94. Gorter and Casimir (note 85), 966.
95. Ibid.
96. Meissner to Schachenmeier, 2 July 1934 (WM, Archive N45, book III). Schachenmeier, R., "Zur Elektronentheorie der Supraleitung," *PZ*, **35** (1934), 966–969.
97. Schachenmeier, R., "Wellenmechanische Vorstudien zu einer Theorie der Supraleitung," *ZP*, **74** (1932), 503–546 (submitted for publication in December of 1931). Schachenmeier acknowledges stimulating discussions with Laue.
98. Casimir, H., *Haphazard Reality* (chap. 1, note 21), 131.
99. Quoted in Hoddeson, Baym, and Eckert, "The development" (chap. 6, note 38), 318.
100. Frenkel, J., "Beitrag zur electrischen Theorie der festen Körper," *ZP*, **25** (1924), 1–30.
101. Frenkel, J., "On a possible explanation of superconductivity," *PR*, **43** (1933), 908 (submitted 27 December 1932).
102. Ibid., 911.
103. Bethe, H., and H. Fröhlich, "Magnetische Wechselwirkung der Metallelektronen. Zur Kritik der Theorie der Supraleitung von Frenkel," *ZP*, **85** (1933), 392; Hoddeson et al., "The development" (chap. 6, note 38), 41.
104. Hoddeson et al. (chap. 6, note 38).
105. Gavroglu and Goudaroulis, as noted earlier in Section 9.1, dwell at length on methodology and phenomenology in the elucidation of superconductivity and superfluidity. Zero resistance, according to them, did not, strictly speaking, establish the "paradoxical situation" required to break out of the confines of the existing theoretical framework, since it could conceivably be treated within existing models. The same holds true for the Onnes–Tuyn experiment, due to the a priori faith in Lippmann and Maxwell. However, the *diamagnetic* behavior did "establish a paradoxical situation within the dominant theoretical framework since it was that framework which stipulated that lowering the temperature cannot cause the creation of a diamagnetic state. ... With the creation of a paradoxical situation and the subsequent formulation of the 'right' problem, a satisfactory explanation was soon to follow." Gavroglu and Goudaroulis, *Methodological Aspects* (chap. 4, note 41), 122.

Notes to Chapter 11

1. Keeley, T.C., K. Mendelssohn, and J.R. Moore, "Experiments on supraconductors," *NA*, **134** (1934), 773–774.
2. Mendelssohn, K., "Prewar work" (chap. 7, note 71), 9.
3. De Haas, W.J., and J.M. Casimir-Jonker, "Penetration of a magnetic field into supraconductive alloys," *NA*, **135** (1935), 30–31.
4. Ibid., 31. The Toronto group had also found that the alloys, e.g., SnPb, rose metal (Bi-Pb-Sn) "showed no reduction in flux either on cooling [in a field] or removing the field." F.G.A. Tarr and J.O. Wilhelm, "Magnetic effects in superconductors," *Canadian Journal of Research*, **12** (1935), 265–271.
5. Schubnikow, L.W., and W.J. Chotkewitsch, "Spezifische Wärme von supraleitenden Legierungen," *PZS*, **6** (1934), 605–607.
6. Rjabinin, J.N., and L.W. Schubnikow, "Magnetic properties and critical currents of superconducting alloys," *PZS*, **7** (1935), 122–125.
7. Ibid., 122–123.

8. Ibid., 124–125.
9. Ibid., 125.
10. Shoenberg, D. (chap. 3, note 29), 43.
11. Berlincourt, T.G., "Type II superconductivity: quest for understanding," *TM*, **MAG-23** (1987), 403–412. In this paper, Berlincourt, one of the principals in the experimental phase of this struggle for high-field superconductivity, ably covers the complex chain of developments that led to the Ginzburg–Landau–Abrikosov–Gorkov theory of type II superconductivity.
12. Mendelssohn, K., and J.R. Moore, "Supra-conducting alloys," *Supplement to NA*, **135** (1935), 826–827.
13. Mendelssohn, K., "Hysteresis," in "A discussion on supraconductivity and other low temperature phenomena," *PRS*, **A152** (1935), 38–41.
14. McLennan to Kronig, 17 May 1935 (ETH, Hs. 1045:227-241). McLennan was vice president of the Royal Society in 1933–1934. McLennan to Meissner, 17 May 1935 (WM, Archive N45, book IV).
15. Armstrong (chap. 8, note 48), 88.
16. McLennan to Kronig, 18 September 1935 (ETH, Hs. 1045:227–241).
17. McLennan to Kronig, 8 October 1935 (ETH, Hs. 1045:225).
18. Meissner, W., "The magnetic effects occurring on transition to the supraconducting state," in "A discussion on supraconductivity," *PRS*, **A152N** (1935), 13–15.
19. Meissner to Heidenreich, 7 May 1935 (WM, Archive N45, book IV).
20. Meissner to Heidenreich, 7 June 1935 (WM, Archive N45, book IV).
21. Mendelssohn, K., "A discussion" (note 13), 38–41.
22. Mendelssohn, K., and J.D. Babbitt, "Magnetic behaviour of supraconducting tin spheres," *PRS*, **A151** (1935), 316–333.
23. Gorter, C.J., "Superconductivity until 1940" (chap. 8, note 1), 9.
24. Referred to in Mendelssohn to Meissner, 15 February 1936 (WM, Archive N45, book IV).
25. Ibid.
26. Mendelssohn and Babbitt, "Magnetic behaviour" (note 22), 329.
27. Meissner to Mendelssohn, 20 February 1936 (WM, Archive N45, book IV).
28. The paper was finally submitted on 12 June 1936, and Ochsenfeld did not share the authorship; Meissner and F. Heidenreich, "Ueber die Änderung der Stromverteilung" (chap. 10, note 46). Robert Ochsenfeld's papers were destroyed during World War II. *Guide to Sources for History of Solid State Physics*, compiled by J. Warnow-Blewett and J. Teichmann, Report no. 7 of International Catalog of Sources for History of Physics and Allied Sciences (draft, October 1987).
29. Crowther, J.G., *The Cavendish Laboratory 1874–1974* (New York, 1974).
30. Andronikashvili, E.L., *Reflections on Liquid Helium* (New York, 1989). Elveter L. Andronikashvili did experimental low-temperature work for his doctorate under Kapitza in Moscow.
31. Mendelssohn (note 13), 41.
32. Gorter, C.J., "Note on the supraconductivity of alloys," *PH*, **2** (1935), 449–452.
33. Ibid., 452.
34. Gorter, C.J., "Superconductivity until 1940" (chap. 8, note 1), 6.
35. London, H., "Phase-equilibrium of supraconductors in a magnetic field," *PRS*, **A152** (1935), 650–663.
36. Ibid., note on 650–651.
37. Everitt, C.W.F., and W.M. Fairbank, "Fritz London," *DSB*, vol. **VIII** (New York, 1973), 473–479, on 473.

38. Everitt, C.W.F., and W.M. Fairbank, "Heinz London," *DSB*, vol. **VIII** (New York, 1973), 479–483, on 479.
39. Shoenberg, D., "Heinz London, 1907–1970," *Biographical Memoirs of the Royal Society*, **17** (1971), 440–461.
40. Everitt and Fairbank, (note 38), 483.
41. Pippard, A.B., in H. Kamerlingh Onnes Symposium on the Origins of Applied Superconductivity–75th Anniversary of the Discovery of Superconductivity at 1986 Applied Superconductivity Conference, Baltimore, MD, 28 September–3 October 1986.
42. Shoenberg (note 39), 455.
43. London, F., and H. London, "The electromagnetic equations of the supraconductor," *PRS*, **A149** (1935), 71–88, on 71. Communicated by Lindemann to the Royal Society on 24 October 1934.
44. London, H., "An experimental examination of the electrostatic behavior of supraconductors," *PRS*, **155** (1936), 102–110, on 110.
45. London, F., and H. London, "The electromagnetic equations" (note 43), 88.
46. London, H., (note 35).
47. Ibid., 658.
48. Berlincourt, T.G., "Type II superconductivity," *RMP*, **36** (1964), 19–26, on 19–20.
49. Mendelssohn, K., "On different types of superconductivity," *RMP*, **36** (1964) 50–51, on 50.
50. Mendelssohn, K., and J.R. Moore, "Experiments on supraconductive tantalum," *PM*, **21** (1936), 532–544.
51. Ibid., 543.
52. Berlincourt (note 11), 404.
53. Daunt, J.G., and K. Mendelssohn, "Equilibrium curve and entropy difference between the supraconductive and the normal state in Pb, Hg, Sn, Ta, and Nb," *PRS*, **A160** (1937), 127–136.
54. Schubnikow, L.W., W.I. Chotkewitsch, J.D. Schepelew, and J.N. Rjabinin, "Magetische Eigenschaften supraleitender Metalle und Legierungen, *PZS*, **10** (1936), 165–192.
55. Ibid., 189.
56. Ibid., 190.
57. Berlincourt, T.G. (note 11).
58. Balabekyan, O.I., "Lev Vasil'evich Shubnikov," *Soviet Physics Uspekhi*, **9** (1966), 455–459, on 455; translated from *Usp. Fiz. Nauk.*, **89** (1966), 321–325.
59. Mendelssohn to Berlincourt, 1963, quoted in Berlincourt (note 11), 405.

Notes to Chapter 12

1. Misener, A.D., H.G. Smith, and J.O. Wilhelm, "Effect of magnetic fields on the supraconductivity of thin films of tin," Royal Society of Canada, *Transactions*, **29** (1934), 13–21; A.D. Misener, "Magnetic effects and current sensitivity of superconducting films," *Canadian Journal of Research*, **14** (1936), 25–37.
2. Pontius, R.B., "Supraconductors of small dimensions," *NA*, **139** (1937), 1065–1066; "Threshold values of supraconductors of small dimensions," *PM*, **24** (1937), 787–796.
3. Shalnikov, A., "Superconducting thin films," *NA*, **142** (1938), 74.
4. Appleyard, E.T.S., and A.D. Misener, "Superconductivity of thin films of mercury," *NA*, **142** (1938), 474; Appleyard, E.T.S., J.R. Bristow, and H. London,

"Variation of field penetration with temperature in a superconductor," *NA*, **143** (1939), 433–434.

5. Appleyard, E.T.S., J.R. Bristow, H. London, and A.D. Misener, "Superconductivity of thin films. I. Mercury," *PRS*, **A172** (1939), 540–558, on 541–542.
6. Appleyard, Bristow, and London (note 4), 434.
7. Quoted in Appleyard et al. (note 5), 558, apparently referring to an early edition of Shoenberg's famous book.
8. Laue, M. von, "Zur thermodynamik der Supraleitung," *AP*, **32** (1938), 253–258.
9. Shoenberg, D., "Superconducting colloidal mercury," *NA*, **143** (1939), 434–435; "Properties of superconducting colloids and emulsions," *PRS*, **A175** (1940), 49–70. In one technique, Shoenberg laboriously ground mercury in a mortar with lard used to disperse the droplets.
10. London, F., "Zur Theorie magnetischer Felder im Supraleiter," *PH*, **3** (1936), 450–462; Shoenberg (chap. 3, note 29), 234.
11. Gorter, C.J., and H.B.G. Casimir (chap. 10, note 85).
12. Shoenberg (chap. 3, note 29), on 197; Daunt, J.G., A.R. Miller, A.B. Pippard, and D. Shoenberg, "Temperature dependence of penetration depth of a magnetic field in superconductors," *PR*, **74** (1948), 842.
13. Pippard, A.B., "Early superconductivity research (except Leiden)," *TM*, **MAG-23** (1987), 372.
14. Daunt, J.G., "The magnetic threshold curves of superconductors," *PR*, **72** (1947), 89–90.
15. Daunt, J.G., A.R. Miller, A.B. Pippard, and D. Shoenberg, "Temperature dependence ..." (note 12).
16. Casimir, H.B.G., "On the variation with temperature of the surface layer of superconducting mercury," *PH*, **7** (1940), 887–896.
17. Sommerfeld, A. (chap. 10, note 90).
18. Discussion following Keesom (chap. 10, note 83), 944.
19. Keesom (chap. 10, note 83), 943.
20. Kok, J.A., "Supraconductivity and Fermi–Dirac statistics," *NA*, **134** (1934), 532–533; "Some remarks on supraconductivity and Fermi–Dirac statistics," *PH*, **1** (1934), 1103–1106.
21. Gorter (chap. 8, note 1), 4.
22. Shoenberg (chap. 3, note 29), 64.
23. Daunt, J.G., and K. Mendelssohn, "Equilibrium curve and entropy difference between the supraconductive and the normal state in Pb, Hg, Sn, Ta, and Nb," *PRS*, **A160** (1937), 127–136.
24. Ibid., 133–134.
25. Sommerfeld, A., "Zur spezifischen Wärme der Metallelektronen," *AP*, **28** (1937), 1–10.
26. Daunt and Mendelssohn, "Equilibrium curve" (note 23), 133.
27. Mendelssohn to Sommerfeld, 8 September 1937, SON.
28. Sommerfeld (note 25), 6.
29. Mendelssohn (chap. 7, note 71), 11.
30. Daunt, J.G., and K. Mendelssohn, "Thomson effect of supraconductive lead," *NA*, **141** (1938), 116.
31. Daunt, J.G., and K. Mendelssohn, "An experiment on the mechanism of superconductivity," *PRS*, **A185** (1946), 225–239.
32. Ibid., 239.
33. London, F., "Macroscopical interpretation of supraconductivity," *PRS*, **A152** (1935), 31.

34. Welker, H., "Ueber ein elektronentheoretisches Modell des Supraleiters," *Zeitschrift für technische Physik*, **19** (1938), 606–611.
35. Ibid., 611.
36. Daunt, J.G., T.C. Keeley, and K. Mendelssohn, "Absorption of infrared light in supraconductors," *PM*, **23** (1937), 264–271.
37. Ibid., 270.
38. London, H., "The high-frequency resistance of superconducting tin," *PRS*, **176** (1940), 522–533.
39. Ibid., 532.
40. Ibid.
41. Serin (chap. 9, note 15), 248.
42. Keesom, W.H., and P.H. van Laer, "Measurements of the atomic heats of tin in the superconductive and in the non-superconductive state," *PH*, **5** (1938), 193–201 [*PLC*, **252b**].
43. Keesom, W.H., and M. Desirant, "The specific heats of tantalum in the normal and in the superconductive state," *PH*, **8** (1941), 273–288 [*PLC*, **275b**].
44. Serin (chap. 9, note 15), 232.
45. Brown, A., M.W. Zemansky, and H.A. Boorse, "Behavior of the heat capacity of superconducting niobium below 4.5 °K," *PR*, **86** (1952), 135; "The superconducting and normal heat capacities of niobium," *PR*, **92** (1953), 52, 57.
46. Aside from appearing to rule out a complicated relationship then in vogue, proposed in 1947 by H. Koppe on the basis of Heisenberg's contemporary theory of superconductivity.
47. Corak, W.B., B.B. Goodman, C.B. Satterthwaite, and A. Wexler, "Exponential temperature dependence of the electronic specific heat of superconducting vanadium," *PR*, **96** (1954), 1442–1444.
48. Bardeen, J., "Development of concepts in superconductivity," *Proceedings of Eighth International Conference on Low Temperature Physics*, ed. R.O. Davies (Washington, D.C., 1963), 3–8.
49. Corak et al., "Exponential temperature" (note 47), 1444. The niobium data referred to here is that of Brown et al. The tin measurements refer to measurements taken the previous year by Goodman on the thermal conductivity of superconducting tin. The electronic contribution to the thermal conductivity of the superconductor is related to its electronic specific heat; it too was found by Goodman to vary in an exponential manner well represented by Eq. (12-7). Goodman, B.B., "The thermal conductivity of superconducting tin below 1 °K," *PRS*, **A66** (1953), 217–227.
50. Pippard (note 13), 371.
51. Bardeen (note 48), 3.
52. London, F., *Superfluids*, vol. 1 (New York, 2nd ed., 1961).
53. Ibid., 145, 150.
54. Ibid., 152. The earliest prediction of flux quantization is in F. London's "On the problem of the molecular theory of superconductivity," *PR*, **74** (1948), 570.
55. Pippard, "The surface impedance of superconductors and normal metals at high frequencies. I. Resistance of superconducting tin and mercury at 1200 Mcyc./sec.," *PRS*, **A191** (1947), 370–384; "III. The relation between impedance and superconducting penetration depth," 399–415.
56. Pippard (note 13), 372.
57. Pippard, "The surface impedance of superconductors and normal metals at high frequencies. II. The anomalous skin effect in normal metals," *PRS*, **A191** (1947), 385–399.

58. Ibid. The difference between λ' and λ was found to be much more pronounced at 9400 Mc, as expected since the effect of the normal state should be correspondingly enhanced at the higher frequency. Pippard, "The surface impedance of superconductors and normal metals at high frequencies. IV. Impedance at 9400 Mc./sec. of single crystals of normal and superconducting tin," *PRS*, **A203** (1950), 98–118.
59. H. London to Pippard, private communications; Pippard (note 13), 373.
60. Ibid. Pippard's recollection is that the seed for this investigation was planted in a New York hotel on his way home from the conference, where he pondered the problems associated with a sharp transition at T_c.
61. Pippard, "Field variations of the superconducting penetration depth," *PRS*, **A203** (1950), 210–223.
62. Ibid., 219.
63. Ibid., 220.
64. Pippard (note 13), 373.
65. Pippard, "An experimental and theoretical study of the relation between magnetic field and current in a superconductor," *PRS*, **A216** (1953), 547–568.
66. Pippard, "Long range order in superconductors," in *Proceedings of International Conference on Low Temperature Physics*, ed. R. Bowers (Oxford, 1951), 118.
67. Ibid.
68. Pippard (note 65).
69. Reuter, G.E.H., and E.H. Sondheimer, "The theory of the anomalous skin effects in metals," *PRS*, **A195** (1948), 336.
70. Pippard (note 65).
71. Pippard (note 13), 372.
72. Ibid., 373; Pippard, Ph.D. thesis (University of Cambridge, 1949).
73. Ginzburg, V.L., and L.D. Landau, "On the theory of superconductivity," *Zhurnal Eksperimental'noi i Teoreticheskoi Fiziki*, **20** (1950), 1064–1082 [in Russian].
74. Quoted in Berlincourt (chap. 11, note 11), 406.
75. Pippard (note 13), 373.
76. Pippard, A.B., "The historical context of Josephson's discovery," *Superconductor Applications: Squids and Machines*, ed. by B.B. Schwartz and S. Foner (New York and London, 1976), 8.
77. Zavaritskii, N.V., "Investigation of superconducting properties of thallium and tin films deposited at low temperatures," *Doklady Akademii Nauk, SSSR*, **86** (1952), 501–504 [in Russian].
78. Abrikosov, A.A., "Fritz London Award Address," *Low Temperature Physics-LT-13*, ed. by K.D. Timmerhaus, W.J. O'Sullivan, and E.F. Hammel (New York and London, 1974), 2.
79. Ibid., 2.
80. Abrikosov, A.A., "Influence of size on the critical field of superconductors of the second group," *Doklady Akademii Nauk, SSSR*, **86** (1952), 489–492 [in Russian].
81. Abrikosov (note 78), 2.
82. The concept of superfluidity, it may be recalled, had its origins in Onnes's observation, in 1911, of a density maximum in liquid helium slightly above 2 °K, confirmed by J.D.A. Boks in 1924 as a sharp maximum with a discontinuity in slope. In the last year of his life (1926), on the basis of ongoing Leiden measurements on the surface tension, latent heat, and specific heats, Onnes speculated on the existence of two states of liquid helium—states termed helium I and helium II by Keesom and M. Wolfke the following year. In 1930 Keesom and van den Ende uncovered an enormous drop in viscosity below 2.19 °K—a property subsequently found to be more complicated than simply a matter of magnitude at Toronto and

Leiden. In 1932 Keesom, with K. Clusius, rediscovered the sharp specific heat maximum at the helium I-II transition point, a point named, on Ehrenfest's suggestion, the "lambda point" by Keesom and his daughter, Anna Petronella, in their continued pursuit of the subject. The years 1936–1937 brought new surprises: the Keesoms, with J. Allen, Peierls and Z. Uddim hard on their heels at Cambridge, found not only an enormous thermal conductivity in helium II, but it increased with decreasing temperature gradient—unconventional behavior at its best! Equally unconventional proved to be the viscosity of helium II, revisited by Kapitza in Moscow and Allen and D. Misener in Cambridge in early 1938. Below the lambda point it fell to an immeasurably small value and became essentially independent of pressure gradient—highly nonclassical behavior for which Kapitza coined the term *superfluidity*.

83. Landau, L.D., "The theory of superfluidity of helium II," *Journal of Physics (USSR)*, **5** (1941), 71–90; reprinted in Galasiewicz, Z.M., *Helium 4* (Oxford, 1971).
84. Onsager, L., Remark at a Low Temperature Physics Conference on Shelter Island, New York, in 1948, published in *Nuovo Cimento*, supplement, **6** (1949), 249; Feynman, R., and M. Cohen, "Energy spectrum of the excitations in liquid helium," *PR*, **102** (1956), 1189–1204.
85. Ibid., 4.
86. Ibid., 5.
87. Ibid., 4.
88. Abrikosov, A.A., "On the magnetic properties of superconductors of the second group," *Soviet Physics JETP*, **5** (1957), 1174–1182 [English translation from *Zhurnal eksperimental'noi i teoreticheskoi fiziki*, **32** (1957), 1442–1452].
89. Abrikosov (note 78), 5. Pippard credits B.B. Goodman for first drawing attention in the West to Abrikosov's paper at the IBM conference in 1961. Pippard (note 76), 9; Goodman, B.B., "The magnetic behavior of superconductors of negative surface energy," *IBM Journal of Research and Development*, **6** (1962), 63.
90. The apparent similarities between the intermediate state in a type I superconductor and the mixed state in a type II superconductor are superficial at best. In the former state, the size of the macroscopic lamellae (sometimes a fraction of a millimeter) is determined by the interplay between the magnetic energy that increases as the size of the lamellae increases and thus favors small lamellae, and the positive interface surface energy that decreases as the size of the lamellae increases. In the latter state, where the supernormal interface is negative, it is energetically favorable for the regions of flux penetration to be as small as possible. The size of the flux tubes is typically 0.01 to 0.1 μm. Their minimum size is limited only by a fundamental quantum constraint, according to which each tube contains exactly one quantum of magnetic flux, $\phi_0 = hc/2e = 2.07 \times 10^{-15}$ Wb.
91. Reynolds, C.A., B. Serin, W.H. Wright, and L.B. Nesbitt, "Superconductivity of isotopes of mercury," *PR*, **78** (1950), 487; Maxwell, E., "Isotope effect in the superconductivity of mercury," *PR*, **78** (1950), 477.
92. Fröhlich, H., "Theory of the superconducting state. I. The ground state at the absolute zero of temperature," *PR*, **79** (1950), 845–856.
93. Fröhlich, H., "Isotope effect in superconductivity," *PRS*, **63A** (1950), 778.
94. Huby, R., "Physics at Liverpool," *NA*, **166** (1950), 552–555.
95. Pippard (note 13), 374.
96. Allen, W.D., R.H. Dawton, J.M. Lock, A.B. Pippard, and D.H. Shoenberg, "Superconductivity of tin isotopes," *NA*, **166** (1950), 1071; Allen, W.D., R.H. Dawton, M. Bär, K. Mendelssohn, and J.L. Olsen, ibid., 1071–1072.

97. Bardeen (note 48), 5.
98. Bloch, F., "Heisenberg and the early days of quantum mechanics," *PT*, **29** (December, 1976), 23–27; Allen, P.B., and W.H. Butler, "Electrical conduction in metals," *PT*, **31** (December, 1978), 44–49.
99. Ibid., 5.
100. Ibid., 6.
101. Cooper, L.N., "Origins of the theory of superconductivity," *TM*, **MAG-23** (1987), 376–379.
102. Ibid., 377.
103. Ibid.
104. Schrieffer, J.R., "Macroscopic quantum phenomena from pairing in superconductors," in *les Prix Nobel, en 1972* (Nobel Foundation, 1972); reprinted in *PT*, **26** (July, 1973), 23–20.
105. Bardeen, J., L.N. Cooper, and J.R. Schrieffer, "Microscopic theory of superconductivity," *PR* (letter), **106** (1957), 162–164.
106. Cooper (note 101), 378.
107. Bardeen, J., L.N. Cooper, and J.R. Schrieffer, "Theory of superconductivity," *PR*, **108** (1957), 1175–1204. A succinct summary of the theoretical state of the art in superconductivity in the period just prior to the advent of the mature BCS formulation is found in the proceedings of the Symposium on the Many Body Problem held in 1957 at Stevens Institute of Technology.
108. Hulm, J.K., J.E. Kunzler, and B.T. Matthias, "The road to superconducting materials," *PT*, **34**, no. 1 (1981), 34–43.
109. Shoenberg (chap. 3, note 29). The first edition of this venerable monograph was written in Russian during Shoenberg's Moscow tenure in the late 1930s.
110. Ibid., 40.
111. Hulm, J.K., et al. (note 108), 37. Ferroelectrics are materials with a permanent electrical polarity within a certain temperature range. Hulm's thesis was a dual one on ferroelectrics and on the thermal conductivity of superconductors (the latter topic partially on the suggestion of Brian Pippard, fellow graduate student under Shoenberg at the Mond).
112. Aschermann, G., E. Friederich, E. Justi, and J. Kramer, "Supraleitfähige Verbindungen mit extrem hohen Sprungtemperaturen (NbH und NbN)," *PZ*, **42** (1941), 349–360.
113. Geballe, T.H., B.T. Matthias, G.W. Hull, Jr., and E. Corenzwit, "Absence of an isotope effect in superconducting rutheneum," *PRL*, **6** (1961), 275–277; Matthias, B.T., T.H. Geballe, and V.B. Compton, "Superconductivity," *RMP*, **35** (1963), 1–22; Matthias, B.T., "Superconductivity and its relation to transition elements," *Proceedings of the Eighth International Conference on Low Temperature Physics*, ed. R.O. Davis (Washington, D.C., 1963), 135–138.
114. Hulm, J.K., et al. (note 108), 37.
115. Yntema, G.B.. "Niobium superconducting magnets," *TM*, **MAG-23** (1987), 390–395, on 390.
116. Yntema, G.B., "Superconducting winding for electromagnet," *PR*, **98** (1955), 1197.
117. Wilson, M.N., *Superconducting Magnets* (Oxford, 1983).
118. Autler, S.H., "Superconducting electromagnetics," *Review of Scientific Instruments*, **31** (1960), 369–373.
119. Berlincourt, T.G., "Emergence of Nb-Ti as supermagnet material," *CR*, **27** (1987), 283–289.
120. Kunzler, J.E., E. Buehler, F.S.L. Hsu, B.T. Matthias, and C. Wahl, "Production of

magnetic fields exceeding 15 kilogauss by a superconducting solenoid," *JAP*, **32** (1961), 325–326.
121. Kunzler, J.E., "Recollections of events" (chap. 7, note 77).
122. Kunzler, J.E., E. Buehler, F.S.L. Hsu, and J.H. Wernick, "Superconductivity in Nb_3Sn at high current density in a magnetic field of 88 kGauss," *PRL*, **6** (1961), 89–91; Kunzler, J.E., "Recollection of events" (chap. 7, note 77), 398.
123. Hulm, J.K., "Superconductivity research in the good old days," *TM*, **MAG-19** (1983), 161–166, on 165.
124. Kunzler, J.E., "Superconductivity in high magnetic fields at high current densities," *Proceedings International Conference on High Magnetic Fields*, ed. H. Kolm, B. Lax, F. Bitter, and R. Mills (New York and London, 1962).
125. Bozorth, R.M., H.J. Williams, and D.D. Davis, "Critical field for superconductivity in niobium-tin," *PRL*, **5** (1960), 148.
126. Hulm, Kunzler, and Matthias (note 108), 43.
127. Morin, F.J., J.P. Maita, H.J. Williams, R.C. Sherwood, J.H. Wernick, and J.E. Kunzler, "Heat capacity evidence for a large degree of superconductivity in V_3Ga in high magnetic fields," *PRL*, **8** (1962), 275.
128. Goodman, B.B., "The magnetic behavior" (note 89); Gorkov, L.P., "Theory of superconducting alloys in a strong magnetic field near the critical temperature," *Soviet Physics JETP*, **10** (1960), 998–1004.
129. Anderson, P.W., "Theory of flux creep in hard superconductors," *PRL*, **9** (1962), 309–311; Kim, Y.B., C.F. Hempstead, and A.R. Strnad, "Critical persistent currents in hard superconductors," *PRL*, **9** (1962), 306–309; "Flux creep in hard superconductors," *PR*, **131** (1963), 2486–2495.
130. Wilson, M.N., C.R. Walters, J.D. Levin, and P.F. Smith, "Experimental and theoretical studies of filamentary superconducting composites," *British Journal of Applied Physics D*, **3** (1970), 1515–1585.
131. Stekly, Z.J.J. and J.L. Zar, "Stable superconducting coils," *TNS*, **23** (1965), 367–372; Laverick, C., "The performance of large superconducting coils," *Advances in Cryogenic Engineering*, vol. II (New York, 1965).
132. Wilson et al. (note 130).

Notes to Chapter 13

1. A cursory summary of a few prototype devices is nevertheless in order to provide some sense of the breadth of industrial involvement, of the interrelationship between government, industry, and academia, and of the international scope of applied superconductivity.

   Representative of superconducting electrical machinery is the 3250-hp Fawley homopolar motor built by the International Research and Development Co., Ltd. and tested under full load at the Fawley power station in 1971. It was followed by a worldwide succession of prototype synchronous machines developed by most of the electrical corporate giants. The Massachusetts Institute of Technology has been a leading force in this effort. It peaked out when a joint Electric Power Research Institute–Westinghouse project organized to construct a 300-MVA machine for TVA was abandoned in 1984. Magnetohydrodynamic (MHD) generators for the direct conversion of thermal energy to electrical energy must rely on superconducting magnets if they are to produce more power than they consume. One such (dipole) magnet was constructed at Argonne National Laboratory in the mid-1970s and flown to the USSR for successful deployment in a collaborative MHD program at the High Temperature Institute in Moscow.

Superconducting magnets will be essential for plasma confinement in controlled thermonuclear reactors, because of the need for very high fields over large volumes. A major engineering achievement in this area was the operation of the coil system for Lawrence Livermore Laboratory's Mirror Fusion Test Facility. Employing 42 magnets of various configurations, including a pair of "yin yang" mirror field coils (in which the current path resembles the seam in a baseball) and weighing over 1200 tons in all, the system underwent final tests in 1986. Not to be outdone is the Large Coil Task project centered at Oak Ridge National Laboratory. It employs 6 very large D-shaped coils in a toroidal array. Three were constructed in the U.S. (General Dynamics, General Electric, and Westinghouse), two in Europe (Euratom, Swiss Institute for Nuclear Research/Brown Boveri), and one came from Japan (Japan Atomic Research Institute/Hitachi). Each coil was funded by the Government of the country where it was produced.

Worthy of note is Brookhaven National Laboratory's 150-m long superconducting power transmission line, which reached full power in 1982. It is the only facility of its kind designed to test superconducting cables with simultaneous voltage and current excitation at the 1000-MVA level.

Finally, there is magnetically levitated rail transportation, long under on again-off again development in Japan and Germany, and splendidly represented by Japanese National Railway's ML 500 test vehicle, which holds the world rail speed record of 577 km h^{-1}. As if this were not enough, another mode of transportation exploiting superconductivity (pioneered by the United States Navy) was ushered in with much fanfare in the launching in 1990 of Japan's Yamato I, a 280-ton 30-m long ship using superconducting thrusters developed by a joint venture of Mitsubishi, Toshiba, and Kobe Steel.

Birmingham, B.W., and C.N. Smith, "A survey of large scale applications of superconductivity in the US," *CR*, **16** (1976), 59–71; Komarek, P., "Superconducting magnets in the world of energy, especially in fusion power," *CR*, **16** (1976), 131–142; Hulm, J.K., and C. Laverick, "International cooperative-collaborative perspectives, superconductive science & technology," *TM*, **MAG-23** (1987), 423–426.

2. Reardon, P.J., "High energy physics and applied superconductivity," *TM*, **MAG-13** (1977), 704–718. An excellent summary of the state of superconductivity for HEP ca. 1977.
3. Purcell, J.R., 1968 summer study on superconducting devices and accelerators, *Proceedings*, ed. J.P. Blewett, H. Hahn, and A.G. Prodell (Brookhaven National Laboratory, report 50155, Upton, New York, 1969). The 1968 3-week summer study at BNL, organized by John Blewett, was an important forum for airing state-of-the-art concepts in applied superconductivity. A technical highlight of the study was the concept of intrinsic stability introduced by P.F. Smith and his colleagues at the Rutherford Laboratory.
4. Purcell, J.R., and H. Desportes, "The NAL bubble chamber magnet," *Proceedings of 1972 Applied Superconductivity Conference*, (Annapolis, MD, 1–3 Mar. 1972), 246–249. Giger, U., P. Pagani, and C. Trepp, "The low temperature plant for the big European bubble chamber BEBC," *CR*, **11** (1971), 451–455.
5. Wilson, M.N. (chap. 12, note 117).
6. Allinger, J., et al., "8° superconducting bending magnet in a primary proton beam," International Conference on High Energy Accelerators, IX, *Proceedings* (Stanford, 1974); Allinger, J., G. Danby, and J.J. Jackson, "High field supercon-

ducting magnets for accelerators and particle beams," *TM*, **MAG-11** (1975), 463–466.

7. Reardon (note 2), 708–710.
8. Aggus, J., et al., "Superconducting magnet system for the AGS high energy unseparated beam," *TNS*, **NS-22** (1975), 1164–1167.
9. Adams, J.B., "The European 300 GeV programme," Conference on High Energy Accelerators, *Proceedings*, ed. M.H. Blewett and N. Vogt-Nilson (Geneva, 1971), 29.
10. Turowski, P., J.H. Coupland, and J. Perot, "Pulsed superconducting dipole magnets of the GESSS collaboration," International Conference on High Energy Accelerators, IX, *Proceedings* (Stanford, 1974), 174–178.
11. Avery, R., T. Elioff, A. Garren, W. Gilbert, M. Green, H. Grunder, E. Hartwig, D. Hopkins, G. Lambertson, E. Lofgren, K. Lou, R. Meuser, R. Peters, L. Smith, J. Staples, R. Thomas, and R. Wolgast, *Experimental Superconducting Accelerator Ring (ESCAR)*, (Lawrence Berkeley Laboratory, report 2603, May 1974); Ninth International Conference on High Energy Accelerators, *Proceedings* (Stanford, 1975), 179–183.
12. Gilbert, W.S., R.B. Meuser, W.L. Pope, and M.A. Green, "ESCAR superconducting magnet system," *TNS*, **NS-22** (1975), 1129–1132.
13. Reardon (note 2), 710.
14. A common criticism of ESCAR concerned its relatively modest ring size: As a test vehicle of the new generation of superconducting accelerators (both ISABELLE and the Fermilab Saver/Doubler were in advanced stages of design by mid-1970), but five orders of magnitude smaller in circumference, ESCAR could quite possibly give rise to more irrelevant machine-operational problems and issues than it could clarify. Moreover, being a cold-bore machine it seemed likely to have all kinds of operational problems because particles scattering off the walls of the beam tube could cause vacuum instabilities. LBL's Glen Lambertson retorted that one of the functions of ESCAR was precisely to investigate such problems in an accelerator small enough that major modifications would be economically feasible. Metz, W.D., "Two superconducting accelerators: Physics spurs technology," *SC*, **200** (1978), 188–191.
15. Lambertson, G., A. Borden, J. Cox, W. Eaton, W. Gilbert, J. Holl, E. Knight, R. Main, R. Meuser, J. Rechen, R. Schafer, F. Toby, and F. Voelker, *Design, Construction, and Operation of 12 ESCAR Bending Magnets* (Lawrence Berkeley Laboratory, report 8210, September, 1978).
16. Lambertson, G.R., W.S. Gilbert, and J.B. Rechen, *Final Report on the Experimental Superconducting Synchrotron (ESCAR)* (Lawrence Berkeley Laboratory, report 8211, March 1979).
17. *Report of the Panel on High Energy Accelerator Physics of the General Advisory Committee to the Atomic Energy Commission and the President's Science Advisory Committee*, TID-18636, 26 April 1963.
18. *Design Study of a Cold Magnet Synchrotron (CMS)*, ed. A. van Steenbergen (Brookhaven National Laboratory, report 15430, 1970).
19. *200-GeV Intersecting Storage Accelerators ISABELLE, A Preliminary Design Study*, ed. J.P. Blewett and H. Hahn (Brookhaven National Laboratory, report 16716, 1972). John Blewett named the facility after his sailboat.
20. *A Proposal for Construction of a Proton–Proton Storage Accelerator Facility ISABELLE*, ed. H. Hahn and M. Plotkin (Brookhaven National Laboratory, report 18891, 1974; revised report 20161, 1975).
21. Luminosity is a term used in specifying the probability of interaction of particle

beams in collision. The higher the luminosity the greater the rate of collisions. It is expressed in cm^{-2} sec^{-1}.

22. Dahl, P.F., *Superconducting Devices at Brookhaven National Laboratory* (Brookhaven National Laboratory, report 50498R, 1978).
23. The importance of intrinsic stability was first impressed on the American superconducting magnet builders by P.F. Smith and colleagues of the Rutherford Laboratory during Brookhaven's highly successful 1968 summer study, as noted earlier (note 3).
24. McInturff, A.D., P.F. Dahl, and W.B. Sampson, "Pulsed field losses in metal-filled superconducting multifilamentary braids," *JAP*, **43** (1972), 3546–3551.
25. ERDA, High Energy Physics Advisory Panel, Subpanel on New Facilities, *Report* (Washington, D.C., June 1977).
26. Chasman, R., and R.L. Gluckstern, "Extrapolation of the ISABELLE design to 400x400 GeV" (Brookhaven National Laboratory, report 19062, June 1974).
27. *Report of the Subpanel on New Facilities of the High Energy Physics Advisory Panel to the Atomic Energy Commission* (U.S. Atomic Energy Commission, July 1974).
28. Ibid., 19.
29. Van Steenbergen, A., "Design of a 400x400 GeV version of ISABELLE," International Conference on High Energy Accelerators, *Proceedings* (Protvino, USSR, July 1977), 150–159. Technical issues of concern included the so-called transition energy of the large 400-GeV ring and an associated beam instability.
30. Ibid., 156.
31. Hoddeson, L., "The first large-scale application of superconductivity: The Fermilab Energy Doubler, 1972–1983," *HSPBS*, **18** (1987), 25–54. The following account of the Fermilab program relies largely on Lillian Hoddeson's excellent article.
32. Westfall, C.L., "The site contest for Fermilab," *PT*, **42** (1989), 44–52.
33. R.R. Wilson to NAL Users, 8 December 1972, quoted in Hoddeson (note 31), 32.
34. Quoted in Hoddeson (note 31), 33. The first published reference to a Doubler appears in R.R. Wilson's "Future options at NAL Batavia," International Conference on High Energy Accelerators, *Proceedings* (Yerevan, USSR, 1970), 103–105.
35. Sheldon, R., and B. Strauss, "0.5 meter prototype energy doubler-quadrupole magnet," July 1971 (FN-235, Fermilab History Collection).
36. G.H. Morgan to author, 12 February 1991.
37. Hoddeson (note 31), 42, quoting J. Richie Orr.
38. Wilson to Secretary of Energy J.R. Schlesinger, 22 October 1977, quoted in Hoddeson (note 31), 44.
39. Hoddeson (note 31), 45, quoting P. Livdahl.
40. *Fermilab Proposals*, vol. 486–492 and vol. 493–515.
41. Necessitated by the addition of correction magnets to the dipoles.
42. Hoddeson (note 31), 48.
43. Dahl, P.F., *The ISABELLE Magnets: A Brief Description* (Brookhaven National Laboratory, report 51508, 1982). Leon Lederman first suggested using Fermilab magnets for ISABELLE to the chairman of the Committee on ISABELLE Magnet R&D, Eric Forsyth, during the summer of 1980. Palmer's first suggestion to use the modified Tevatron magnet appears in an internal BNL Report, "Proposal for a test of cable instead of braid in a dipole magnet," dated 23 September 1980. In December of 1980, he received permission to proceed with a 1.8-m model dipole, a stock of four-year-old cable having been located at Fermilab.
44. Bleser, E.J., et al., "Superconducting magnets for the CBA project," *Nuclear Instruments and Methods in Physics Research*, **A235** (1985), 435–463. An alterna-

tive CBA magnet design was devised (and successfully tested) as a possible cost-cutting option in the course of the project. It has subsequently become known as the "two-in-one" approach in the accelerator magnet community. In this design, the adjacent dipole coils that belong to the two counter-rotating particle beams of the same electrical charge are assembled in a common iron yoke and cryostat. The concept was first suggested by J.P. Blewett in 1971. See also "2-in-1 Magnets," *CBA Newsletter No. 2*, ed. N.V. Baggett, P.F. Dahl, and H.L. McNally (Brookhaven National Laboratory, November 1982).

45. Machine projects under way at that time included conversion of CERN's 450-GeV fixed-target Super Proton Synchrotron to a proton–antiproton collider, the previously described Fermilab Doubler (with its expected proton–antiproton upgrade) and Brookhaven's ISABELLE, as well as several electron–positron colliders nearing completion. Projects not yet approved but in the R&D stage included two additional major electron–positron colliders: Japan's Tristan and CERN's Large Electron–Positron collider (LEP).
46. *Report of the 1983 HEPAP Subpanel on New Facilities for the U.S. High Energy Physics Program* (U.S. Department of Energy, Office of Energy Research, report DOE/ER-0169, July 1983).
47. Ibid., 4.
48. J. Sandweiss to A.W. Trivelpiece, 12 July 1983, reproduced in *Report of 1983 HEPAP Subpanel* (note 46). Considerable debate concerning the merits of the SSC soon surfaced in various quarters, scientific and political—a debate spawned by a long-standing undercurrent of opposition to what many members of the scientific community saw as an inordinate expenditure on high-energy facilities out of proportion to the scientific returns. In the context of the SSC the debate found common expression in terms of "big" science vs. "small" or "table-top" science. Spokespersons for small science included prominent members of the solid state and superconductivity communities (in contradistinction to the superconducting magnet subcommunity which, after all, made the SSC possible). Three factors, in particular, fueled the debate: the consistent anti-SSC stance of the editorial page of the *New York Times*, the authority of Nobel laureate P.W. Anderson, and the concomitant promise apparent of the high-temperature superconductors (Chapter 14).
49. Only in proton–proton colliders are 2-in-1 magnets feasible since two apertures are required—one with the field vector pointing up and the other with the vector pointing down. In these magnets the field lines from one aperture are returned, not through the midplane, but across the other magnetic aperture located beside the first. The total weight of iron is about the same as would be required for each of the single magnets; another advantage is the compactness of the design, which renders it practical in cramped accelerator tunnels. Important drawbacks of the scheme are greater complexity in magnet assembly and loss of flexibility in machine operation. See also Bleser, "Superconducting Magnets" (note 44).
50. *Report of the Reference Designs Study Group of the Superconducting Super Collider*, draft II (May 1984).
51. Reardon, P., "Cold iron cos θ magnet option for the SSC," *TNS*, **NS-32** (1985), 3466–3470.
52. Fisk, H.E., et al., "The ironless cos θ magnet option for the SSC," *TNS*, **NS-32** (1985), 3456–3461.
53. Wilson, R.R., "Superferric magnets for 20 TeV," 1982 DPF Summer Study on Elementary Particle Physics and Future Facilities, *Proceedings*, ed. R. Donaldson, R. Gustavson, and F. Paige (Snowmass, 1982), 330–332, on 330 and Huson, R., et al., "20 TeV colliding beam facilities: New, low-cost approaches," 315–321.

54. Allinger, J., G. Danby, B. Devito, S. Hsieh, J. Jackson, and A. Prodell, "Studies of performance and field reproducibility of a precision 40 kG superconducting dipole magnet," *TNS*, **NS-20** (1973), 678–682, on 678.
55. Wilson, "Superferric magnets," and Huson, "20 TeV colliding beam" (note 53).
56. Huson, R., et al., "Superferric magnet option for the SSC," *TNS*, **NS-32** (1985), 3462–3465.
57. Reardon (note 51), 3466.
58. Ibid.
59. *SSC Magnet Selection Advisory Panel Report to the Director of the Central Design Group* (9 September 1985).
60. Ibid., 4–5.
61. Memorandum by Maury Tigner, dated 13 September 1985.
62. Maury Tigner to P.J. Reardon, K. Berkner, and R.A. Lundy, 31 January 1986.
63. *SSC Program FY 86: Laboratory Task Specifications* (SSC Central Design Group, 16 December 1985).
64. Willen, E., et al., *Performance of Six 4.5m SSC Dipole Model Magnets* (Brookhaven National Laboratory, report 38813, 1986).
65. Author's files.
66. F.R. Huson and P.M. McIntyre to W. Wallenmeyer, 7 April 1986.
67. *Report of the HEPAP Subpanel to Review Recent Information on Superferric Magnets* (May 1986).
68. *Conceptual Design of the Superconducting Super Collider*, ed. J.D. Jackson (SSC Report SR-2020, March 1986).
69. *Report of the DOE Review Committee on the Conceptual Design of the Superconducting Super Collider* (DOE/ER-0267, May 1986). Tigner to "all hands" (CDG staff), 19 May 1986, puts it simply: "In brief, we got high marks."

Notes to Chapter 14

1. Gavaler, J.R, "Superconductivity in Nb-Ge films above 22 K," *Applied Physics Letters*, **23** (1973), 480–482.
2. Ibid., 482.
3. Edelsack, E.A., D.U. Gubjer, and S.A. Wolf, "The rocky road to high temperature superconductivity," in *Novel Superconductivity*, ed. S.A. Wolf and V.Z. Kresin (New York, 1987).
4. Müller, K.A., and J.G. Bednorz, "The discovery of a class of high-temperature superconductors," *SC*, **237** (1987), 1133–1139. This provides a good review of the discovery and technical status of the subject as of September 1987.
5. Perovskites (named after the Russian mineralogist Count L.A. von Perovski) are natural minerals with a particular atomic structure; they comprise the Earth's most abundant minerals, and also include approximately 150 synthetic compounds. The mineral perovskite is ideally calcium titanate ($CaTiO_3$), but extensive substitutes exist. Perovskites, principally barium titanate, form the basis of a $20 billion per annum electroceramics industry (e.g., capacitors, voltage-surge protectors, piezoelectric transducers). R.M. Hazen, "Perovskites," *Scientific American*, **258** (June 1988), 74–81.
6. London, F., "Supraconductivity in aromatic compounds," *Journal of Chemical Physics*, **5** (1937), 837–838.
7. Cohen, M.L., "The existence of a superconducting state in semiconductors," *RMP*, **36** (1964), 240–243; Scholey, J.F., W.R. Hosler, and M.L. Cohen, "Superconduc-

tivity in semiconducting $SrTiO_3$," *PRL*, **12** (1964), 474–475. John Hulm obtained some evidence for superconductivity in GeTl.

8. Rusakov, A.P., S.G. Grigoryan, A.V. Omel'chenko, and A.E. Kadyshevich, "Isomorphic phase transitions in CuCl at high pressure," *Soviet Physics JETP*, **45** (1977), 380–384.
9. Müller and Bednorz (note 4).
10. Er-Rakho, L., C. Michel, J. Provost, and B. Raveau, "A series of oxygen-defect perovskites containing Cu^{II} and Cu^{III}: The oxides $La_{3-x}Ln_xBa_3[Cu^{II}_{5-2y}Cu^{III}_{1+2y}]O_{14+y}$," *Journal of Solid State Chemistry*, **37** (1981), 151.
11. Müller and Bednorz (note 4), 1136.
12. Bednorz, J.G., and K.A. Müller, "Possible high T_c superconductivity in the Ba-La-Cu-O system," *ZP*, **64** (1986), 189–193.
13. Tanaka, S., "Research on high-T_c superconductivity in Japan," *PT* (December, 1987), 53–57.
14. Chu, C.W., P.H. Hor, R.L. Meng, L. Gao, Z.J. Huang, and Y.Q. Wang, "Evidence for superconductivity above 40 K in the La-Ba-Cu-O compound system," *PRL*, **58** (1987), 405–407.
15. Cava, R.J., R.B. van Dover, B. Batlogg, and E.A. Rietman, "Bulk superconductivity at 36 K in $La_{1.8}Sr_{0.2}CuO_4$," *PRL*, **58** (1987), 408–410.
16. Pool, R., "Superconductor credits bypass Alabama," *SC*, **241** (1988), 655–657.
17. Ibid., 656.
18. Ibid.
19. Wu, M.K., J.R. Ashburn, C.J. Torng, P.H. Hor, R.L. Meng, L. Gao, Z.J. Huang, Y.Q. Wang, and C.W. Chu, "Superconductivity at 93 K in a new mixed-phase Y-Ba-Cu-O compound system at ambient pressure," *PRL*, **58** (1987), 908–910, on 908 and Hor, P.H., L. Gao, R.L. Meng, Z.J. Huang, Y.Q. Wang, K. Forster, J. Vassilious, and C.W. Chu, "High-pressure study of the new Y-Ba-Cu-O superconducting compound system," 911–912. Robert Hazen, in his account of these events, dwells on two sets of errors in the original typescript, which were only corrected at the 11th hour as the paper went to press. One was the consistent substitution of the symbol Yb (for the element ytterbium) for Y (yttrium), inexplicably igniting a widespread rumor in advance of publication by unknown parties despite the seeming confidentiality of the publication process adhered to by *Physical Review Letters*. A second was a systematic substitution of the numerical coefficient "4" for "1" in the chemical formula for the perovskite. Were the errors deliberate, an attempt by Chu to protect his discovery? According to Hazen, Chu refuses to comment on the episode. Hazen, Robert M., *The Breakthrough: The Race for the Superconductor* (New York, 1988), 58–65.
20. Hazen (note 5).
21. Dahl, P.F., "James Dewar, Walther Nernst, and Heike Kamerlingh Onnes in events leading to superconductivity in 1911"; Baym, G., "Quantum theories of superconductivity, 1929–1933"; Schrieffer, R., "The development of the microscopic theory of superconductivity"; Anderson, P.W., "Its not over 'til the fat lady sings." Abstracts of papers in *Bulletin of the American Physical Society*, **32**, No. 3 (March 1987), 619–620. [The abstract of Dahl's paper appears on 619–620; those of the remaining papers appear on p. 620.]
22. Khurana, A., "Superconductivity seen above the boiling point of nitrogen," *PT* (April 1987), 17–23, on 20. By August 1987, over 400 manuscripts on high-T_c superconductivity had been submitted to the American Physical Society, and more than 300 to *Physical Review Letters* alone. This prompted drastic changes in procedural handling of manuscripts at APS's editorial offices. *High-Temperature Su-*

perconductivity: Reprints from Physical Review Letters and Physical Review B, January–June, 1987 (American Institute of Physics, New York, 1987).

23. Pool, Robert, "New superconductors answer some questions," *SC*, **240** (1988), 146–147, on 146.
24. Pool, Robert, "Feud flares over thallium superconductor," *SC*, **247** (1990), 1029.
25. The depairing critical current density, according to BCS theory, is reached when the kinetic energy of the electron pair is so high that it is energetically unfavorable to maintain the pairing. It is given by

$$J_{cd} = \frac{n_s e \Delta}{P_f}$$

where n_s is the number density of superconducting electron pairs, e is the electronic charge, Δ is the energy gap, and P_f is the Fermi momentum of electrons in the given material. For the oxide materials it has been estimated that J_{cd} is approximately 10^5 A/mm^2 at 77 K. Rabinowitz, M., "The science and technology of superconductivity as applied to the electric power industry," EPRI Workshop on High-Temperature Superconductivity, *Proceedings* (EPRI, Palo Alto, Ca, 1988), 2–3 to 2–32.
26. Ibid., 2–8.
27. Geballe, T.H., and J.K. Hulm, "Superconductivity—the state that came in from the cold," *SC*, **239** (1988), 367–374. Potentially a new class of high-temperature superconductors *not* involving a copper oxide and exhibiting three-dimensional isotropic superconductivity may have been unearthed with R. Cava and B. Batlogg's 1988 report of superconductivity near 30 K in a Ba-K-Bi-O perovskite. Cava, R.J., and B. Batlogg et al., "Superconductivity near 30 K without copper: $Ba_{0.6}K_{0.4}BiO_3$ perovskite," *NA*, **332** (1988), 814–816.
28. Khurana, A., "High-temperature superconductors may not be forever," *PT*, **42** (March, 1989), 17–21. The title refers to the time dependence of the magnetization associated with the high flux creep of the oxide superconductors.
29. Yeshurun, Y., and A.P. Malozemoff, "Giant flux creep and irreversibility in an Y-Ba-Cu-O crystal: An alternative to the superconducting-glass model," *PRL*, **60** (1988), 2202–2205.
30. Pool, R., "Superconductivity: Is the party over?" *SC*, **244** (1989), 914–916; Gammel, P.L., L.F. Schneemeyer, J.V. Waszczak, and D.J. Bishop, "Evidence from mechanical measurements for flux-lattice melting in single crystal $YBa_2Cu_3O_7$ and $Bi_{2.2}Sr_2Ca_{0.8}Cu_2O_8$," *PRL*, **61** (1988), 1666–1669.
31. Gammel, P.L., et al., "Evidence from mechanical measurements" (note 30), 1668–1669.

Select Bibliography

The following bibliography lists many, but not all, books, journal articles, and reports cited in the notes. Thus, references to Leiden papers generally refer to the *Proceedings* of the Amsterdam Akademie der Wetenschappen, not to the less accessible *Communications from the Physical Laboratory of the University of Leiden* and their *Supplements*. Abbreviations are those used in the notes.

Abrikosov, A.A., "Influence of size on the critical field of superconductors of the second group," *Doklady Akademii Nauk, SSSR*, **86** (1952), 489–492 [in Russian].

—, "On the magnetic properties of superconductors of the second group," *Soviet Physics JETP*, **5** (1957), 1174–1182.

—, "Fritz London Award Address," *Low Temperature Physics-LT 13*, ed. K.D. Timmerhaus, W.J. O'Sullivan, and E.F. Hammel (New York and London, 1974).

Adams, J.B., "The European 300 GeV programme," Conference on High Energy Accelerators, *Proceedings*, ed. M.H. Blewett and N. Vogt-Nilson (Geneva, 1971).

Allen, P.B., and W.H. Butler, "Electrical conduction in metals," *PT*, **31** (December, 1978), 44–49.

Allen, W.D., R.H. Dawton, M. Bär, K. Mendelssohn, and J.L. Olsen, "Superconductivity of tin isotopes," *NA*, **166** (1950), 1071–1072.

Allen, W.D., R.H. Dawton, J.M. Lock, A.B. Pippard, and D.H. Shoenberg, "Superconductivity of tin isotopes," *NA*, **166** (1950), 1071.

Allinger, J., G. Danby, B. Devito, S. Hsieh, J. Jackson, and A. Prodell, "Studies of performance and field reproducibility of a precision 40 kG superconducting dipole magnet," *TNS*, **NS-20** (1973), 678–682.

Allinger, J., G. Danby, and J.J. Jackson, "High field superconducting magnets for accelerators and particle beams," *TM*, **MAG-11** (1975), 463–466.

Anderson, P.W., "Theory of flux creep in hard superconductors," *PRL*, **9** (1962), 309–311.

Andronikashvili, E.L., *Reflections on Liquid Helium* (New York, 1989).

Appleyard, E.T.S., and A.D. Misener, "Superconductivity of thin films of mercury," *NA*, **142** (1938), 474.

Appleyard, E.T.S., J.R. Bristow, and H. London, "Variation of field penetration with temperature in a superconductor," *NA*, **143** (1939), 433–434.

Appleyard, E.T.S., J.R. Bristow, H. London, and A.D. Misener, "Superconductivity of thin films. I. Mercury," *PRS*, **A172** (1939), 540–558.

A Proposal for Construction of a Proton-Proton Storage Accelerator Facility ISA-BELLE, ed. H. Hahn and M. Plotkin (Brookhaven National Laboratory, report 18891, 1974; revised report 20161, 1975).

Arms, N., *A Prophet in Two Centuries: The Life of F.E. Simon* (Oxford, 1966).

Aschermann, G., E. Friedrich, E. Justi, and J. Kramer, "Supraleitfähige Verbindungen mit extrem hohen Sprungtemperaturen (NbH und NbN)," *PZ*, **42** (1941), 349–360.

Autler, S.H., "Superconducting electromagnetics," *Review of Scientific Instruments*, **31** (1960), 369–373.

Avery, R., T. Elioff, A. Garren, W. Gilbert, M. Green, H. Grunder, E. Hartwig, D. Hopkins, G. Lambertson, E. Lofgren, K. Lou, R. Meuser, R. Peters, L. Smith, J. Staples, R. Thomas, and R. Wolgast, *Experimental Superconducting Accelerator Ring (ESCAR)*, Ninth International Conference on High Energy Accelerators, *Proceedings* (Stanford, 1974), 179–183.

Balabekyan, O.I., "Lev Vasil'evich Shubnikov," *Soviet Physics Uspekhi*, **9** (1966), 455–459.

Bardeen, J., "Developments of concepts in superconductivity," *PT*, **16** (1963), 19–28.

—, "History of superconductivity research," *Impact of Basic Research on Technology*, ed. B. Kursunoglu and A. Perlmutter (New York, 1973), 15–57.

Bardeen, J., L.N. Cooper, and J.R. Schrieffer, "Microscopic theory of superconductivity," *PR* (letter), **106** (1957), 162–164.

Bardeen, J., L.N. Cooper, and J.R. Schrieffer, "Theory of superconductivity," *PR*, **108** (1957), 1175–1204.

Becker, R., G. Heller, and F. Sauter, "Ueber die Stromverteilung in einer supraleitenden Kugel," *ZP*, **85** (1933), 772–787.

Bednorz, J.G., and K.A. Müller, "Possible high T_c superconductivity in the Ba-La-Cu-O system," *ZP*, **64** (1986), 189–193.

Berlincourt, T.G., "Type II superconductivity," *RMP*, **36** (1964), 19–26.

—, "Type II superconductivity: quest for understanding," *TM*, **MAG-23** (1987), 403–412.

—, "Emergence of Nb-Ti as supermagnet material," *CR*, **27** (1987), 283–289.

Bethe, H., and H. Fröhlich, "Magnetische Wechselwirkung der Metallelektronen. Zur Kritik der Theorie der Supraleitung von Frenkel," *ZP*, **85** (1933), 389–397.

Birkenhead, The Earl of, *The Professor and the Prime Minister: the Official Life of Professor F.A. Lindemann, Viscount Cherwell* (Boston, 1962).

Bleser, E.J., J.G. Cottingham, P.F. Dahl, R.J. Engelmann, R.C. Fernow, M. Garber, A.K. Gosh, C.L. Goodzeit, A.F. Greene, J.C. Herrera, S.A. Kahn, J. Kaugerts, E.R. Kelly, H.G. Kirk, R.J. Leroy, G.H. Morgan, R.B. Palmer, A.G. Prodell, D.C. Rahm, W.B. Sampson, R.P. Shutt, A.J. Stevens, M.J. Tannenbaum, P.A. Thompson, P.J. Wanderer, E.H. Willen, "Superconducting magnets for the CBA project," *Nuclear Instruments and Methods in Physics Research*, **A235** (1985), 435–463.

Bloch, F., "Heisenberg and the early days of quantum mechanics," *PT*, **29** (December, 1976), 23–27.

—, "Memories of electrons in crystals," *PRS*, **371A** (1980), 24–27.

Bohr, N., *Studier over metallernes elektrontheori* (Copenhagen, 1911); reprinted with

English translation in *Niels Bohr, Collected Works*, vol. 1, ed. L. Rosenfeld and J.R. Nielsen (Amsterdam, 1971).

Bozorth, R.M., H.J. Williams, and D.D. Davis, "Critical field for superconductivity in niobium-tin," *PRL*, **5** (1960), 148.

Braunbek, W., "Die Ausbreitung elektromagnetischer Wellen in einem Supraleiter," *ZP*, **87** (1934), 470–483.

Breit, G., "Transients of magnetic field in supraconductors," *RN*, **26** (1923), 529–541.

Bridgman, P.W., "Theoretical considerations on the nature of metallic resistance with especial regard to the pressure effect," *PR*, **9** (1917), 269–289.

—, "The discontinuity of resistance preceding supraconductivity," Washington Academy of Sciences, *Journal*, **11** (1921), 455–459.

—, "Rapport sur les phénomènes de conductibilité dans les métaux et leur explication théorique," *Conductibilité électrique des métaux et problèmes connexes*, rapports et discussion du 4 ème Conseil Solvay, Avril 1924 (Paris, 1927), 67–114.

Brown, A., M.W. Zemansky, and H.A. Boorse, "Behavior of the heat capacity of superconducting niobium below 4.5 °K," *PR*, **86** (1952), 134–135.

Brown, A., M.W. Zemansky, and H.A. Boorse, "The superconducting and normal heat capacities of niobium," *PR*, **92** (1953), 52–58.

Buchwald, J.Z., *From Maxwell to Microphysics: Aspects of Electromagnetic Theory in the Last Quarter of the Nineteenth Century* (Chicago, 1985).

Burton, E.F., J.O. Wilhelm, and F.G.A. Tarr, "Magnetic properties of supraconductors," *NA*, **133** (1934), 684.

Callendar, H.L., "Notes on platinum thermometry," *PM*, **47** (1899), 191–222.

Casimir, H.B.G., "On the variation with temperature of the surface layer of superconducting mercury," *PH*, **7** (1940), 887–896.

—, "Superconductivity and superfluidity," in *The Physicist's Conception of Nature*, ed. Jagdish Mehra (Amsterdam and Boston, 1973).

—, "Superconductivity" in *History of Twentieth Century Physics*, Proceedings of the International School of Physics "Enrico Fermi," ed. C. Weiner (New York, 1977).

—, "Development of solid-state physics," in *History of Twentieth Century Physics*, Proceedings of the International School of Physics "Enrico Fermi," ed. C. Weiner (New York, 1977).

—, *Haphazard Reality: Half a Century of Science* (New York and London, 1983).

Cava, R.J., R.B. van Dover, B. Batlogg, and E.A. Rietman, "Bulk superconductivity at 36K in $La_{1.8}Sr_{0.2}CuO_4$," *PRL*, **58** (1987), 408–410.

Chalmers, T.W., "Conduction of electricity through gases," Chapter 10 in *Historic Researches: Chapters in the History of Physical and Chemical Discovery* (New York, 1952).

Cherwell, Viscount, and F. Simon, "Walther Nernst," Royal Society of London, *Obituary Notices*, 4 Nov. 1942, 100–111.

Chu, C.W., P.H. Hor, R.L. Meng, L. Gao, Z.J. Huang, and Y.Q. Wang, "Evidence for superconductivity above 40 K in the La-Ba-Cu-O compound system," *PRL*, **58** (1987), 405–407.

Clausius, R., "Ueber die Zunahme des elektrischen Leitungswiderstandes der einfachen Metalle mit der Temperatur," *APC*, **104** (1858), 650–651.

Clusius, K., "Zwei Vorlesungsversuche mit flüssigem Wasserstoff," *PZ*, **35** (1934), 929–930.

Cohen, E., "Kamerlingh Onnes memorial lecture," Chemical Society, *Journal*, **1** (1927), 1193–1209.

Cohen, M.L., "The existence of a superconducting state in semiconductors," *RMP*, **36** (1964), 240–243.

Conceptual Design of the Superconducting Super Collider, ed. J.D. Jackson (SSC report SR-2020, March 1986).

Cooper, L.N., "Origins of the theory of superconductivity," *TM*, **MAG-23** (1987), 376–379.

Corak, W.B., B.B. Goodman, C.B. Satterthwaite, and A. Wexler, "Exponential temperature dependence of the electronic specific heat of superconducting vanadium," *PR*, **96** (1954), 1442–1444.

Crawford, E., *The Beginnings of the Nobel Institution: The Science Prizes, 1901–1915* (Cambridge, 1984).

Crowther, J.G., *The Cavendish Laboratory 1874–1974* (New York, 1974).

Dahl, P.F., *Superconducting Devices at Brookhaven National Laboratory* (Brookhaven National Laboratory, report 50498R, 1978).

—, "Kamerlingh Onnes and the discovery of superconductivity: The Leyden years 1911–1914," *HSPS*, **15**, no. 1 (1984), 1–37.

—, "Superconductivity after World War I and circumstances surrounding the discovery of a state B = 0," *HSPBS*, **16**, no. 1 (1986), 1–58.

Daunt, J.G., "The magnetic threshold curves of superconductors," *PR*, **72** (1947), 89–90.

Daunt, J.G., and K. Mendelssohn, "Equilibrium curve and entropy difference between the supraconductive and the normal state in Pb, Hg, Sn, Ta, and Nb," *PRS*, **A160** (1937), 127–136.

Daunt, J.G., T.C. Keeley, and K. Mendelssohn, "Absorption of infrared light in supraconductors," *PM*, **23** (1937), 264–271.

Daunt, J.G., and K. Mendelssohn, "Thomson effect of supraconductive lead," *NA*, **141** (1938), 116.

Daunt, J.G., and K. Mendelssohn, "An experiment on the mechanism of superconductivity," *PRS*, **A185** (1946), 225–239.

Daunt, J.G., A.R. Miller, A.B. Pippard, and D. Shoenberg, "Temperature dependence of penetration depth of a magnetic field in superconductors," *PR*, **74** (1948), 842.

DeBruyn Ouboter, R., "Superconductivity: Discoveries during the early years of low temperature research at Leiden 1908–1914," *TM*, **MAG-23** (1987), 355–370.

Debye, P., "Die magnetische Methode zur Erzeugung tiefster Temperaturen," *PZ*, **35** (1934), 923–928.

De Haas, W.J., "Supraleiter im Magnetfeld," *Leipziger Vorträge 1933, Magnetismus*, ed. P. Debye (Leipzig, 1933), 59–73.

De Haas, W.J., G.J. Sizoo, and H. Kamerlingh Onnes, "On the magnetic disturbance of the supraconductvity with mercury, I and II," *RN*, **29** (1926), 233–249, 250–263.

De Haas, W.J., and G.J. Sizoo, "Further measurements on magnetic disturbance of the superconductivity with tin and mercury," *RN*, **29** (1926), 947–963.

De Haas, W.J., G.J. Sizoo, and J. Vogd, "Research about the question whether grey tin becomes supraconductive or not," *RN*, **31** (1928), 350–352.

De Haas, W.J., E. Van Aubel, and J. Voogd, "A superconductor, consisting of two non-superconductors," *RN*, **32** (1929), 724–730.

De Haas, W.J., and J. Voogd, "On the resistance-hysteresis phenomena of tin, lead, indium and thallium at the temperature of liquid helium," *RN*, **32** (1929), 206–213.

De Haas, W.J., and J. Voogd, "On the superconductivity of gallium," *RN*, **32** (1929), 214–217.

De Haas, W.J., and J. Voogd, "The resistance of alloys at the temperature of liquid hydrogen and helium," *RN*, **32** (1929), 715–723.

De Haas, W.J., and J. Voogd, "On the superconductivity of gallium," *RN*, **32** (1929), 733–734.

De Haas, W.J., and J. Voogd, "Disturbance of the superconductivity of the compound

Bi_5Tl_3 and the alloys Sn-Sb and Sn-Cd by magnetic fields," *RN*, **32** (1929), 874–882.

De Haas, W.J., E. Van Aubel, and J. Voogd, "On the supra-conductivity of alloys," *RN*, **33** (1930), 258–261.

De Haas, W.J., and J. Voogd, "The influence of magnetic fields on supraconductors," *RN*, **33** (1930), 262–272.

De Haas, W.J., S. Ayoama, and H. Bremmer, "Thermal conductivity of tin at low temperatures," *RN*, **34** (1931), 75–77.

De Haas, W.J., and H. Bremmer, "Conduction of heat of lead and tin at low temperatures," *RN*, **34** (1931), 325–338.

De Haas, W.J., and J. Voogd, "Measurements on the electrical resistance of pure indium, thallium, and gallium at low temperatures and of the magnetic disturbance of the supraconductivity of thallium," *RN*, **34** (1931), 51–55.

De Haas, W.J., and J. Voogd, "Further investigations on the magnetic disturbance of the supraconducting state of alloys," *RN*, **34** (1931), 56–62.

De Haas, W.J., and J. Voogd, "The magnetic disturbance of the supraconductivity of single-crystal wires of tin," *RN*, **34** (1931), 63–69.

De Haas, W.J., and J. Voogd, "On the steepness of the transition curve of supraconductors," *RN*, **34** (1931), 192–203.

De Haas, W.J., and H. Bremmer, "Thermal conductivity of indium at low temperatures," *RN*, **35** (1932), 131–136.

De Haas, W.J., and J.M. Casimir-Jonker, "Unterschungen über den Verlauf des Eindringens transversalen Magnetfeldes in einem Supraleiter," *PH*, **1** (1934), 291–296.

De Haas, W.J., J. Voogd, and J.M. Jonker, "Quantitative Untersuchung über einen möglichen Einfluss der Achsenorientierung auf die magnetische übergangsfigur," *PH*, **1** (1934), 281–290.

De Haas, W.J., and J.M. Casimir-Jonker, "Penetration of a magnetic field into supraconductive alloys," *NA*, **135** (1935), 30–31.

De Haas-Lorentz, G.L., "Iets over he mechanisme van inductiever schijnselen," *PH*, **5** (1925), 384–388.

Dewar, J., "The nadir of temperature and allied problems" (Bakerian lecture, Royal Society, June 13, 1901), *PRS*, **68** (1901), 360–366.

—, "Electrical resistance thermometry," *PRS*, **73** (1904), 244–251.

Donnelly, R.J., "Leo Dana: Cryogenic science and technology," *PT*, **40** (April 1987), 38–44.

Drude, P., "Zur Elektronentheorie der Metalle," *AP*, **1** (1900), 566–613.

Edelsack, E.A., D.U. Gubjer, and S.A. Wolf, "The rocky road to high temperature superconductivity," in *Novel Superconductivity*, ed. S.A. Wolf and V.Z. Kresin (New York, 1987).

Ehrenfest, P., "Phasenumwandlungen in üblichen und erweiterten Sinn, classifiziert nach den entsprechenden Singularitäten des thermodynamischen Potentiales," *RN*, **36** (1933), 153–157.

Einstein, A., "Die Plancksche Theorie der Strahlung und die Theorie der spezifischen Wärme," *AP*, **22** (1907), 180–190.

—, "Theoretische Bemerkungen zur Supraleitung der Metalle," *Het natuurkundig laboratorium der Rijks-Universiteit te Leiden in de jaren 1904–1922* (Leiden, 1922), 429–435.

Everitt, C.W.F., and W.M. Fairbank, "Fritz London," *DSB*, vol. **VIII** (New York, 1973), 473–479.

Everitt, C.W.F., and W.M. Fairbank, "Heinz London," *DSB*, vol. **VIII** (New York, 1973), 479–483.

Feffer, S.M., "Arthur Schuster, J.J. Thomson, and the discovery of the electron," *HSPBS*, **20**, no. 1 (1989), 33–61.

Feynman, R., and M. Cohen, "Energy spectrum of the excitations in liquid helium," *PR*, **102** (1956), 1189–1204.

Fisk, H.E., B.C. Brown, J.A. Carson, D.A. Edwards, H.T. Edwards, N.H. Engler, J.D. Gonezy, R.W. Hanft, K.P. Koepke, M. Kuchnir, R.A. Lundy, P.M. Mantsch, P.O. Mazur, A.D. McInturff, T.H. Nicol, R.C. Niemann, R.J. Powers, R.H. Remsbottom, C.H. Rode, E.E. Schmidt, and A. Szymulanski, "The ironless cos θ magnet option for the SSC," *TNS*, **NS-32** (1985), 3456–3461.

Frenkel, J., "Beitrag zur electrischen Theorie der festen Körper," *ZP*, **25** (1924), 1–30.

—, "On a possible explanation of superconductivity," *PR*, **43** (1933), 907–912.

Fröhlich, H., "Theory of the superconducting state. I. The ground state at the absolute zero of temperature," *PR*, **79** (1950), 845–858.

—, "Isotope effect in superconductivity," *PRS*, **63A** (1950), 778.

Gammel, P.L., L.F. Schneemeyer, J.V. Waszczak, and D.J. Bishop, "Evidence from mechanical measurements for flux-lattice melting in single crystal $YBa_2Cu_3O_7$ and $Bi_{2.2}Sr_2Ca_{0.8}Cu_2O_8$," *PRL*, **61** (1988), 1666–1669.

Gavaler, J.R., "Superconductivity in Nb-Ge films above 22 K," *Applied Physics Letters*, **23** (1973), 480–482.

Gavroglu, K., and Y. Goudaroulis, *Methodological Aspects of the Development of Low Temperature Physics 1881–1956: Concepts Out of Context(s)* (Dordrecht, 1989).

Geballe, T.H., and J.K. Hulm, "Superconductivity—the state that came in from the cold," *SC*, **239** (1988), 367–374.

Ginzburg, V.L., and L.D. Landau, "On the theory of superconductivity," *Zhurnal Eksperimental'noi i Teoreticheskoi Fiziki*, **20** (1950), 1064–1082 [in Russian].

Goodman, B.B., "The thermal conductivity of superconducting tin below 1 °K," *PRS*, **AGG** (1953), 217–227.

—, "The magnetic behavior of superconductors of negative surface energy," *IBM Journal of Research and Development*, **6** (1962), 63–67.

Gorkov, L.P., "Theory of superconducting alloys in a strong magnetic field near the critical temperature," *Soviet Physics JETP*, **10** (1960), 998–1004.

Gorter, C.J., "Some remarks on the thermodynamics of supraconductivity," Musée Teyler, *Archives*, **7** (1933), 378–386.

—, "Theory of supraconductivity," *NA*, **132** (1933), 931.

—, "Note on the supraconductivity of alloys," *PH*, **2** (1935), 449–452.

—, "The two fluid model for superconductors and helium II," *Progress in Low Temperature Physics*, vol. I (New York and Amsterdam, 1955).

—, *Lectures on Magnetism and Superconductivity* (Tata Institute of Fundamental Research, Bombay, 1963).

—, "Superconductivity until 1940 in Leiden and as seen from there," *RMP*, **36** (1964), 3–7.

Gorter, C.J., and H.B.G. Casimir, "On supraconductivity I," *PH*, **1** (1934), 306–320.

Gorter, C.J., and H.B.G. Casimir, "Zur Thermodynamik des supraleitenden Zustandes," *PZ*, **35** (1934), 963–966.

Guide to Sources for History of Solid State Physics, compiled by J. Warnow-Blewett and J. Teichmann, report no. 7 of international catalog of sources for history of physics and allied sciences (draft, October 1987).

Hazen, R.M., *The Breakthrough: The Race for the Superconductor* (New York, 1988).

Heilbron, J.L., "Physics at McGill in Rutherford's time," in *Rutherford and Physics at the Turn of the Century*, ed. M. Bunge and W.R. Shea (New York, 1979).

—, *The Dilemmas of an Upright Man: Max Planck as Spokesman for German Science* (Berkeley and Los Angeles, 1986).

Hermann, A., "Max von Laue," in *DSB*, vol. **VIII** (New York, 1975), 50–53.

—, "Johannes Stark," in *DSB*, vol. **3** (New York, 1975), 613–616.

Het natuurkundig laboratorium der Rijksuniversiteit te Leiden in de jaren 1904–1922. Gedenkboek aangeboden aan H. Kamerlingh Onnes, Directeur van het laboratorium, bij gelegenheid van zijn veertiwarig professoraat op 11 November 1922 (Leiden, 1922).

High-Temperature Superconductivity: Reprints from Physical Review Letters and Physical Review B, January–June, 1987 (American Institute of Physics, New York, 1987).

Hoddeson, L., "The first large-scale application of superconductivity: The Fermilab Energy Doubler, 1972–1983," *HSPBS*, **18**, no. 1 (1987), 25–54.

Hoddeson, L., and G. Baym, "The development of the quantum mechanical electron theory of metals: 1900–28," *PRS*, **A371** (1980), 8–23.

Hoddeson, L., G. Baym, and M. Eckert, "The development of the quantum mechanical electron theory of metals: 1928-1933," *RMP*, **59** (1987), 287–327.

Hor, P.H., L. Gao, R.L. Meng, Z.J. Huang, Y.Q. Wang, K. Forster, J. Vassilious, and C.W. Chu, "High-pressure study of the new Y-Ba-Cu-O superconducting compound system," *PRL*, **58** (1987), 911–912.

Huby, R., "Physics at Liverpool," *NA*, **166** (1950), 552–555.

Hulm, J.K., "The thermal conductivity of tin, mercury, indium and tantalum at liquid helium temperatures," *PRS*, **204** (1950), 98–123.

—, "Superconductivity research in the good old days," *TM*, **MAG-19** (1983), 161–166.

Hulm, J.K., J.E. Kunzler, and B.T. Matthias, "The road to superconducting materials," *PT*, **34**, no. 1 (January 1981), 34–43.

Hulm, J.K., and C. Laverick, "International cooperative-collaborative perspective, superconductive science & technology," *TM*, **MAG-23** (1987), 423–426.

Hume-Rothery, W.J., *The Metallic State* (Oxford, 1931).

Huson, F.R., H. Bingham, J. Colvin, M. Davidson, J. Greenough, S. Heifets, H. Hinterberger, M. Kobayashi, K. Lau, G. Lopez, P. McIntyre, W. MacKay, D. Neuffer, G. Phillips, S. Pissanetsky, P. Rajan, D. Raparia, R. Rocha, W. Schmidt, R. Stegman, D. Swenson, T. Tominaka, P. VanderArend, R. Weinstein, W. Wenzel, R. Wolgast, and J. Zeigler, "Superferric magnet option for the SSC," *TNS*, **NS-32** (1985), 3462–3465.

Institut International de Physique Solvay, *La théorie du rayonnement et les quanta, rapports et discussions de la réunion tenue à Bruxelles, du 30 octobre au 3 novembre 1911 sous les auspices de M.E. Solvay*, ed. P. Langevin and M. de Broglie (Paris, 1912).

Joffe, A.F., "Twenty years of Soviet physics," *PZS*, **12** (1937), 494–505.

Justi, E., and M. von Laue, "Phasengleichgewichte dritter Art," *PZ*, **35** (1934), 945–963.

Kaiser, W., "Early theories of the electron gas," *HSPBS*, **17**, Part 2 (1987), 271–297.

Kamerlingh Onnes, H., "The liquefaction of helium," *RN*, **11** (1909), 168–185.

—, "Further experiments with liquid helium. B. On the change in the resistance of pure metals at very low temperatures, etc. III. The resistance of platinum at helium temperatures," *RN*, **13** (1911), 1093–1112.

—, "Further experiments with liquid helium. C. On the change of electric resistance of pure metals at very low temperatures, etc. IV. The resistance of pure mercury at helium temperatures," *RN*, **13** (1911), 1274–1276.

—, "Further experiments with liquid helium. D. On the change of the electrical resistance of pure metals at very low temperatures, etc. V. The disappearance of the resistance of mercury," *RN*, **14** (1912), 113–115.

—, "Further experiments with liquid helium. G. On the electrical resistance of pure metals, etc. VI. On the sudden change in the rate at which the resistance of mercury disappears," *RN*, **14** (1912), 818–821.

—, "Report on the researches made in the Leiden cryogenics laboratory between the second and third international congress of refrigeration: Superconductivity," *PLS*, **34b** (1913), 55–70.

—, "Untersuchungen über die Eigenschaften der Körper bei niedrigen Temperaturen, welche Untersuchungen unter anderem auch zur Herstellung von flüssigem Helium geführt haben," *PLS*, **35** (1913), 1–36.

—, Further experiments with liquid helium. H. On the electrical resistance of pure metals, etc. VII. The potential difference necessary for the electric current through mercury below 4.19 °K *RN*, **15** (1913), 1406–1427, 1427–1430; *RN*, **16** (1914), 113–124.

—, "Further experiments with liquid helium. H. On the electrical resistance, etc. VIII. The sudden disappearance of the ordinary resistance of tin, and the super-conductive state of lead," *RN*, **16** (1914), 673–688.

—, "Further experiments with liquid helium. I. The Hall-effect, and the magnetic change in resistance at low temperatures. IX. The appearance of galvanic resistance in supra-conductors, which are brought into a magnetic field, at a threshold value of the field," *RN*, **16** (1914), 987–992.

—, "Further experiments with liquid helium. J. The imitation of an Ampere molecular current or of a permanent magnet by means of a supra-conductor," *RN*, **17** (1915), 12–20, 278–283.

—, "Further experiments with liquid helium. L. The persistence of currents without electro-motive force in supra-conducting circuits," *RN*, **17** (1915), 514–519.

—, "Demonstration of liquid helium," *PLS*, **43c** (1919), 13–19; translated from Nederlandisch Natur-en Geneeskundig Congress, XVII (1919), *Handelingen*.

—, "On the lowest temperature yet obtained," Faraday Society, *Transactions*, **18** (1922), 145–174.

—, "Les supraconducteurs et le modèle de l'atome Rutherford–Bohr," *Atomes et électrons* (Paris, 1923).

—, "Nouvelles expériences avec les supraconductors," Institut International de Physique Solvay, *Conductibilité électrique des métaux et problèmes connexes* (Paris, 1927), 251–281.

Kamerlingh Onnes, H., and J. Clay, "On the change of resistance of the metals at very low temperatures and the influence exerted on it by small amounts of admixtures," *RN*, **10** (1908), 207–215.

Kamerlingh Onnes, H., and G. Holst, "Further experiments with liquid helium. M. Preliminary determination of the specific heat and of the thermal conductivity of mercury at temperatures obtainable with liquid helium, besides some measurements of thermoelectric forces and resistances for the purpose of these investigations," *RN*, **17** (1915), 760–767.

Kamerlingh Onnes, H., and W. Tuyn, "Further experiments with liquid helium. Q. On the electric resistance of pure metals, etc. X. Measurements concerning the electrical resistance of thallium in the temperature field of liquid helium," *RN*, **25** (1923), 443–450.

Kamerlingh Onnes, H., and W. Tuyn, "Further experiments with liquid helium. R. On the electric resistance of pure metals, etc. XI. Measurements concerning the electric resistance of ordinary lead and of uranium lead below 14 °K," *RN*, **25** (1923), 451–457.

Kangro, H., "Wilhelm Carl Werner Otto Fritz Franz Wien," *DSB*, vol. **XIV** (New York, 1975), 337–342.

Kapitza, P., "The study of the specific resistance of bismuth crystals and its change in strong magnetic fields and some allied problems," *PRS*, **119A** (1928), 358–442.

Keeley, T.C., K. Mendelssohn, and J.R. Moore, "Experiments on supraconductors," *NA*, **134** (1934), 773–774.

Keesom, W.H., "On the equation of state of an ideal monatomic gas according to the quantum-theory," *RN*, **16** (1914), 227–236.

—, "On the theory of free electrons in metals," *RN*, **16** (1914), 236–245.

—, "Prof. Dr. H. Kamerlingh Onnes. His life-work, the founding of the cryogenic laboratory," *PLS*, **57** (1926), 2–21.

—, "Methods and apparatus used in the cryogenic laboratory. XXII. A cryostat for temperatures below 1 °K," *RN*, **32** (1929), 710–714.

—, "Das kalorische Verhalten von Metallen bei den tiefsten Temperaturen," *PZ*, **35** (1934), 939–944.

—, "On the disturbance of supraconductivity of an alloy by an electric current," *PH*, **2** (1935), 35–36.

—, *Helium* (Amsterdam, London, New York, 1942).

Keesom, W.H., and H. Kamerlingh Onnes, "On the question of the possibility of a polymorphic change at the point of transition into the supraconductive state," *PLC*, **174b** (1924), 43–45.

Keesom, W.H., and J.N. van den Ende, "The specific heat of solid substances at the temperatures obtainable with the aid of liquid helium. II. Measurements of the atomic heats of lead and of bismuth," *RN*, **33** (1930), 243–254; A correction, *RN*, **34** (1931), 210–211.

Keesom, W.H., and J.N. van den Ende, "The specific heat of solids at temperatures obtainable with liquid helium. IV. Measurements of the atomic heats of tin and zinc," *RN*, **35** (1932), 143–155.

Keesom, W.H., and J.A. Kok, "On the change of the specific heat of tin when becoming supraconductive," *RN*, **35** (1932), 743–748.

Keesom, W.H., and J.A. Kok, "Measurements of the specific heat of thallium at liquid helium temperatures," *PH*, **1** (1934), 175–181.

Keesom, W.H., and J.A. Kok, "Measurements of the latent heat of thallium connected with the transition, in a constant external magnetic field, from the supraconductive to the non-supraconductive state," *PH*, **1** (1934), 503–512.

Keesom, W.H., and J.A. Kok, "Further calorimetric experiments on thallium," *PH*, **1** (1934), 595–608.

Keesom, W.H., and P.H. van Laer, "Measurements of the atomic heats of tin in the superconductive and in the non-superconductive state," *PH*, **5** (1938), 193–201.

Keesom, W.H., and M. Desirant, "The specific heats of tantalum in the normal and in the superconductive state," *PH*, **8** (1941), 273–288.

Khurana, A., "High-temperature superconductors may not be forever," *PT*, **42** (March, 1989), 17–21.

Kim, Y.B., C.F. Hempstead, and A.R. Strnad, "Critical persistent currents in hard superconductors," *PRL*, **9** (1962), 306–309.

Kim, Y.B., C.F. Hempstead, and A.R. Strnad, "Flux creep in hard superconductors," *PR*, **131** (1963), 2486–2495.

Klein, M.J., "Einstein, specific heats, and the early quantum theory," *SC*, **148** (1965), 173–180.

—, *Paul Ehrenfest, Volume I: The Making of a Theoretical Physicist* (Amsterdam, 1970).

Kok, J.A., "Supraconductivity and Fermi–Dirac statistics," *NA*, **134** (1934), 532–533.

—, "Some remarks on supraconductivity and Fermi–Dirac statistics," *PH*, **1** (1934), 1103–1106.

Kramers, H.A., *Between Tradition and Revolution* (New York, 1987).

Kronig, R. de L., "Zur Theorie der Supraleitfähigkeit," *ZP*, **78** (1932), 744–750.

—, "Zur Theorie der Supraleitfähigkeit II," *ZP*, **80** (1933), 203–216.

Kuhn, T.S., *Black-Body Theory and the Quantum Discontinuity 1894–1912* (Oxford, 1978).

Kunzler, J.E., "Superconductivity in high magnetic fields at high current densities," *RMP*, **33** (1961), 501–509.

—, "Superconductivity in high magnetic fields at high current densities," *Proceedings of International Conference on High Magnetic Fields*, ed. H. Kolm, B. Lax, F. Bitter, and R. Mills (New York and London, 1962).

—, "Recollections of events associated with the discovery of high field-high current superconductivity," *TM*, **MAG-23** (1987), 396–402.

Kunzler, J.E., E. Buehler, F.S.L. Hsu, B.T. Matthias, and C. Wahl, "Production of magnetic fields exceeding 15 kilogauss by a superconducting solenoid," *JAP*, **32** (1961), 325–326.

Kunzler, J.E., E. Buehler, F.S.L. Hsu, and J.H. Wernick, "Superconductivity in Nb_3Sn at high current density in a magnetic field of 88 kGauss," *PRL*, **6** (1961), 89–91.

Kurti, N., Obituary for F.E. Simon in *Biographical Memoirs of Fellows of the Royal Society* (1958), 224–256.

—, "Cryomagnetic research at the Clarendon Laboratory," *Search and Research*, ed. J.P. Wilson (London, 1971).

Lambertson, G., A. Borden, J. Cox, W. Eaton, W. Gilbert, J. Holl, E. Knight, R. Main, R. Meuser, J. Rechen, R. Schafer, F. Toby, and F. Voelker, *Design, Construction, and Operation of 12 ESCAR Bending Magnets* (Lawrence Berkeley Laboratory, report 8210, September 1978).

Lambertson, G.R., W.S. Gilbert, and J.B. Rechen, *Final Report on the Experimental Superconducting Synchrotron ESCAR* (Lawrence Berkeley Laboratory, report 8211, March 1979).

Landau, L.D., "Zur Theories der Supraleitfähigkeit," *PZS*, **4** (1933), 43–49.

—, "The theory of superfluidity of helium II," *Journal of Physics (USSR)*, **5** (1941), 71–90.

Langton, H.H., *Sir John Cunningham McLennan: A Memoir* (Toronto, 1939).

Laue, M. von, "Zur Deutung einiger Versuche über Supraleitung," *PZ*, **33** (1932), 793–795.

—, "Zur Thermodynamik der Supraleitung," *AP*, **32** (1938), 253–258.

Laue, M. von, and F. Möglich, "Ueber das magnetische Feld in der Umgebung von Supraleitern," Preussischen Akademie der Wissenschaften, *Sitzungberichte*, **16** (1933), 544–565.

Laverick, C., "The performance of large superconducting coils," *Advances in Cryogenic Engineering*, vol. II (New York, 1965).

Lindemann, F.A., "Ueber die Berechnung molekularer Eigenfrequenzen," *PZ*, **11** (1910), 609–612.

—, "Untersuchungen über die spezifische Wärme bei tiefen Temperaturen. IV," *SB* (1911), 316–321.

—, "Note on the metallic state," *PM*, **29** (1915), 127–140.

Lippmann, G., "Sur les propriétés des circuits électriques dénués de résistance," Académie des Sciences, *Comptes rendus*, **168** (1919), 73–78.

London, F., "Macroscopical interpretation of supraconductivity," *PRS*, **A152** (1935), 24–33.

—, "Zur Theorie magnetischer Felder im Supraleiter," *PH*, **3** (1936), 450–462.

—, "Supraconductivity in aromatic compounds," *Journal of Chemical Physics*, **5** (1937), 837–838.

—, "On the problem of the molecular theory of superconductivity," *PR*, **74** (1948), 562–573.

—, *Superfluids*, vol. 1 (New York, 2nd ed., 1961).
London, F., and H. London, "The electromagnetic equations of the supraconductor," *PRS*, **A149** (1935), 71–88.
London, H., "Phase-equilibrium of supraconductors in a magnetic field," *PRS*, **A152** (1935), 650–663.
—, "An experimental examination of the electrostatic behavior of supraconductors," *PRS*, **155** (1936), 102–110.
—, "The high-frequency resistance of superconducting tin," *PRS*, **176** (1940), 522–533.
Lorentz, H.A., "The motion of electrons in metallic bodies, I, II and III," *RN*, **7** (1905), 438–453, 585–593, 684–691.
—, "Application de la théorie des électrons aux propriétés des métaux," Institut International de Physique Solvay, *Conductibilité électrique des métaux et problèmes connexes* (Paris, 1927), 1–45.
Martin, T., *The Royal Institution* (London, 1961).
Matthias, B.T., "Superconductivity and its relation to transition elements," *Proceedings of the Eighth International Conference on Low Temperature Physics*, ed. R.O. Davis (Washington, D.C., 1963).
Maxwell, E., "Isotope effect in the superconductivity of mercury," *PR*, **78** (1950), 477.
McInturff, A.D., P.F. Dahl, and W.B. Sampson, "Pulsed field losses in metal-filled superconducting multifilamentary braids," *JAP*, **43** (1972), 3546–3551.
McLennan, J.C., "The cryogenic laboratory of the University of Toronto," *NA*, **112** (1923), 135–139.
—, "Electric supra-conduction in metals," *Supplement to Nature*, **130** (1932), 879–886.
—, "Electrical phenomena at extremely low temperatures," *Reports on Progress in Physics*, **1** (1934), 198–227.
—, "Opening address in a discussion on supraconductivity and other low temperature phenomena," *PRS*, **A152** (1935), 1–8.
—, "Electrical conductivity of metals at the lowest temperatures," *The Royal Institution Library of Science: Physical Sciences*, vol. 9, ed. W.L. Bragg and G. Porter (Amsterdam, 1970).
McLennan, J.C., L.E. Howlett, and J.O. Wilhelm, "On the electrical conductivity of certain metals at low temperatures," Royal Society of Canada, *Transactions*, **23** (1929), 287–300.
Mehra, J., and H. Rechenberg, *The Historical Development of Quantum Theory* (4 vols., New York, 1982).
Meissner, W., "Thermische und elektrische Leitfähigkeit einiger Metalle zwischen 20 und 373° abs.," *AP*, **47** (1915), 1001–1058.
—, "Ueber die Heliumsverflüssigungsanlage der Physikalisch-Technische Reichsanstalt und einige Messungen mit Hilfe von flüssigem Helium," *PZ*, **26** (1925), 689–694.
—, "Das neue Kaltelaboratorium der Physikalisch-Technische Reichsanstalt in Berlin," *PZ*, **29** (1928), 610–623.
—, "Messungen mit Hilfe von flüssigem Helium. II. Widerstand von Metallen. Supraleitfähigkeit von Tantal. Beiträge zur Erklarung der Supraleitfähigkeit. Spezifische Wärme des gasformigen Heliums," *PZ*, **29** (1928), 897–904.
—, "Supraleitfähigkeit von Thorium," *Die Naturwissenschaften*, **17** (1929), 390–391.
—, "Messungen mit Hilfe von flüssigem Helium. V. Supraleitfähigkeit von Kupfersulfid," Physikalisch-Technische Reichsanstalt, *Mitteilung* (1929), 570–572.
—, "Arbeiten des Charlottenburger Kältelaboratorium über Supraleitfähigkeit und über das Wasserstoff-Isotop," *Helvetica physica acta*, **6** (1933), 414–418.
—, "Bericht über neuere Arbeiten zur Supraleitfähigkeit," *PZ*, **35** (1934), 931–944.

—, "The magnetic effects occurring on transition to the supraconducting state," in "A discussion on supraconductivity and other low temperature phenomena," *PRS*, **A152** (1935), 1–46.

—, "Max von Laue als Wissenschaftler und Mensch," Bayerische Akademie der Wissenschaften, *Sitzungsberichte*, **9** (1960), 101–121.

Meissner, W., and H. Franz, "Messungen mit Hilfe von flüssigem Helium. VIII. Supraleitfähigkeit von Niobium," Physikalisch-Technische Reichsanstalt, *Mitteilung* (1930), 558–559.

Meissner, W., and B. Voigt, "Messungen mit Hilfe von flüssigem Helium. XI. Widerstand der reinen Metalle in tiefen Temperaturen," *AP*, **7** (1930), 761–797.

Meissner, W., and R. Ochsenfeld, "Ein neuer Effect bei Eintritt der Supraleitfähigkeit," *Die Naturwissenschaften*, **21** (1933), 787–788.

Meissner, W., with R. Ochsenfeld and F. Heidenreich, "Magnetische Effecte bei Eintritt der Supraleitfähigkeit," *Zeitschrift für die gesamte Kälte-Industrie*, **41** (1934), 125–130.

Meissner, M., and F. Heidenreich, "Ueber die Änderung der Stromverteilung und der magnetischen Induktion beim Eintritt der Supraleitfähigkeit," *PZ*, **37** (1936), 449–470.

Mendelssohn, K., "Production of high magnetic fields at low temperatures," *NA*, **132** (1933), 602.

—, "Hysteresis," in "A discussion on superconductivity and other low temperature phenomena," *PRS*, **A152** (1935), 1–46.

—, "Heat conduction in superconductors," *Progress in Low Temperature Physics*, I, ed. C.J. Gorter (New York and Amsterdam, 1955), 184–201.

—, "Prewar work on superconductivity as seen from Oxford," *RMP*, **36** (1964), 7–12.

—, "On different types of superconductivity," *RMP*, **36** (1964), 50–51.

—, "Walther Nernst: An appreciation," *CR*, **4** (1964), 129–135.

—, "The Clarendon Laboratory, Oxford," *CR*, **6** (1966), 129–140.

—, *The Quest for Absolute Zero* (New York, 1966).

—, *The World of Walther Nernst: The Rise and Fall of German Science* (Edinburgh, 1973).

Mendelssohn, K., and J.D. Babbitt, "Persistent currents in supraconductors," *NA*, **133** (1934), 459–460.

Mendelssohn, K., and J.D. Babbitt, "Magnetic behaviour of supraconducting tin spheres," *PRS*, **A151** (1935), 316–333.

Mendelssohn, K., and J.R. Moore, "Supra-conducting alloys," *Supplement to NA*, **135** (1935), 826–827.

Mendelssohn, K., and J.R. Moore, "Experiments on supraconductive tantalum," *PM*, **21** (1936), 532–544.

Misener, A.D., "Magnetic effects and current sensitivity of superconducting films," *Canadian Journal of Research*, **14** (1936), 25–37.

Misener, A.D., H.G. Smith, and J.O. Wilhelm, "Effect of magnetic fields on the supraconductivity of thin films of tin," Royal Society of Canada, *Transactions*, **29** (1934), 13–21.

Morin, F.J., J.P. Maita, H.J. Williams, R.C. Sherwood, J.H. Wernick, and J.E. Kunzler, "Heat capacity evidence for a large degree of superconductivity in V_3Ga in high magnetic fields," *PRL*, **8** (1962), 275–277.

Müller, K.A., and J.G. Bednorz, "The discovery of a class of high-temperature superconductors," *SC*, **237** (1987), 1133–1139.

Nernst, W., "Untersuchungen zur spezifischen Wärme bei tiefen Temperaturen. I, II and III," *SB* (1910), 247–261, 262–282, *SB* (1911), 306–315.

Nernst, W., and F.A. Lindemann, "Spezifische Wärme und Quantentheorie," *Zeitschrift für Elektrochemie*, **17** (1911), 817–827.

Obreimow, I., and L. Schubnikow, "Eine Methode zur herstellung einkristalliger Metalle," *ZP*, **25** (1924), 31–36.

Peierls, R., "Zur kinetischen Theorie der Wärmeleitung in Kristallen," *AP*, **4** (1929), 1055–1101.

—, "Zur Theorie der elektrischen und thermischen Leitfähigkeit von Metallen," *AP*, **5** (1930), 121–148.

—, "Zwei Bemerkungen zur Theorie der Leitfähigkeit," *AP*, **5** (1930), 244–246.

Pippard, A.B., "The surface impedance of superconductors and normal metals at high frequencies. I. Resistance of superconducting tin and mercury at 1200 Mcyc./sec.," *PRS*, **A191** (1947), 370–384; "II. The anomalous skin effect in normal metals," 385–399; "III. The relation between impedance and superconducting penetration depth," 399–415; "IV. Impedance at 9400 Mc./sec. of single crystals of normal and superconducting tin," *PRS*, **A203** (1950), 98–118.

—, "Field variations of the superconducting penetration depth," *PRS*, **A203** (1950), 210–223.

—, "An experimental and theoretical study of the relation between magnetic field and current in a superconductor," *PRS*, **A216** (1953), 547–568.

—, "The historical context of Josephson's discovery," *Superconductor Applications: Squids and Machines*, ed., B.B. Schwartz and S. Foner (New York and London, 1976).

—, "Early superconductivity research (except Leiden)," *TM*, **MAG-23** (1987), 371–375.

Pontius, R.B., "Superconductors of small dimensions," *NA*, **139** (1937), 1065–1066.

—, "Threshold values of supraconductors of small dimensions," *PM*, **24** (1937), 787–796.

Pool, R., "New superconductors answer some questions," *SC*, **240** (1988), 146–147.

—, "Superconductor credits bypass Alabama," *SC*, **241** (1988), 655–657.

—, "Superconductivity: Is the party over?" *SC*, **244** (1989), 914–916.

—, "Feud flares over thallium superconductor," *SC*, **247** (1990), 1029.

Reardon, P.J., "High energy physics and applied superconductivity," *TM*, **MAG-13** (1977), 704–718.

—, "Cold iron cos θ magnet option for the SSC," *TNS*, **NS-32** (1985), 3466–3470.

Reinganum, M., "Theoretische Bestimmung des Verhältnisses von Wärme-und Elektricitätsleitung der Metalle aus der Drude'schen Elektronentheorie," *AP*, **2** (1900), 398–403.

Report of the 1983 HEPAP Subpanel on New Facilities for the U.S. High Energy Physics Program (U.S. Department of Energy, Office of Energy Research, report DOE/ER-0169, July 1983).

Report of the HEPAP Subpanel to Review Recent Information on Superferric Magnets (May 1986).

Report of the Reference Designs Study Group of the Superconducting Super Collider, draft II (May 1984).

Reynolds, C.A., B. Serin, W.H. Wright, and L.B. Nesbitt, "Superconductivity of isotopes of mercury," *PR*, **78** (1950), 487.

Riecke, E., "Zur Theorie des Galvanismus und der Wärme," *APC*, **66** (1898), 353–389, 545–581.

—, "Ueber das Verhaltnis der Leitfähigkeiten der Metalle für Wärme und für Elektricität," *AP*, **2** (1900), 835–842.

—, "Die jetzigen Anschaugen über das Wesen des metallischen Zustandes," *PZ*, **10** (1909), 508–519.

Rjabinin, J.N., and L.W. Schubnikow, "Dependence of magnetic induction on the magnetic field in supraconducting lead," *NA*, **134** (1934), 286–287.
Rjabinin, J.N., and L.W. Schubnikow, "Verhalten eines Supraleiters in magnetischen Feld," *PZS*, **5** (1934), 641–643.
Rjabinin, J.N., and L.W. Schubnikow, "Magnetic properties and critical currents of superconducting alloys," *PZS*, **7** (1935), 122–125.
Rose-Innes, A.C., and E.H. Rhoderick, *Introduction to Superconductivity* (New York, 1969).
Ruhemann, M. and B., *Low Temperature Physics* (Cambridge, 1937).
Rutgers, A.J., "Note on supraconductivity," *PH*, **1** (1934), 1055–1058.
Schachenmeier, R., "Wellenmechanische Vorstudien zu einer Theorie der Supraleitung," *ZP*, **74** (1932), 503–546.
—, "Zur Elektronentheorie der Supraleitung," *PZ*, **35** (1934), 966–969.
Schopman, J., "Industrious science: Semiconductor research at the N.V. Phillips' Gloeilampenfabrieken, 1930–1957," *HSPBS*, **19** (1988), 137–172.
Schrieffer, J.R., "Macroscopic quantum phenomena from pairing in superconductors," *PT*, **26** (July, 1973), 23–28.
Schubnikow, L.W., "Ueber die herstellung von Wismutheinkristallen," *RN*, **33** (1930), 327–331.
Schubnikow, L.W., and W.J. de Haas, "Magnetische widerstandsvergrösserung in Einkristallen von Wismut bei tiefen Temperaturen," *RN*, **33** (1930), 130–133.
Schubnikow, L.W., and W.J. de Haas, "Die Abhängigkeit des elektrischen Widerstandes von Wismutheinkristallen von der Reinheit des Metalles," *RN*, **33** (1930), 350–362.
Schubnikow, L.W., and W.J. de Haas, "Neue Erscheinungen bei der Widerstandsanderung von Wismuthkristallen im Magnetfeld bei der Temperatur von flüssigem Wasserstoff," *RN*, **33** (1930), 363–378.
Schubnikow, L.W., and W.J. de Haas, "A new phenomenon in the change of resistance in a magnetic field of single crystals of bismuth," *NA*, **126** (1930), 500.
Schubnikow, L.W., and W.J. Chotkewitsch, "Spezifische Wärme von supraleitenden Legierungen," *PZS*, **6** (1934), 605–607.
Schubnikow, L.W., W.J. Chotkewitsch, J.D. Schepelew, and J.N. Rjabinin, "Magnetische Eigenschaften supraleitender Metalle und Legierungen," *PZS*, **10** (1936), 165–192.
Schuster, A., "Experiments on the discharge of electricity through gases. Sketch of a theory," *PRS*, **37** (1884), 317–339.
Serin, B., "Superconductivity. Experimental part," *Encyclopedia of Physics*, ed. S. Flügge, vol. **XV** (Berlin, 1956), 210–273.
Shalnikov, A., "Superconducting thin films," *NA*, **142** (1938), 74.
Shoenberg, D., "Superconducting colloidal mercury," *NA*, **143** (1939), 434–435.
—, *Superconductivity* (Cambridge, 1965).
—, "Heinz London, 1907–1970," *Biographical Memoirs of the Royal Society*, **17** (1971), 440–461.
Silsbee, F.B., "A note on electrical conduction in metals at low temperatures," Washington Academy of Sciences, *Journal*, **6** (1916), 597–602.
—, "Note on electrical conduction in metals at low temperatures," Bureau of Standards, *Bulletin* (1917), 301–306.
—, "Current distribution in supraconductors," National Academy of Sciences, *Proceedings*, **13** (1927), 516–518.
Sizoo, G.J., "Onderzoekingen over den suprageleidenden toestand van metallen," Doctoral Dissertation, University of Leiden, 1926.

Sizoo, G.J., and H. Kamerlingh Onnes, "Further experiments with liquid helium. X. On the electric resistance of pure metals, etc. XIV. Influence of elastic deformation on the supraconductivity of tin and indium," *RN*, **28** (1925), 656–666.

Sizoo, G.J., W.J. de Haas, and H. Kamerlingh Onnes, "Influence of elastic deformation on the magnetic disturbance of the superconductivity with tin. Hysteresis phenomena," *RN*, **29** (1926), 221–223.

Sommerfeld, A., "Zur Elektronentheorie der Metalle," *Die Naturwissenschaften*, **15** (1927), 825–832.

—, "Zur spezifischen Wärme der Metallelektronen," *AP*, **28** (1937), 1–10.

SSC Magnet Selection Advisory Panel Report to the Director of the Central Design Group (9 September 1985).

Stark, J., "Ueber elektrische und mechanische Schussflächen in Metallen," *PZ*, **13** (1912), 585–589.

—, "Bemerkung über den Zustand der Elektronen in der Supraleitung," *PZ*, **36** (1935), 515–516.

Steiner, K., and P. Grassmann, *Supraleitung* (Braunschweig, 1937).

Stekly, Z.J.J., and J.L. Zar, "Stable superconducting coils," *TNS*, **23** (1965), 367–372.

Summer study on superconducting devices and accelerators, 1968, *Proceedings*, ed. J.P. Blewett, H. Hahn, and A.G. Prodell (Brookhaven National Laboratory, report 50155, Upton, New York, 1969).

Tanaka, S., "Research on high-T_c superconductivity in Japan," *PT* (December, 1987), 53–57.

Tarr, F.G.A., and J.O. Wilhelm, "Magnetic effects in superconductors," *Canadian Journal of Research*, **12** (1935), 265–271.

Thomson, J.J., *The Corpuscular Theory of Matter* (London, 1907).

—, "Conduction of electricity through metals," *PM*, **30** (1915), 192–202.

—, *Recollections and Reflections* (New York, 1975).

Turowski, P., J.H. Coupland, and J. Perot, "Pulsed superconducting dipole magnets of the GESSS collaboration," International Conference on High Energy Accelerators, IX, *Proceedings* (Stanford, 1974), 174–178.

Tuyn, W., "Measurements on the disturbance of the superconductivity of thallium by magnetic fields," *RN*, **31** (1928), 687–691.

Tuyn, W., and H. Kamerlingh Onnes, "Measurements concerning the electric resistance of indium in the temperature field of liquid helium," *RN*, **26** (1923), 504–509.

Tuyn, W., and H. Kamerlingh Onnes, "Further experiments with liquid helium. AA. The disturbance of supraconductivity by magnetic fields and currents. The hypothesis of Silsbee," *PLC*, **174a** (1926), 13–39; Franklin Institute, *Journal*, **201** (1926), 379–410.

Van Aubel, E., W.J. de Haas, and J. Voogd, "New super-conductors," *RN*, **32** (1929), 218–225.

Van Proosdij, B.A., "Some letters from Albert Einstein to Heike Kamerlingh Onnes," *Janus*, **XLVII** (1959), 133.

Welker, H., "Über ein elektronentheoretisches Modell des Supraleiters," *Zeitschrift für technische Physik*, **19** (1938), 606–611.

Westfall, C.L., "The site contest for Fermilab," *PT*, **42** (1989), 44–52.

Wien, W., "Zur Theorie der elektrischen Leitung in Metallen," *SB*, **5** (1913), 184–200.

Wien, W., "On the laws of thermal radiation," *Nobel Lectures including Presentation Speeches and Laureate's Biographies: Physics 1901–1921* (Amsterdam, London, and New York, 1967), 275–286.

Willen, E., *Performance of Six 4.5m SSC Dipole Model Magnets* (Brookhaven National Laboratory, report 38813, 1986).

Wilson, M.N., *Superconducting Magnets* (Oxford, 1983).

Wilson, M.N., C.R. Walters, J.D. Levin, and P.F. Smith, "Experimental and theoretical studies of filamentary superconducting composites," *British Journal of Applied Physics D*, **3** (1970), 1515–1585.

Wilson, R.R., "Superferric magnets for 20 TeV," 1982 DPF summer study on elementary particle physics and future facilities, *Proceedings*, ed. R. Donaldson, R. Gustavson, and F. Paige (Snowmass, 1982).

Wu, M.K., J.R. Ashburn, C.J. Torng, P.H. Hor, R.L. Meng, L. Gao, Z.J. Huang, Y.Q. Wang, and C.W. Chu, "Superconductivity at 93 K in a new mixed-phase Y-Ba-Cu-O compound system at ambient pressure," *PRL*, **58** (1987), 908–910.

Yeshurun, Y., and A.P. Malozemoff, "Giant flux creep and irreversibility in an Y-Ba-Cu-O crystal: An alternative to the superconducting-glass model," *PRL*, **60** (1988), 2202–2205.

Yntema, G.B., "Superconducting winding for electromagnet," *PR*, **98** (1955), 1197.

—, "Niobium superconducting magnets," *TM*, **MAG-23** (1987), 390–395.

Zavaritskii, N.V., "Investigation of superconducting properties of thallium and tin films deposited at low temperatures," *Doklady Akademii Nauk SSSR*, **86** (1952), 501–504 [in Russian].

Name Index

Subject Index

Printed in the United States
By Bookmasters